CONTENTS

HEAT TREATMENT '87

Proceedings of the
international conference
included in
Materials '87
held at
the Royal Lancaster Hotel
London
11-15 May 1987

The Institute of Metals
London
1988

Book 402
published by
The Institute of Metals
1 Carlton House Terrace
London SW1Y 5DB

Distributed in North America by
The Institute of Metals
North American Publications Center
Old Post Road
Brookfield VT05036 USA

British Library Cataloguing in Publication Data

Heat treatment '87.
1. Materials. Heat treatment
I. Institute of Metals, *1985, CRC Unit*
II. Materials '87
620.1'121

ISBN 0-904357-96-1

Library of Congress Cataloging in Publication Data

Heat treatment '87.

1. Metals--Heat treatment--Congresses. I. Institute of Metals.
TN672.H365 1988 671.3'6 88-8833
ISBN 9-0943-5796-1

Compiled by the Institute's CRC Unit
from original typescripts and illustrations
provided by the authors

Printed and made in England by
Henry Ling Ltd, The Dorset Press, Dorchester

Tribology of materials

W H ROBERTS

The author is Manager of the National Centre of Tribology, United Kingdom Atomic Energy Authority, Risley.

SYNOPSIS

The correct selection of materials, incorporated in a sound design, is an essential prerequisite for achieving reliability, predictable performance and economically acceptable lifetimes for such mechanical engineering components (tribo-components) as bearings, seals and gears. Lubricants, whether wet or dry, are also "materials" to be considered as crucial elements in any "tribo-system".

The paper briefly recapitulates on some tribology fundamentals, covering wear, friction, lubrication (including boundary lubrication); and indicates how these are encountered in real (engineering) situations. Types of wear and the importance of environmental factors in determining tribological performance are discussed. Lubrication in the "difficult" environments of nuclear power reactors and spacecraft is mentioned. The concept of "surface engineering" in tribology is introduced and, an outline given of a recently completed, £1 million per annum, industry DTI co-sponsored 3-year project on plasma-assisted coating technology (PACT). Finally, mention is made of a new DTI/industry 3-year programme, structured around eight generic research projects aimed at assisting industry in utilising modern (and established) tribological surface modification techniques to good effect.

INTRODUCTION

In tribology, "the science and technology of interacting surfaces in relative motion", the correct selection of materials incorporated in a sound design is an essential prerequisite for achieving reliability, predictable performance and economically acceptable lifetimes for mechanical engineering components, such as bearings, gears and seals ("tribo-components"). Note that lubricants, wet or dry, are also tribo-materials, and therefore crucial elements in any tribo-system.

In this paper some aspects of tribology fundamentals - wear, friction, lubrication (including boundary lubrication) will be introduced, and related to real (engineering) situations. The materials approach to solving some practical (engineering) problems will be discussed - including self-lubricating materials (especially polymers), dry lubrication and also lubrication in "difficult environments" such as in nuclear power reactors and in spacecraft. Finally, the concept of "surface engineering" will be introduced and mention made of two substantial "surface coatings/surface treatments" projects, co-sponsored by industry and DTI, and managed by the National Centre of Tribology, at Risley.

SURFACES

However smooth or polished they may appear, all surfaces on a microscopic scale are rough. Thus when two surfaces are brought together they will contact at surface high spots (so-called "asperities"). The number and sizes of these real areas of contact depend on the applied load (Fig.1). Relative lateral movement of the surfaces will lead to asperity interaction and in general to disruption and degradation of the surface. In engineering terms this will manifest itself as surface damage and wear. The asperity interaction will also give rise to a resistance to relative movement of the surfaces, hence friction.

THE LUBRICANT

The function of a lubricant, be it solid, liquid or indeed gaseous, interposed between the two juxtaposed surfaces, is to minimise interaction between the opposing asperities, and hence reduce wear and friction. If the lubricant film thickness is much greater than the combined roughnesses of the two surfaces ($h \gg R_a$) (Fig.2), then we have the situation of hydrodynamic, or "full film", lubrication giving "zero" wear of the two bearing materials, and a frictional resistance to relative motion which is a function of the viscous shear properties of the lubricant film.

In practice, "zero" wear can be achieved provided that system conditions exclude foreign bodies from bridging the lubricant gap, and that shock loads or vibration do not give rise to intermittent (damaging) asperity contact between the surfaces. Thus the reliability and predictable performance of such tribo-components as gears, bearings, cams, tappets and seals is largely determined by:

(a) suitable choice of lubricant, (conventionally mineral oil) and especially selecting the correct oil viscosity, and

(b) having in the oil necessary additives, such as oxidation and corrosion inhibitors, and extreme pressure (EP) (strictly, extreme temperature) additives which help the surfaces to survive potentially damaging asperity encounters.

Conventional bearing materials in such "ideal" situations cover a wide range of alloys - copper alloys (brass, bronzes); cast irons; steels; soft bearing alloys, (lead, tin) - as well as polymers.

BOUNDARY LUBRICATION

When the separation of the surfaces is not effective ($h \rightarrow 0$) (Fig.2) we have the condition of boundary lubrication where both friction and wear are significantly higher than for full film lubrication. It will be readily apparent that there can be complex chemical interactions between surfaces and lubricant (and also lubricant additives), as well as the environment. This aspect of tribology has been, and continues to be, the subject of much research [1, 2].

Figure 3 shows that a 7-fold variation in friction coefficient for a bronze-steel combination in sliding contact, under different regimes of boundary lubrication leads to a very large (some 4 orders of magnitude) variation in the wear rate of the bronze.

WEAR

Wear is a complex process and material selection plays an important role in wear control[3]. In a practical, engineering system the variables in the wear process are many. They include (Fig.4) not only the lubricant, as already mentioned, but also temperature, load, motion, relative velocity and, most importantly, environment.

Types of Wear

Several types of wear have been categorised - scuffing, adhesive, abrasive, corrosive, surface fatigue, fretting and erosion. In practice, more than one type will be operative, with one mode predominating.

It is probably true that abrasive wear accounts for more than half the wear encountered in industrial situations. Mineral extraction is an obvious example of abrasive wear - the cutting teeth of a digging tool are subject to "two-body" abrasive wear. "Three-body" abrasive wear arises when the abrasive is trapped between two contacting surfaces. In such cases, the solution is primarily a design (sealing) solution; that is, taking steps to exclude abrasive wear particles from bearings and gear boxes so as to reduce the chances of premature failure.

The solution to corrosive wear problems, for example in offshore plant and equipment [4], is not only efficient sealing (ie design), but also choice of suitable corrosion-resistant bearing materials, and also water-resistant lubricants (including greases).

Interactive Effects

The data in Table 1 show the million-fold spread of wear rates, coupled with a 7.5-fold variation in friction coefficient for six materials combinations, in dry sliding, in air, at ambient temperature. Such data illustrates the importance of material selection in determining wear rates, even under "benign" operating conditions.

Temperature/speed

Figure 5 shows, in simplified form, how the breakdown of a beneficial surface film can lead to dramatic changes in adhesive wear rate (surface damage). The resultant wear rate will be determined by the relative rate of formation of the beneficial film, and the rate of destruction/ attrition of the film by the wear process. A balance of rate processes of this kind is a feature of boundary lubrication.

In practice the "mild" (oxidative) and "severe" adhesive wear rate of some steels can vary by a factor of 100, depending upon load and speed (Fig.6). On the right hand side of the figure, oxidative (mild) wear has been re-established. The conditions for such mild/severe wear transitions with steels will depend upon their composition. Surface treatments, such as nitriding, can also suppress mild-to-severe wear transitions in steels [5]. Figure 7 shows the mild/severe wear behaviour of brass sliding on steel (in dry carbon dioxide gas) at elevated temperatures up to 300°C. At 300°C mild wear increases progressively with temperature (probably due to softening), but there is no transition to "severe" wear due to the formation of beneficial oxide films.

Figure 8 illustrates the dramatic combined effects of two variables, namely temperature and sliding speed, on the wear of brass. A large variation in wear rate is exhibited (x250), with lowest wear at 400°C.

Environment

From what has been said, it is not surprising that different materials combinations in the same environment give different wear behaviour. Conversely, the same material combination in different environments also gives widely different wear rates [6]. Figures 9a and 9b indicate the sharp difference in magnitude and in variation of wear with temperature for the same materials in two different gaseous environments - air and carbon dioxide respectively [7].

For components working immersed in high temperature molten sodium, surface films formed by reaction of elemental constituents, such as molybdenum, aluminium and chromium in bearing alloys, with the liquid metal (and also any oxygen impurity) can provide effective boundary lubrication [8].

DRY LUBRICATION

Dry and self-lubricating materials can be categorised under three headings:

Lamellar solids: of which graphite and molybdenum disulphide (MoS_2) are probably best known.

Materials with low adhesive forces: a category which embraces polymers as a class of materials.

Soft materials: that is, low shear strength solids, of which lead is of great importance in the lubrication of bearings in spacecraft.

Self-lubricating Materials

The wear rates of polymers sliding on smooth steel counterfaces can vary by over 4 orders of magnitude; with several orders of magnitude variation even for a given class of polymer. The data in Figure 10 have been extracted from NCT's Polymer Bearing Design Guide which will be re-issued, in expanded form, in a new Design Data Item by the Engineering Sciences Data Unit (ESDU) in November 1987, as Item 87007. Polymers have a wide range of application as bearing materials in industry; they generally complement those for ceramics and hard metals (Table 2).

Three well-known bioengineering applications for polymers, operating under boundary-lubricated conditions, are hip and knee joint prostheses utilizing ultra-high molecular weight polyethylene, with sinovial fluid giving "marginal" lubrication; and total replacement heart valves in polypropylene (where the lubricant is blood).

Lubrication in Vacuum

Reliable, maintenance-free, operation of commercial satellites (for up to 15 years), depends upon effective lubrication of moving parts [9]. These include de-spin mechanisms, solar array drive mechanisms and antenna pointing mechanisms. Surface films of lead, of less than 1 micron thickness, applied by vacuum sputtering, have been shown to be extremely effective and in NCT's European Space Tribology Laboratory (ESTL), applying such films to bearings for spacecraft use is a "space approved" process [10].

The recent successful encounter of the GIOTTO space probe with Halley's comet (13/14 March 1986) owed its success in large part to the lead film lubrication of the antenna de-spin mechanism bearings. The probe was launched some 9 months before the encounter and the bearings are still operating successfully. The requirement here was for a steady bearing torque (friction), over a wide temperature range of -55°C to +80°C [11].

SURFACE ENGINEERING

Whilst improved plastics and ceramics in bulk form will almost certainly continue to have an important role in tribology, what cannot be doubted is that the most significant trend in the technology is that towards "designing" engineering component surfaces so as to achieve the requisite enhancement of friction and anti-wear properties over those of the bulk (structural) substrate material. Thus we have the concept of "surface engineering". In addition to the well-established localised thermal hardening techniques (induction, flame, spark, and electron beam), laser heating has emerged as a significant process both for transformation hardening and for alloying. Current developments of thermochemical treatments involving carbon and/or nitrogen are in the direction of utilising plasma-assisted techniques [12]. Modifying surfaces by implanting energetic ions into them (ion implantation) is moving out of the laboratory into industrial applications [13].

Surface engineering is concerned both with surface modification and coating treatments, singly and in combination ("hybrid" systems). Just as importantly, it is also concerned with the substrate so that the benefits of the surface modifications can be fully exploited.

It is not appropriate to go into detail on the very extensive range of surface coatings and treatments which are available to improve surface tribological properties. The recently issued HMSO [14] and ESDU [15] guides are excellent documents which also cover the process of selecting surface treatments and coatings. The effective depth of treatment/coating can vary between several millimetres for weld overlays; down to thicknesses of only a few microns for vapour deposited films (PVD and CVD); and of 0.5 micron, and less, for ion implantation. Hardness of PVD and CVD coatings are very high, in the range 1500-4000HV. Transmission electron microscope (TEM) studies of such films indicate that these high hardness values are associated with dense dislocation networks [16].

PLASMA-ASSISTED COATING TECHNOLOGY (PACT)

It is well-established that TiN coatings applied by (PVD) to high-speed steel drills of conventional geometry give significant improvements in performance (metal cutting rates and life) [17, 18]. However, such wear-resistant coatings have potential for much wider application in engineering. A 3-year project aimed at introducing the new technology of ion plating and sputter ion plating to many parts of industry for which plasma-assisted coatings could be of benefit was recently completed. This £1 million per annum, industry/DTI co-sponsored project, managed and co-ordinated by the National Centre of Tribology, provided scientific support to the 110 companies involved, which covered a broad spectrum of engineering. The undoubted attraction of the project for industry, and its success, stemmed in large measure from the effective transfer of the technology. NCT established a close working relationship with industry - both coaters and the potential users of coatings - and injected scientific resources into the programme, guiding companies into the new technologies and providing assistance in technological assessment.

COATING AND SURFACING TECHNOLOGY (CAST)

As a result of the success of the previous collaborative PACT project, NCT was encouraged by DTI to seek industrial support and finance for a new, more broadly based, 3-year generic research programme on coating and surfacing technology (CAST) for tribological applications. This programme is now underway.

With experience gained from the PACT programme attention was given to the most effective ways of meeting the likely demands of the large number of potential users in many different areas of technology. As a result the new programme has been structured around eight research projects aimed at assisting industry in utilising to good effect a wide range of modern (and established) surface modification techniques (Table 3). The eight groups are:

1. Friction Control Through Surface Engineering.
2. Reducing Wear in High-Stress Abrasive and Corrosive Environments.
3. Reducing Wear in Low-Stress Abrasive Environments.
4. Optimisation of Substrate Heat Treatment for PVD Coating.
5. Improvements in Tooling for Difficult-to-Handle Materials.
6. Surface Treatments and Coatings for Reducing Friction and Wear at Elevated Temperatures.
7. Research and Development of Quality Control Techniques and Specifications for PVD Coatings.
8. Preservation of Sharp Edges.

Each group will be concerned with researching a specific area of tribology relevant to the interests of its membership. The underlying theme in all the group projects is that the research should result in surface modification processes being utilised industrially in innovative ways in order to secure scientific and, in turn, economic and technical advantages in engineering design, manufacturing, production and equipment operation. In particular, it is necessary to understand the scientific aspects of new (and some existing) technology coatings and be able to predict the mechanistic processes involved in their degradation under envisaged operating conditions. This will lead to a rational basis for the selection of coatings and surface treatment for specific technological applications.

CONCLUSION

There is no one, universal, material or coating or treatment that will solve all tribological problems in engineering. Practical situations involve materials interactions under a wide range of conditions which directly influence and determine the system tribology. The materials interactions within a "tribo-system" are generally complex. Only by a proper understanding of the processes involved can we hope to optimise solutions and achieve desired friction and wear parameters, and hence the aimed-for performance and reliability of operation of tribo-components. In practice, materials solutions are, almost inevitably, compromises (very often because of economic considerations). However, with advances in materials science and the technology of materials processing to give surface modification treatments and coatings, singly or in combination, the future for achieving significant improvement in component tribological performance looks both promising and exciting.

REFERENCES

1. CAMERON, A. Lubricant chemistry and tribology chemistry - boundary and extreme pressure lubrication. Proc. I.Mech.E. Intl. Conf. Tribology - Friction, Lubrication and Wear Fifty Years On, Vol. 1 (1987), 355-364.

2. LANSDOWN, A.R. Lubricants. idem, 365-378.

3. WEAR CONTROL HANDBOOK. Eds. M B Peterson and W O Winer, (1980). ASME, New York.

4. HALL, P. Tribology problems offshore. Proc. I.Mech.E. Seminar Tribology Offshore, (1984), 1-4.

5. FARROW, M. and GLEAVE, C. Wear resistant coatings. Trans. Inst. Metal Finishing, Vol.62, Part 2, (1984), 74-80.

6. ROBERTS, W.H. Tribology in nuclear power generation. Tribology Intl., Vol.14, No.1 (1981), 17-28.

7. CLARK, W.T., PRITCHARD, C. and MIDGLEY, J.W. Mild wear of unlubricated hard steels in air and carbon dioxide. I.Mech.E Proceedings, Vol.182, Part 3N, (1967/68), 97-106.

8. CAMPBELL, C.S. and LEWIS, M.W.J. Some aspects of the tribological behaviour of materials in sodium. Proc. 2nd Intl. Conf. Liqud Metal Technol. in Energy Prodn., (1980), 3/27-34.

9. ROBERTS, W.H. Some recent trends in tribology in the UK and Europe. Tribology Intl., Vol.19, No.6, (1985), 295-311.

10. TODD, M.J. and ROBBINS, E.J. Ion-plated lead as a film lubricant for bearings in vacuum. In "Selected ESTL papers on ball bearings for satellites", ESA TRIB/1, (1980), 1-13. European Space Agency, Paris.

11. TODD, M.J. and PARKER, K. Giotto's antenna de-spin mechanism: its lubrication and thermal vacuum performance. Proc. 21st Aerospace Mechs. Symp., NASA Conf. Publn. 2470, (1987), 295-314.

12. BELL, T. and DEARNLEY, P.A. Plasma surface engineering. Proc. IFHT Intl. Seminar Plasma Heat Treatments, (1987), 13-53.

13. DEARNALEY, G. Ion implantation and ion assisted coatings for wear resistance. Surface Engineering, Vol.2, No.3, (1986), 213-221.

14. DEPARTMENT OF TRADE AND INDUSTRY. Wear resistant surfaces in engineering (a guide to their production, properties and selection), (1986). HMSO, London.

15. ESDU DESIGN ITEM 86040. Selection of surface treatments and coatings for combating wear of load-bearing surfaces, (1986). Engineering Sciences Data Unit, London.

16. STEVENS, K.T. and DOUGLAS, A. Physical vapour deposition coatings - their properties and potential applications. Proc. I.Mech.E Intl. Conf. Tribology - Friction, Lubrication and Wear Fifty Years On, Vol.2, (1987), 629-636.

17. MATTHEWS, A. Titanium nitride PVD coating technology. Surface Engineering, Vol.1, No.2, (1985), 93-104.

18. THOMAS, A and THOMSON, P. Performance and wear of TiN coated twist drills. Proc. 6th Intl. Conf. Ion and Plasma Assisted Techniques (IPAT '87), (1987), 201-206.

TABLE 1

WEAR AND FRICTION

	Friction Coefficient	Wear Factor
PTFE vs Steel	0.08	10^6
Steel vs Steel	0.30	10^5
Brass vs Steel	0.60	10^4
Stellite vs Steel	0.60	10^3
Chrome - Plate vs Steel	0.40	10^2
Tungsten Carbide vs Itself	0.35	1

TABLE 2

CONDITIONS FAVOURING USE OF PARTICULAR TYPES OF NON-METALLIC BEARING MATERIALS

PROBLEM AREA	BEST CANDIDATE	
	PLASTICS	CERAMICS OR HARD METALS
Dry Operation	X	X
Marginal Lubrication with Oils or grease	X	
Lubrication with Water, Fuel, etc	X	X
Very high Temperatures		X
Very low Temperatures	X	X
Conditions of Possible Misalignment	X	
Bearings with High Stiffness and Dimensional Stability		X
Corrosive Fluids	X	X
Application where Damage to Journal must be Avoided in the Event of Lubricant Interruption	X	
High Wear Resistance		X
Bearings Resistant to Shock or Vibration	X	

TABLE 3

COATING AND SURFACING TECHNOLOGY (CAST)

THE SURFACE MODIFICATION PROCESSES TO BE INCORPORATED IN THE PROGRAMME

Physical vapour deposition (PVD)

Chemical vapour deposition (CVD)

Ion implantation

Thermochemical diffusion treatments including plasma-assisted processes.

Chemical and electrochemical processes.

Welding, metal spraying, plasma and high-velocity coating techniques.

Friction surfacing.

Laser and electron beam modification techniques.

Dip and polymer coating processes.

Mechanical surface modifications and phase transformation processes.

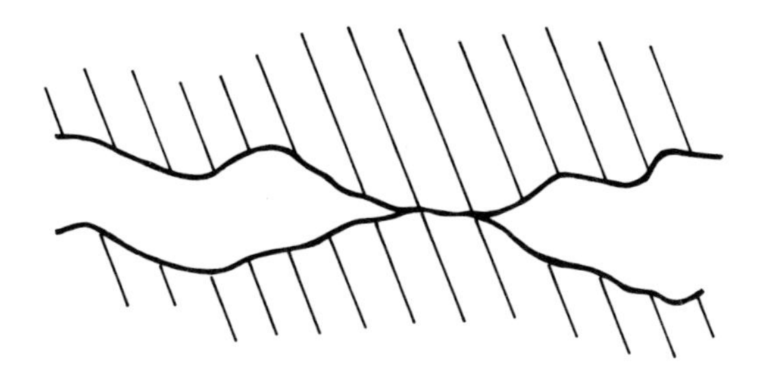

Fig.1: Schematic of an asperity contact.

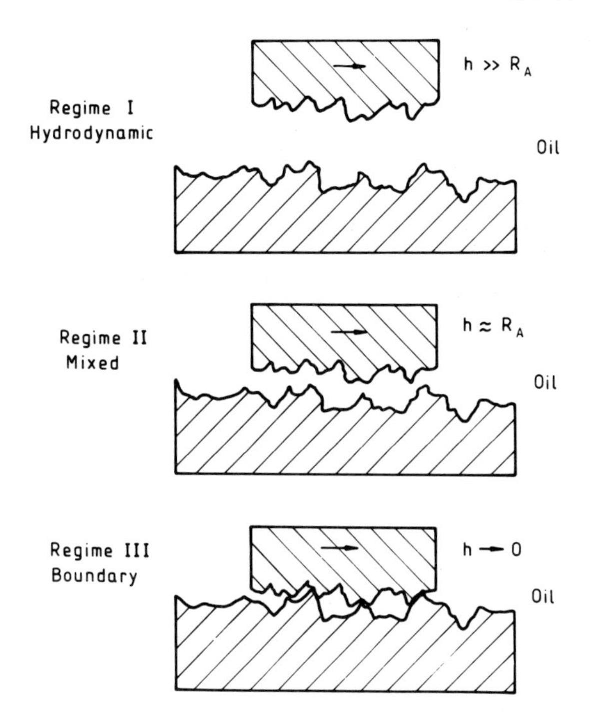

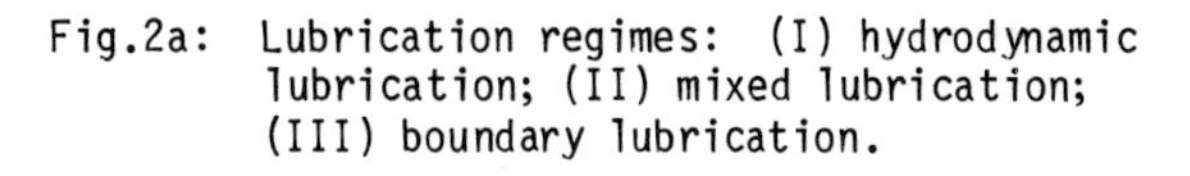

Fig.2a: Lubrication regimes: (I) hydrodynamic lubrication; (II) mixed lubrication; (III) boundary lubrication.

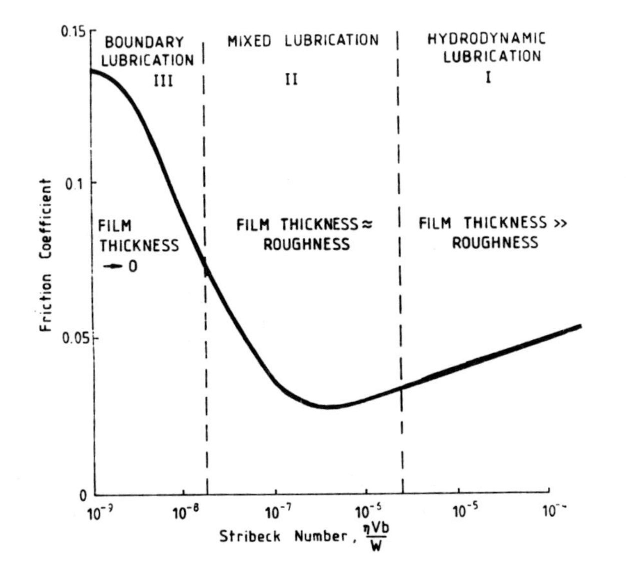

Fig.2b: Idealised Stribeck curve showing friction coefficient as a function of (viscosity x velocity/specific loading)

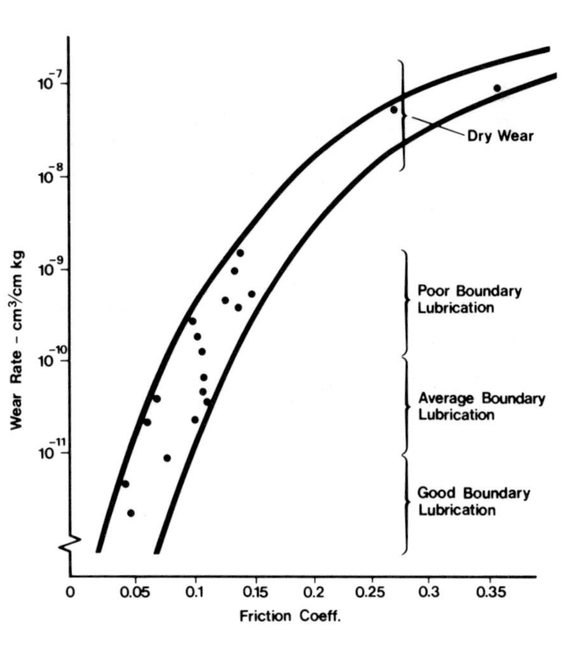

Fig.3: Experimental results for bronze under various environmental conditions.

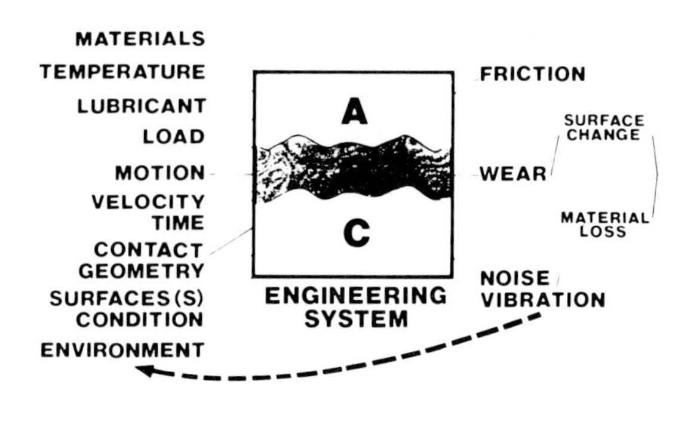

Fig.4: Variables in the wear process.

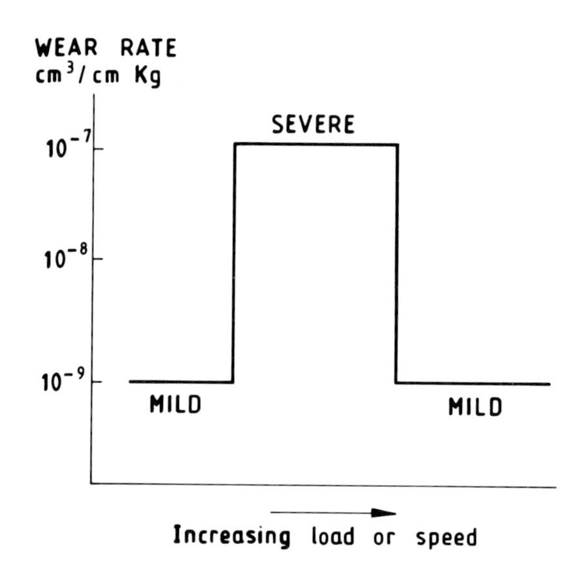

Fig.6: Adhesive wear: mild-to-severe transitions (low- or medium-carbon steels).

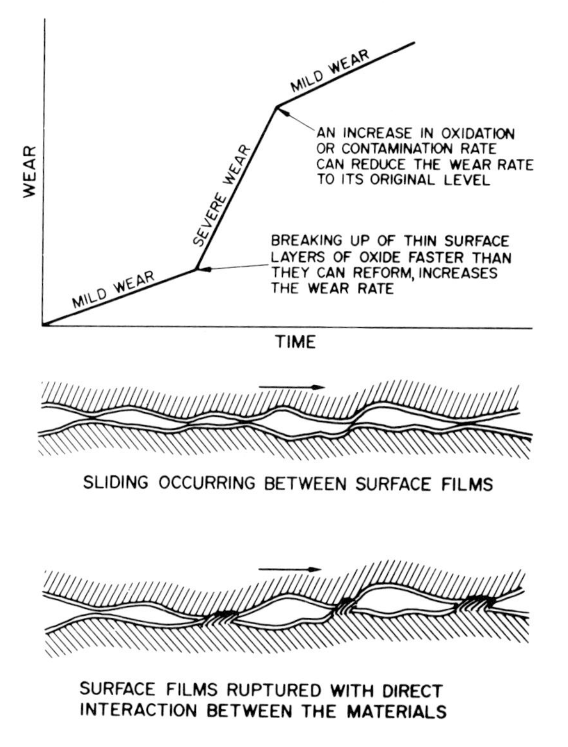

Fig.5: A simplified picture of adhesive wear.

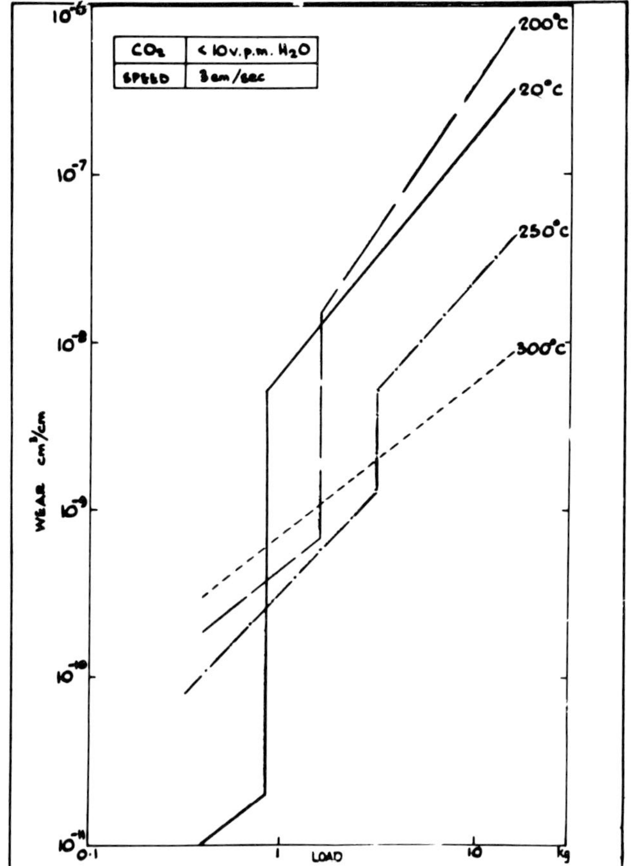

Fig.7: Mild-to-severe wear transitions for brass sliding on steel in dry carbon dioxide.

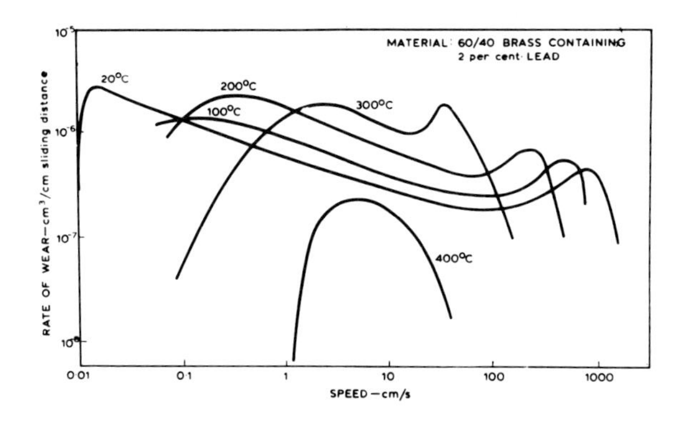

Fig.8: Wear of brass in air as a function of speed and temperature.

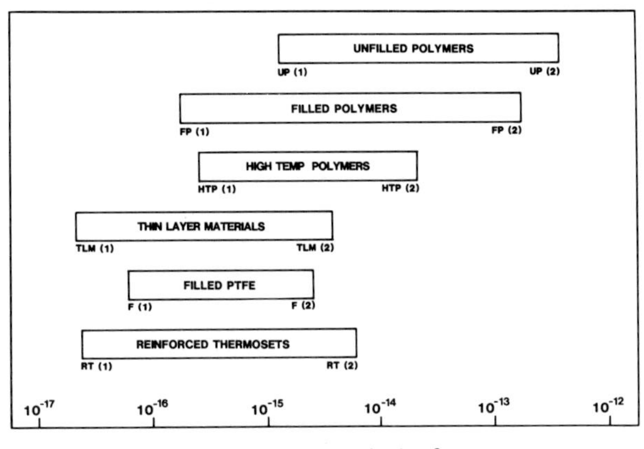

Fig.10: Limiting room temperature specific wear rates for 6 classes of polymer (moderate bearing pressures).

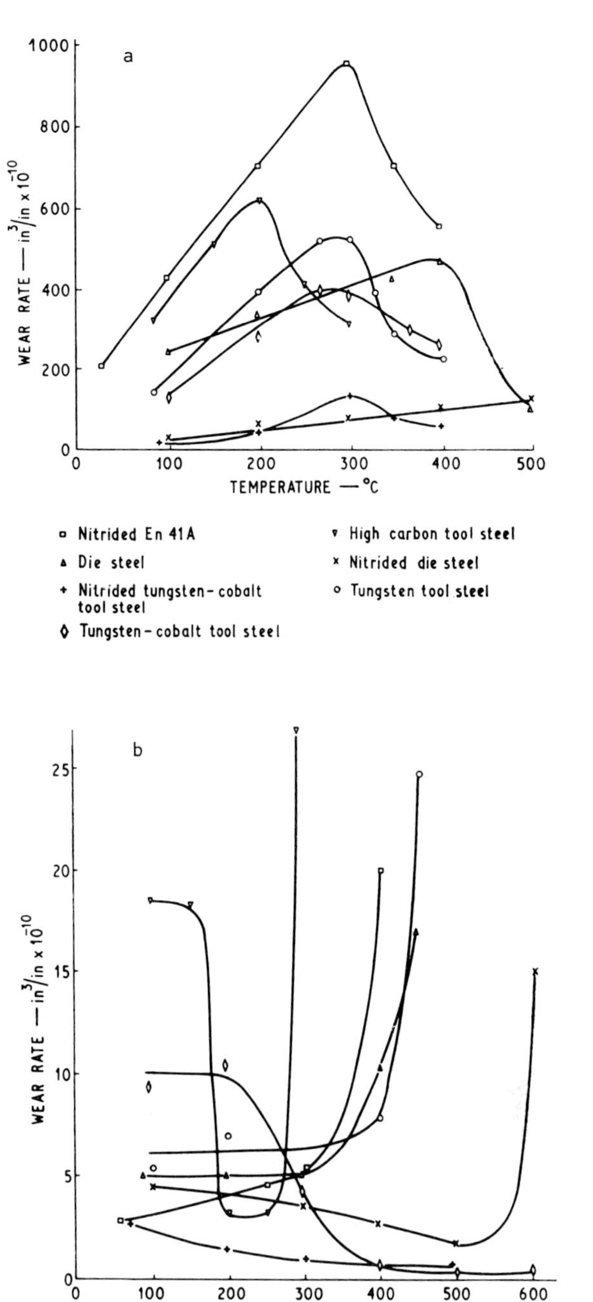

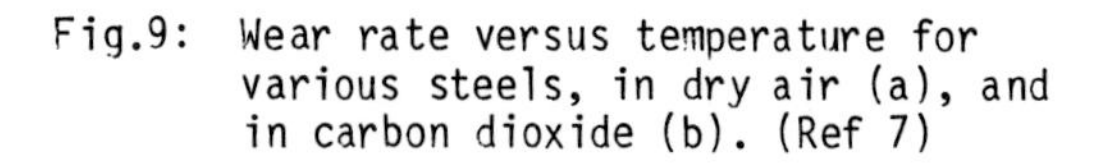

Fig.9: Wear rate versus temperature for various steels, in dry air (a), and in carbon dioxide (b). (Ref 7)

Plasma spraying of ceramic coatings — application to thermal barriers, wear protection and friction

P FAUCHAIS, A VARDELLE, M VARDELLE, J F COUDERT and D BERNARD

The authors are with the University of Limoges, Limoges, France.

SYNOPSIS

This paper attempts to summarize the authors' actual knowledge of the parameters influencing heat and momentum transfer to solid particles introduced in d.c. plasma jets: particle size and injection velocity distribution, plasma gas nature and flowrate, torch design, and particle manufacturing process. This study was carried out both by modelling plasma flow with particles in flight and by measuring the velocity and surface temperature of the particles. The thermophysical properties of some coatings (thermal barriers, wear resistant coatings) were correlated to the molten state and velocity of the particles upon impact together with the substrate and coating cooling while spraying.

INTRODUCTION

The plasma spraying process consists of introducing solid particles into a d.c. plasma jet in order to accelerate and, if possible, melt them to form a coating with the splats resulting in a lamellar structure. The coatings consist of many layers of overlapping thin lenticular particles or splats and almost any material that can be melted without decomposing or evaporating can be used to form the coatings.[1-5] Of all the materials, ceramics seem to have the most promising applications because such coatings can deeply modify the surface properties of metal or plastic materials with their high hardness, strength, wear resistance, thermal insulation, and corrosion resistance. In general ceramic materials have a high melting point and a poor thermal conductivity. On one hand it is necessary to use plasma jets with a high enthalpy and heat transfer capacity (i.e. Ar–H_2 plasma: Ar for momentum transfer and H_2 for heat transfer) to melt the particles and also to design the plasma generator (nozzle, plasma gas injection conditions) in such a way that the longest plasma jets are obtained in order to improve the residence time of the particles.[6-8] But on the other hand, ceramic materials have a low thermal conductivity which, with the high heat transfer from the plasma, results in heat propagation phenomena (the particle surface is melted and even evaporates while the central part is still almost at room temperature). These heat propagation phenomena are enhanced when particulates are manufactured by agglomeration of small particles which are then insufficiently sintered. It is thus of primary importance to choose carefully the powders to be sprayed first for their morphology, but also for their size distribution, that is, with their injection velocity distribution, which is of primary importance for their melting and velocity distributions upon impact. A good coating is generally obtained with particles completely melted and with a sufficient velocity upon impact to accommodate the surface roughness. The impact velocity, as well as the mean temperature and diameter of the particles, seem to control the contact area between the resulting lamellae and the substrate or the previously deposited and solidified layers, i.e. the thermal and mechanical properties of the deposits. These properties depend also on the temperature gradients inside the layers, i.e. on one hand on the powder flowrate (which, if too high, reduces the heat transfer to the particles) and relative velocity between substrate and plasma torch, and on the other hand on the ways substrate and deposits are cooled down while spraying.

This paper presents
- the techniques used to measure the different parameters: plasma jet temperatures and velocities, surface temperature and trajectory distributions of the particles in plasma jet
- the models used to calculate the heat and momentum transfer between the plasma and the particles, taking into account their size and injection velocity distributions
- the influence of the particle morphology on their melting
- the importance of the torch design on the plasma jet lengths and the modifications of these jets by the induced pumping of the surrounding atmosphere as well as the possible chemical reactions of the particles with the pumped air during their flight
- the influence of deposit cooling while spraying

and the importance of these parameters will be illustrated by their influence on the thermal and mechanical properties of coatings used as electrical barriers, thermal barriers (stabilized zirconia), and protection against wear (WC–Co).

COATING FORMATION

The formation of protective coatings by spraying a stream of molten metal or ceramic particles was first developed using combustion flames into which the spray material was fed as powder, wire, or rod. In the 1960s commercial plasma spraying equipment became available in which a d.c. plasma jet was used to melt a powder feed and project the droplets at high velocity against the material to be coated. The powder, directed at a substrate (surface to be coated), forms on impact a coating consisting of many layers of overlapping thin lenticular particles or splats. Almost any material that can be melted without decomposing can be used to form the coating. The major advantage over the flame spraying process is the higher particle velocity obtainable (up to 500 m/s) and the high temperatures achieved in the plasma jet (up to 15000 K) that makes it possible to melt even the most refractory material to produce high quality coatings. Plasma spraying is therefore particularly suitable for the formation of ceramic coatings for wear and thermal and corrosion protection.

In the commercial spraying process, the mean mass flow rate of sprayed powder is about 1–2 kg/h for the 30–50 kW range plasma generators. This flowrate, for the 30μm diameter particles, corresponds to about 10^8 part/m², accounting for the mean velocity of the relative particle-target displacement. During the freezing time of a particle (10^{-7}–10^{-6} s) about 50 to 100 particles arrive in one square meter. During freezing, assuming the ratio of the flattened particle to its initial diameter is six, a fraction of 10^{-6} to 10^{-7} of the substrate is covered. The arrival of the next particle on the same spot takes about 0.01 s, i.e. 10^4 times the freezing time. Therefore the heat content removal from one particle will not be influenced by a new incoming particle. Thus the coating is built up particle after particle and consists of successive layers of material.[1]

Heat transfer calculations show that the freezing of particles occurs in a few microseconds and the complete cooling amounts to 100 μs. The zone of thermal effect in the underlying material is quite small and the temperature gradients reach 10^5 K.cm^{-1}. The coating has a sandwichlike structure, as shown in Fig. 1, and its properties are regarded as resulting from deformation and solidification processes of individual particles and their interaction on contact. The wetting and flow properties of the liquid droplets are of first importance since they will influence porosity of the coating and the substrate interface.

The flow and solidification of molten particles on impact with surfaces is difficult to treat theoretically due to the interaction between heat transfer and crystal growth. The very high velocity at which the various processes occur makes experimental observation extremely difficult. However, as these processes control the quality of coating, numerous studies have been devoted to them.[9–14]

A very recent development by Houben[15] takes into account the propagation of a shock wave into the flattening particles and shows that very high pressures could be reached inside them. A critical radius of the deforming particle was introduced resulting in a lateral outflow of material escaping from the pressurized zone.

The greater the critical radius, the more potential energy will be stored in the pressurized zone, thus providing an explosive situation. These calculations are illustrated by examples with molybdenum particles splatted onto the steel substrate with three types of flattened particles:

- pancake type (Fig. 2a), of rather small particles with not too high velocities for which the adherence seems good over the whole contact surface with microcracks to release stresses when the particle freezes
- flower type (Fig. 2b), corresponding to a central zone adhering well to the substrate within an erosion crater due to the beginning of the lateral flow and with lateral free standing material exhibiting no cracks (the cooling down is much lower due to the absence of contact with the substrate)
- exploded type particles (Fig. 2c), corresponding to overheated particles.

COATING PROPERTIES

The extremely rapid cooling rate from the liquid state, characteristic of plasma sprayed material, may result in the formation of metastable phases or suppress crystallization. The classic example of this effect is provided by alumina coatings which predominantly consist of one or more of the many metastable forms (eta, gamma, delta, theta) rather than α–Al_2O_3, the only stable structure.[13]

The residual stresses in coatings may be divided into microstresses within individual particles and macrostresses within the coating as a whole. Microstresses arise because of restraint due to the thermal contraction of individual particles as they cool in the solid state relative to the underlying material which remains at a constant temperature. These stresses will therefore depend largely upon the expansion coefficient of the coating material and the elastic constants of coating and substrate. They would also be expected to be influenced at the interface by the yield strength of the substrate, plastic deformation of which would allow stress relaxation, and the effectiveness of the particle–substrate bond.

Macroscopic residual stresses will arise after the cooling of coated material to ambient temperature due to the difference in thermal expansion coefficients between coating and substrate and the presence of temperature gradients during coating formation.[1,16–18] High interfacial stresses may lead to peeling of the coating, particularly on smooth substrates, while high tensile stresses may lead to cracking in the coating itself.

The nature of coating formation by impact and solidification of separate droplets results in some porosity which generally lies in the 5–20% range and depends both on spraying conditions and the substrate material. All the particles have the same mean direction when arriving at the substrate[3] and a shadow effect occurs, causing porosity (Fig. 3c). Narrow holes cannot be filled out (Fig. 3b). The valleys in the surface of the substrate or the lamellae can include some air or gas (Fig. 3c) causing formation of pores or adherence failures. The particle may burst during the process of flattening to a lamella (Fig. 3d).

Mechanical anchorage or interlocking is considered to be the most important mechanism for ceramics. Grit blasting provides an ideal surface topology for interlocking and bond strength increases with increasing surface roughness in both shear and tensile tests. This is mainly due, as previously mentioned, to a better contact of the particles with the peaks of the rough surface, the lamella following the peak contour and the contact being greatly improved due to the contraction of the lamella when cooling (Fig. 4).

The adhesion is of course better if the particles have a good plasticity or a low viscosity when they are liquid, good wettability, high velovity upon impact, and a high expansion coefficient. It is also of primary importance to avoid any formation of oxide at the surface, i.e. spraying must be performed a few minutes after sand blasting.[19] The adhesion limit lies between 30 and 50 MPa and depends strongly on surface roughness.[20] The sand blasting of the substrate surface is also a good method of reducing internal stresses by dividing them into smaller components.[20]

Thus the mechanical, electrical, and thermo-physical properties of the deposits depend on the properties of the starting material, the powder, and the end product, i.e. the micro-structure of the deposit. The different defects encountered in the sprayed layers are:[21]

- adhesion and cohesion defects due to successive passages of the torch with incompletely melted particles travelling at the periphery of the jet
- stratification of the coating due to the cracks relaxing the stresses formed at high temperature gradients, especially in ceramic coatings
- microcracks in the lamellae
- open porosities at the surface of the deposit due to bad contacts between the lamellae, or gas inclusion, or poor accommodation of the surface roughness by insufficiently melted particles
- cracks orthogonal to the substrate due to residual tensile stresses acting mainly near the surface of the deposit
- gas inclusions between the lamellae with thicknesses between 0.01 and 0.1 μm[22] and resulting from poor contact between the lamellae
- inclusion of non-melted particles.

This is the result of the different thermal histories undertaken by the particles, of their trajectories through zones of different chemical compositions, and of their different kinetic energies upon impact. The coatings present non-homogeneous grain sizes, crystalline phases, and chemical composition and it is of primary importance to control the momentum and heat transfers between the plasma and the particles on which their velocity and temperature distributions upon impact depend.

MEASUREMENT TECHNIQUES FOR PLASMAS AND PARTICLES IN FLIGHT

In spite of an intensive research effort that has been devoted over the last thirty years to the study of electric arcs, mathematical modelling has developed slowly due to difficulties encountered in the analytical description of the electrode regions.[23] It is still impossible to predict temperature and velocity fields starting from the nozzle dimensions and shape, gas flowrate and nature, arc current intensity and voltage, and thermal efficiency and initial velocity and temperature distributions at the nozzle exit must be assumed. Moreover, experiments[24] show that for given plasma parameters and nozzle design, the design of the arc chamber, and especially the way the gas is introduced into the arc chamber, plays an important role in deciding the length of the isotherms. The nozzle design is also very important for the turbulent mixing of the plasma jet with the surrounding air.[28] Thus, to characterize a plasma torch it is of primary importance to measure the flow velocity and temperature fields, and not only the power level and thermal efficiency.

Although there are certain similarities between ordinary fluids and thermal plasmas, heat and momentum transfers become much more complex due to the presence of charged particles, steep gradients, modified viscous drag, and heat and mass transfer coefficients due to strongly varying plasma properties, vaporization and evaporation, non-continuum effects, radiation, heat propagation inside the particle, particle shape, or a combination of these effects.[25]

The fact that particles have size and injection velocity distributions must also be accounted for and it is thus very difficult to predict what will be the temperature and velocity of particles upon impact on a substrate. Here also measurements are needed.

Plasma jet measurements

All the temperature measurements are performed by using emission spectroscopy and the steep gradients of the emitting plasma jet are accounted for by assuming an axisymmetrical plasma jet and performing Abel's inversion, on line if possible.[26]

Even if equilibrium is generally assumed in thermal plasma jets flowing at atmospheric pressure, much care has to be taken in the choice of thermometers.[27]

Of course it is far easier to choose atomic lines for temperature measurement, especially when working with large slits of the monochromator (intensity measurement with no wavelength scanning), however care is necessary. If the absolute intensity is used equilibrium has to be assumed and the measured population of the excited state can be related to temperature. But a plasma jet flowing in air pumps it very rapidly and thus the composition of the starting plasma is continuously changed. It is very difficult to account for this pumping. For example, Fig. 5[30] shows the argon concentration axial profiles along the axis of a DC plasma torch working with two arc current intensities

(450 and 600 A) and two gas flowrates (23.6 and 47.2 l(STP)/mn). The fast pumping of the surrounding air (more than 50% at 30 mm from the nozzle exit) is enhanced when the current intensity as well as the gas flowrate are increased.

If the intensity ratio of two atomic lines is used, the obtained excitation temperature is not necessarily equal to that of the heavy species (molecules, ions, atoms) which control momentum and heat transfers to the particles.[30] If the equality can be assumed with confidence in the plasma core ($n_e > 10^{22}$ e/cm^3) it is quite different in the fringes or in the plume of the plasma.[26] Moreover, the temperature deduced from atomic lines cannot be obtained below 8000 K due to the fast decrease of emission coefficients below these temperatures.

Thus it is better to use molecular spectra to measure gas temperature from the rotational lines of a well isolated band transition.[26] However it is more complex because about 30 to 40 rotational lines have to be considered simultaneously with the Abel's inversion performed for all of them. This necessitates for example the use of 2D optical multichannel analysers instead of photomultipliers, the data being treated by a computer.

In comparison to temperature, flow velocity measurements are rather poor at atmospheric pressure. Doppler shift cannot be used because with velocities below 1500 m/s the shift is far too small to be measured from lines enlarged by collisions and Stark effects.[26] Thus the only method used is laser anemometry of small ceramic particles (<3 µm) injected with the plasma gas in the arc chamber. But firstly the precision is low because of the weak signal in a region of very intense emission from the plasma and secondly because in the plasma core (T>8000 K) the mean free path of atoms and molecules being of the same order of magnitude as that of the solid particles, their velocity will always be lower than that of the flow.[26]

Figs. 6 and 7 show typical examples of such measurements for Ar–H_2 DC plasma jets. Fig. 7 shows temperature and velocity cartographies of and Ar–H_2 plasma jet at atmospheric pressure. The generator nozzle is a conical convergent followed by a cylindrical pipe 8 mm in diameter and 26 mm in length. The flowrates (STP conditions) were 75 l/mn of argon and 37 l/mn of H_2. It can be seen in this figure firstly that the radial gradients are very high (up to 4000 K/mm and 200 m/s/mm) and secondly that the length of the plasma jet is rather small (the zones where the temperatures are greater than 5000 K are shorter than 50 mm along the plasma jet axis). This is due to the fast pumping of the surrounding air in the plasma jet (see Fig. 5) which cools it very fast because of oxygen dissociation at about 4000 K.[31] With such gradients the particles introduced in the plasma jet will undertake very different thermal treatments depending on their trajectories and their heating rates as well as their accelerations will be high. Fig. 7 illustrates the influence of the power level increase when the current intensity is raised for a given flowrate and gas nature (Ar–H_2 mixture: 40 l/mn Ar, 8 l/mn H_2).[24] The diameter of the hot core of the plasma increases as well as the plasma velocity and momentum. In fact the increased pumping rate of air (Fig. 5) drastically reduces the length increase and a much longer plasma jet should be obtained if for example the pumped surrounding gas was not air but argon.

Measurements on particles in flight

Simultaneous measurements of velocity, number density, and surface temperature of solid particles (10<Ø<100 µm) in flight in DC spraying plasma jets have been developed at Limoges.

The particle concentration is measured in different cross-sections of the plasma jet by counting, during a given time, pulses resulting from light scattered by the particles passing through the focused point of a laser beam. The evolution of the maxima of these radial concentration curves corresponds to the most probable trajectory of the particles.

Along these trajectories the particle velocity and surface temperature are measured simultaneously by laser doppler velocimetry (LDV) and by discrete in-flight colour pyrometry.

In the LDV the measuring volume is in the order of 0.03 mm^3 and the method accuracy of 3%.[29] While for colour pyrometry, the measuring volume corresponds to a chord in the plasma flow of which a cross-section is 1.5 x 0.125 mm^2 and the accuracy about 10%.[30] It must be stressed that in the plasma core (T>8000 K) surface temperature measurements are impossible because the light emitted by the plasma enclosed in the measurement volume is 3 or 4 orders of magnitude higher than that emitted by the particle.

It must be borne in mind that such measurements give only statistical mean values resulting from a great number of particles with different sizes passing through the measurement volume. Of course, as already stressed, the momentum and temperature of the particles will depend on their trajectories in plasmas where gradients are very high. A given trajectory results from the particle mass and injection velocity. The powders are never unisized but have a more or less broad size distribution and acquire their momentum from a carrier gas.

For a given injector, nature, and gas flowrate, this carrier gas has a velocity distribution inside the injector (zero velocity at the wall). And finally, at the injector exit, the particle trajectory distribution is the product of the two distributions (particle size and carrier gas velocity)[31] and it is generally rather broad. For example, when using small alumina particles with a narrow size distribution (-21+15 µm) injected in the Ar–H_2 plasma jet characterized in Fig. 6, it can be seen[29] in Fig. 9 that the normalized trajectory distribution, measured in a cross-section of the plasma jet 20 mm downstream from the nozzle exit, is distributed in more than 1/3 of the hot plasma jet cross-section. Thus it is very easy to understand why the velocity distribution of these particles is rather broad, as shown in Fig. 10, which represents the evolution of the velocity radial distribution along the axis of the plasma jet characterized in Fig. 6. It is to be noted that even if the mean trajectory of the particles is not confused with the plasma jet axis, the maximum velocities are measured along the jet axis where the plasma flow velocity is the highest.

Such measurements also allow a better understanding of the influence of the particle diameters. The velocity of zirconia particles along the jet axis are reported in Fig. 11, and it can be seen that the maximum velocity of the particles increases with decreasing particle size. The largest particles (-80+63 μm) attain a steady state velocity value whereas the velocity of smaller particles diminishes at an increased rate. On one hand the heat transfer to the particles decreases with the residence time, but on the other hand the heat quantity required to melt a particle is lower when its diameter decreases. Thus the smaller the particle, the higher its surface temperature will be even if its residence time is the shortest. This is illustrated in Fig. 12 for two size distributions (-21+15 μm and -52+40μm) of alumina particles injected in the Ar–H_2 plasma jet characterized in Fig. 6. Of course the biggest particles with their highest thermal inertia have the highest temperature in the plasma plume where they cool down.

PARAMETERS CONTROLLING THE THERMOMECHANICAL PROPERTIES OF DEPOSITS

Substrate and coating temperature while spraying

The cooling rates of substrate and coating while spraying are of primary importance for the quality of ceramic coatings as emphasized by Wilms et al.[32] They have shown, by X-ray diffraction analysis, that starting from a steel substrate the first layers of alumina coatings are amorphous. Then the layers successively present very fine, medium, and finally coarse grains. This evolution was attributed to the increase of the coating temperature with its thickness. This is due to the heating of coating and substrate by the hot plasma gas and the poor evacuation of the particle heat content to the water cooled substrate through the previously deposited layers with their low thermal conductivity. Pawlowski et al.[33] have shown that the surface of such alumina deposits can reach 1500°C as soon as their thickness is about 0.3 mm on a water cooled copper substrate. Moreover, such temperature gradients in ceramic coatings, of which the expansion coefficient is lower than that of the substrate, induces stresses first compressive and then tensile[18] and limits the possibility of obtaining thick ceramic coatings. The high stresses are released by cracks, or detachment of the deposit from substrate in extreme cases (detachment at the interface or between successively deposited layers). The temperature control of coating and substrate while spraying is thus of primary importance. This can be achieved by

- cooling down the front face of the deposit while spraying either with a CO_2 jet[18] or an air jet[34] or a liquid gas[35]
- stopping the plasma jet with an air barrier blown orthoganally to it.[34] It is also important to note that such an air barrier allows elimination of the non-melted particles travelling at the periphery of the plasma jet where velocity is low (more than one order of magnitude difference) compared to that of the particles travelling in the hot core of the plasma. This reduces the deposit porosity (non-melted particles cannot accommodate surface roughness)
- increasing the relative velocity of displacement torch substrate in order to reduce the thickness of the Gaussian shape deposited stringer at each torch passage.

When doing so zirconia coatings stabilized with yttria (8% wt) can for example be achieved with thicknesses of up to 5 mm.

Particle melting and quenching

In the control of ceramic particle melting α alumina has been found very useful because, due to the high cooling rates (up to 10^6 K/s), the completely melted particles crystallize in the metastable γ phase and the measurement of the ratio $\alpha/(\alpha+\gamma)$ in the sprayed coatings is a very good indication of the quantity of not completely melted particles arriving at the substrate. To obtain a good melting different conditions must be realized:

- proper injection of the particles to penetrate the hot core of the plasma jet
- use of plasma jets with high heat transfer coefficients
- long enough plasma jet and substrate disposed far enough downstream to insure a sufficient residence time for the particles to be completely melted (to account for the heat propagation phenomenon)
- elimination of the non-melted particles travelling at the periphery of the plasma jet

a) The proper injection velocity very much depends on the particle size and density and, for example in the Ar–H_2 plasma jet characterized in Fig. 6, the injection velocities reported in Table 1 must be achieved for alumina particles of different mean sizes. These velocities have been determined experimentally as those giving the highest velocities and surface temperatures of the particles.[36] It is to be noted here that the largest particles require the lowest injection velocity and thus the lowest carrier gas flowrate which will perturbate the less the plasma flow.

b) The heat transfer coefficient depends strongly on the hydrogen content as illustrated in Fig. 13 showing the radial concentration, velocity, and surface temperature distributions for alumina powders (-21+15 μm) in a cross-section of the plasma jet 75 mm downstream of the nozzle exit. Fig. 13a shows the results obtained with a pure argon plasma where the power level and gas flowrate were adjusted in such a way that the mean enthalpy of the gas (taking into account the thermal efficiency) was the same as that of an Ar–H_2 plasma (Fig. 13b). It can be seen that with pure argon, the momentum of the plasma jet being smaller, the the concentration distribution is not centred as well as that obtained with the Ar–H_2 plasma. In spite of higher velocities with Ar–H_2 plasma the surface temperature of the particles is higher than that obtained with pure argon. Of course, when the power level is raised, a better melting of the particles is obtained in spite of a higher particle velocity.[37] This results in a lower porosity (3% at 29 kW against 9% at

20 kW) and in a ratio $\alpha/(\alpha+\gamma)$ lower at higher power levels (2% at 29 kW against 14% at 20 kW). Similar results are obtained with zirconia coatings. For example, Fig. 14 represents the surface temperature evolution along the plasma jet axis (for two power levels) of ZrO_2 powders stabilized with 8% (wt) Y_2O_3 (-125+35 μm). The increase of the power level from 21 to 29 kW results in an increase of 25% in the Vickers Hardness (measured under 0.3 kg) and a decrease of the monoclinic phase content from 6 to less than 1%.[38]

c) Heat propagation phenomenon has to be accounted for, especially for ceramics in $Ar-H_2$ plasmas (heat fluxes of which may be as high as 5.10^8 W/m² [31]). The substrate must not be disposed at the location of the highest surface temperature but further downstream where the centre temperature starts to reach the melting point.[39] If the highest surface temperature of Al_2O_3 (-21+18 μm) particle is reached at 55 mm downstream of the nozzle exit, at 20 kW complete melting is achieved only at 80 mm, as illustrated by the ratio $\alpha/(\alpha+\gamma)$ of 26% and 10% respectively. When the power level is raised to 29 kW with a lengthening of the isotherms (see Fig. 7) and with a substrate at 80 mm, $\alpha/(\alpha+\gamma)$ is lower than 2%.

d) The elimination of the non-melted particles is also important and with a properly designed air barrier (not too high a cold gas momentum to reduce the hot particle cooling, but one sufficient to eliminate the slow particles) the ratio $\alpha/(\alpha+\gamma)$ with a 29 kW $Ar-H_2$ plasma is reduced from 6 to less than 2%.

Particle splatting

The way the particles splat (Fig. 2) is very important even if its control is not yet understood. This has been demonstrated for example by annealing experiments[38,40] of zirconia sprayed coatings. Annealing these coatings (stabilized with 8%(wt)Y_2O_3) at temperatures higher than 1200°C increases thermal conductivity by a factor of 2–4, while the open porosity of the deposits is reduced only be a few percent. After annealing, the pore sizes below 0.1 μm, corresponding to bad contacts between the lamellae, disappear.

Particle manufacturing process

The ways in which the powders are prepared are numerous and they have a great influence on the deposit properties, as shown by recent experiments.[41-46] Metal particles can be melted under vacuum and then centrifugated or blown.[47,48] Metal and ceramics are also prepared from fine powders by agglomeration with afterwards calcination or sintering to densify them.[48-50] Ceramics can also be made by crushing fused ceramics followed by selective milling and sizing. Fig. 15 shows typical morphologies of ceramic particles. The manufacturing method plays an important role through heat transfer from the plasma and chemical reaction with the plasma gas and/or surrounding gas pumped into the plasma jet.

When using plasma gases containing hydrogen, the heat propagation phenomenon can be enhanced by particle morphology, especially when agglomerated or agglomerated and calcinated powders are sprayed. With high heat transfer plasmas the surface of the particle can be melted and then blown up as a balloon by the gas trapped in the porous particle with the central part of the particle not melted at all. This is illustrated in Fig. 16,[8] showing a partial cross-section of an agglomerated stabilized zirconia particle sprayed with the $Ar-H_2$ plasma depicted in Fig. 6 and then collected in water.

Such agglomerated and agglomerated calcinated stabilized zirconia powders result in very porous deposits (up to 39%) against much denser ones (less than 13%) obtained with melted or sintered powders of the same composition and particle size distribution.[51]

Chemical reactions may also be very important when spraying particles very sensitive to oxygen, such as tungsten carbide grains which react with the pumped oxygen when sprayed at ambient atmosphere. The cobalt coated WC powders and the agglomerated sintered ones experience less decarburization when sprayed in air and WC+20% cobalt coated powder gives the best compromise between coating phase composition and hardness. No decomposition was observed during controlled atmosphere spraying but the formation of mixed carbide Co_6W_6C was seen (carbide dissolution in cobalt). When the spraying chamber pressure was decreased the heat transfer to the particles was reduced due to non-continuum effects,[25] and the length of the plasma jet was increased. This results in dense deposits[42] consisting of carbide grains embedded in a cobalt matrix. This is due to the low heat transfer rate which ensures that the cobalt is molten with a minimum degree of overheat limiting the carbide dissolution in liquid cobalt.

CONCLUSIONS

Plasma spraying has been used in industry for almost 20 years and it has been developed mostly by empirical means due to the great number of macroscopic parameters (more than 45) controlling the process.

However, the number of applications is increasing regularly and plasma sprayed deposits are used more and more in industry. This is due to development of the techniques (plasmagun, powder feeders, vacuum spraying, automated production systems, powder production), but also to a better understanding of the physical phenomena encountered (momentum and heat transfer between plasma and particles, the way the sprayed layers are built up, substrate and coating temperature control while spraying - to control residual stresses). This results in better control of the quality of the sprayed deposits with improved mechanical, thermal, and wear resistance properties.

However, research efforts are still needed for a better understanding of:
- heat and momentum transfers between plasma and solid particles (non-continuum effects

for small particles, evaporation, charging effect, loading effect)
- lamellae formation
- the influence of the cold carrier gas on the plasma and on the trajectory distribution of the particles
- the influence of the way the powders have been made (fine particles agglomerated and densified, fused particles)

and to develop new types of plasma generators better adapted for spraying and allowing higher deposition rates.

References

/1/ RYKALIN N.N., KUDINOV V.V. "Pure and Applied Chemistry" 48, (1976), 229.

/2/ VARDELLE A., FAUCHAIS P., VARDELLE M. - Actualité Chimique 10 (1981) 69.

/3/ ZAAT J.H. - Ann. Rev. Mater. Sci. 13 (1983) 9.

/4/ APELIAN D. "Rapid solidification by plasma deposition, in Mat. Res. Soc. Symp. Proc. vol. 30 (1984) (ed.) North-Holland N.Y., Amsterdam

/5/ FAUCHAIS P., BOURDIN E., COUDERT J.F., Mc PHERSON R. - "High pressure plasma and their application to ceramic technology" in Plasma Chemistry Topics in Current Chemistry (ed.) Venugopalan and S. Veprek, Springer Verlag (Berlin) (1983).

/6/ FAUCHAIS P., COUDERT J.F., VARDELLE M., VARDELLE A., Temperature measurements in atmospheric spraying plasma jets : gas and particles in flight, MRS Meeting Annaheim April (1987), to be published

/7/ VARDELLE M., VARDELLE A., FAUCHAIS P., in Advances in Thermal Spraying, p. 379 (Pergamon Press, Publisher, N.Y. 1986)

/8/ FAUCHAIS P., COUDERT J.F., VARDELLE M., VARDELLE A., GRIMAUD A., ROUMILHAC P., State of the art for the understanding of the physical phenomena involved in plasma spraying at atmospheric pressure, National Thermal Spray Conference, Sept. (1987) Orlando, to be published

/9/ KUIJPERS T.W., ZAAT J.H. - Metals Technology 4 (1974) 142

/10/ KHARLAMOW Y.A., HASSAN M.S., ANDERSON R.N. - Thin Solid Films 63 (1979) 111

/11/ MADEJSKI J. - Int. J. Heat and Mass Transfer 19, (1976) 1003

/12/ MADEJSKI J. - Bull. Acad. Pol. des Sciences 24, n° 1 (1976)

/13/ ZOLTOWSKI P. - Rev. Int. Htes. Temp. et Réfract. 6, (1968) 65

/14/ Von MEYER H. - Ber. Dtsch. Keram. Ges. 41, (1964) 112

/15/ HOUBEN J.M. - Paper presented at the 2nd American Thermal Spraying Conference, 31 Oct - 2 Nov 1984, Long Beach, California

/16/ MARINOWSKI C.W., HALDEN F.A., FARLEY E.P. - Electrochem. Techn. 3, (1965) 109

/17/ Mc PHERSON R., SHAFER B.V. - Thin Solid Films 97 (1982) 201

/18/ STEFFENS H.D., HÖHLE H.M., ERTÜRK E. - Schweissen und Schneiden 33 (1981) 159

/19/ TRONCHE A., FAUCHAIS P. - Mat. Science and Eng. 92 (1987) 133

/20/ Thermal Spraying, Practice, Theory and Application (ed.) American Welding Soc. PO Box 35 1040, Miami, Florida

/21/ PAWLOWSKI L. - "Optimisation des paramètres dans le processus de projection par plasma. Etude des propriétés thermiques des couches projetées" Thèse de Doctorat d'Etat, Université de Limoges, June (1985)

/22/ Mc PHERSON R., SHAFER B.V. - Thin Solid Films 97 (1982) 201

/23/ PFENDER E., "ELectric arcs and arc gas heaters" in Gaseous Electronics (ed.) N.N. Hirsh and H.J. Oskam, Academic Press 1 (1978) 291

/24/ ROUMILHAC Ph., COUDERT J.F., LEGER J.M., GRIMAUD A., FAUCHAIS P., Optical and thermal diagnostics to study Ar, Ar-H_2, H_2 plasma jets produced by a spraying plasma gun, ISPC8 Tokyo (1987) (ed.) Prof. Akashi, Univ. of Tokoyo

/25/ PFENDER E., Pure and Applied Chemistry 57 (9) (1985) 1179

/26/ FAUCHAIS P., COUDERT J.F., VARDELLE M., "Diagnostics in thermal plasma spraying" in Plasma-Material Interaction, to be published Academic Press (1988)

/27/ EDDY T.L., SEDGHINASAB A., Non LTE temperatures in argon arcs at various pressures, IEEE 14th Int. Conf. on Plasma Science, Arlington, Virg. USA June (1987)

/28/ BROSSA M., PFENDER E., Probe measurements in plasma jets, to be published in Plasma Chemistry, Plasma Processing (1987)

/29/ VARDELLE M., Thèse de Doctorat d'Etat, University of Limoges, France, July (1987)

/30/ MISHIN J., VARDELLE M., LESINSKI J., FAUCHAIS P., J. of Physics E. 20 (1987) 620

/31/ VARDELLE A., Thèse de Doctorat d'Etat, University of Limoges, France, July (1987)

/32/ WILMS V., HERMAN H., Thin Solid Films 39 (1976) 251

/33/ PAWLOWSKI L., VARDELLE M., FAUCHAIS P., Thin Solid Films 94 (1982) 307

/34/ GUYONNET J. - Brevet C.N.R.S. 124 111 et additifs 168 044 et 7044678

/35/ Brevet CEA-DAM, CEA BP 12, 91680 Bruyère le Chatel, France

/36/ VARDELLE M., VARDELLE A., FAUCHAIS P., BOULOS M., AIChE Journal 29 (2) (1983) 236

/37/ VARDELLE M., VARDELLE A., BESSON. J.L., FAUCHAIS P., Rev. Phys. Appl. 16 (1981) 425

/38/ GITZHOFER F., LOMBARD D., VARDELLE A., VARDELLE M., MARTIN C., FAUCHAIS P., in Advances in Thermal Spraying, p. 379 (Pergamon Press, Publisher, N.Y. 1986)

/39/ FAUCHAIS P., VARDELLE A., VARDELLE M., COUDERT J.F., LESINSKI J., Thin Solid Films 121 (1984) 303

/40/ PAWLOWSKI L., LOMBARD D., FAUCHAIS P., J. Vac. Sci. Tech. 13 (6) (1985) 2494

/41/ VINAYO M.E., KASSABJI F., GUYONNET J., FAUCHAIS P., J. of Ac. Sci. Tech. A3 (6) (1985) 2483

/42/ VINAYO M.E., KASSABJI F., FAUCHAIS P., in Advances in Thermal Spraying, p. 101, (ed.) Pergamon Press (1986)

/43/ RANGASWAMY S., HERMAN S., p. 101, see /42/

/44/ MAZARS P., MANESSE D., LOPVET C. - see /42/ p. 111

/45/ LUGSHEIDER E., ESCHNAUER H., HAUSER B., AGETHEN R. - see /42/ p. 261

/46/ GITZHOFER F. et al. - see /42/ p. 269

/47/ HOUCK D.L. - Techniques for the production of flame and plasma spray powders - Modern Development in Powder Metallurgy, Vols 12 to 14

/48/ SCHWIER G. - see /42/ p. 277

/49/ KVERNES I., LUGSHEIDER E., FAIRBANKS J. - MRS-Europe, Nov. (1985) p. 13, Editions de Physique, Paris

/50/ KRISMER B. - MRS-Europe, Nov. (1985) p. 33, Editions de Physique, Paris

/51/ LUGSCHEIDER E., ESCHNAUER H., HAUSER B., AGETHEN R. - in "Advances in Thermal Spraying" p. 261, (ed.) Pergamon Press (1986)

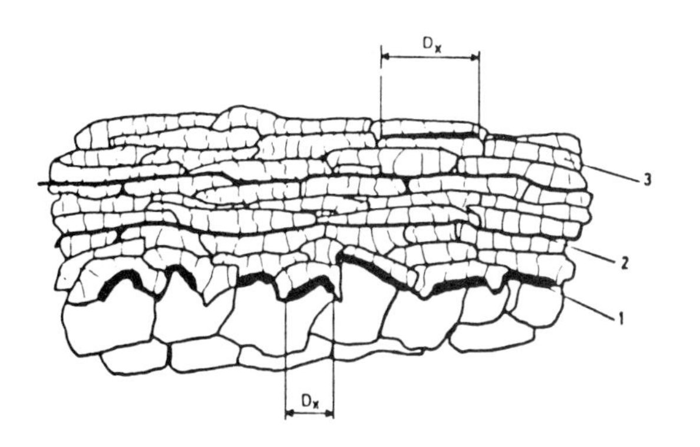

Fig. 1 Structure of plasma sprayed deposits:[1] (1) contact with substrate; (2) contact between solidified flattened particles; (3) cracks in solidified particle

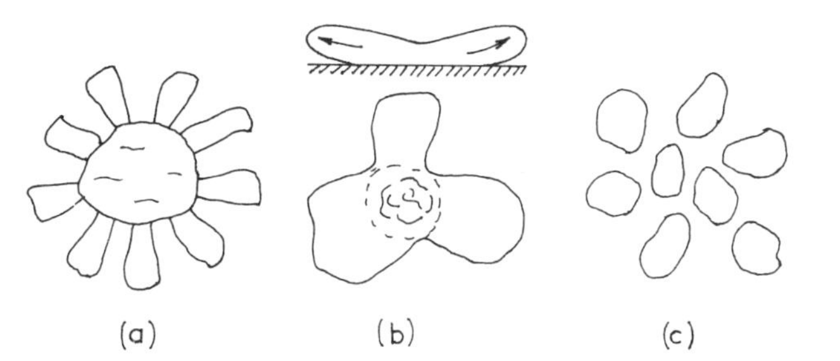

Fig. 2 Three types of flattened particles:[15] (a) pancake; (b) flower; (c) exploded

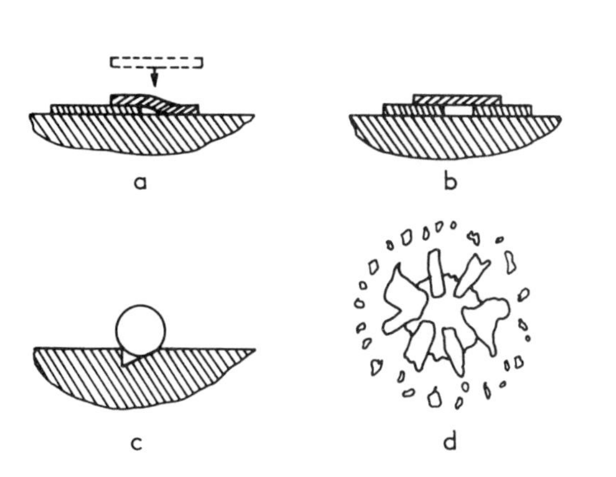

Fig. 3 Shadow effect of impacting particles on substrate[3]

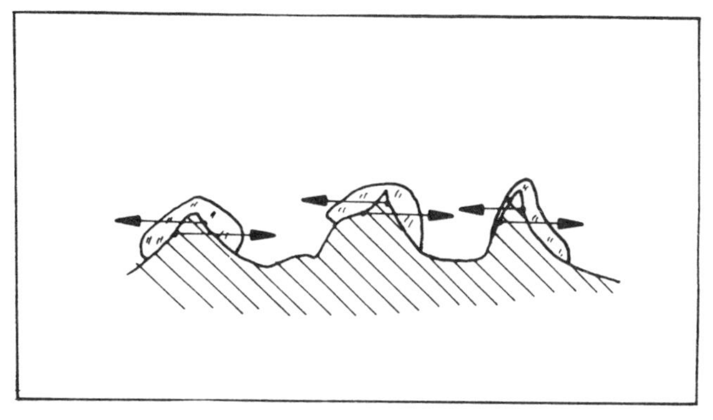

Fig. 4 Particles flattening on surface peaks[21]

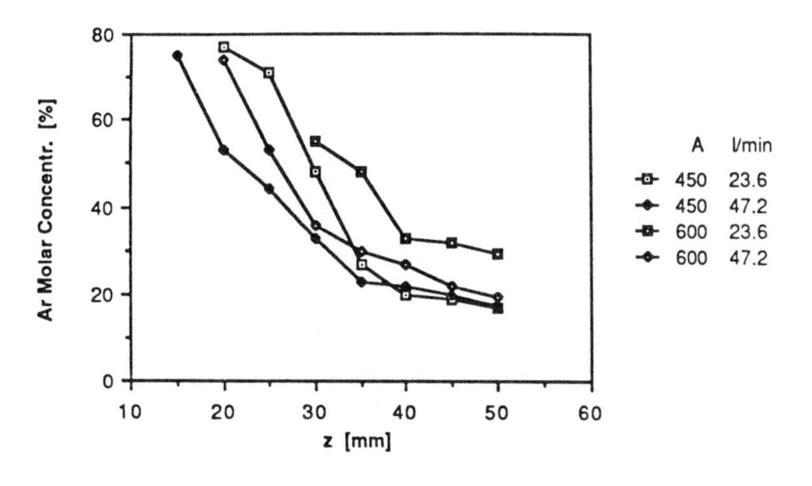

Fig. 5 Ar concentration axial profiles for DC plasma torch[28]

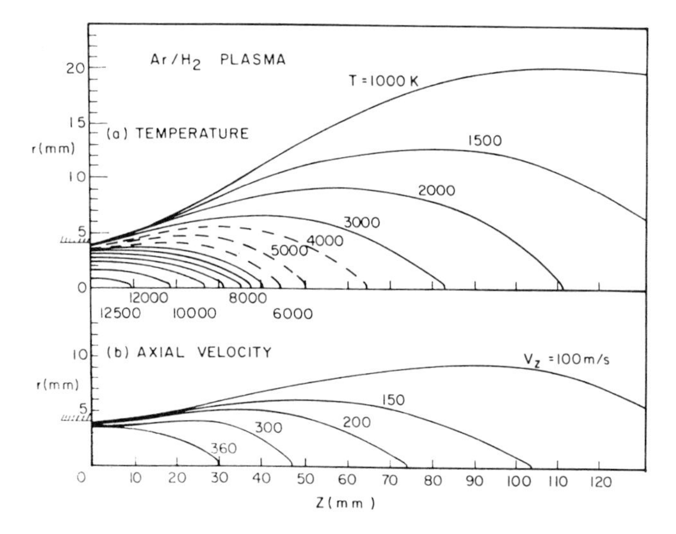

Fig. 6 Isocontours of DC Ar–H_2 plasma jet: (a) temperature (29 kW, D_{Ar}=75 l(STP)/mn, D_{H_2}=37 l(STP)/mn); (b) velocity[29]

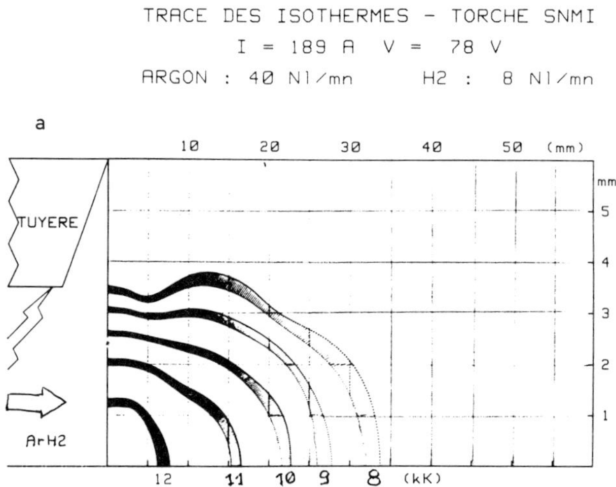

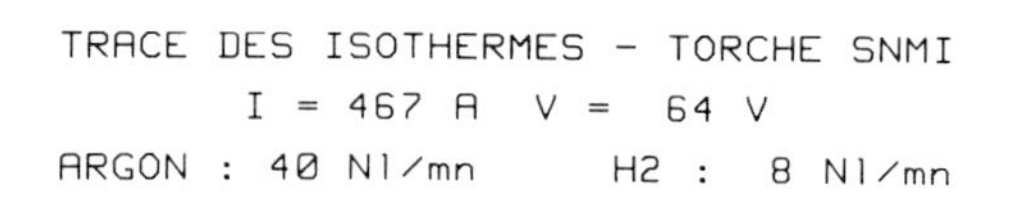

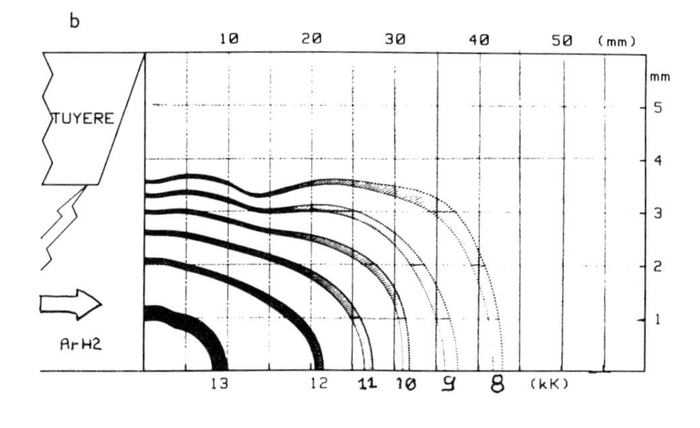

Fig. 7 Ar–H_2 plasma jet temperature isocontours (D_{Ar} = 40 l(STP)/mn, D_{H_2} = 8 l(STP)/mn) (a) for an arc current of 189 A (V = 78V) (b) for an arc current of 467 A (V = 64V)[8]

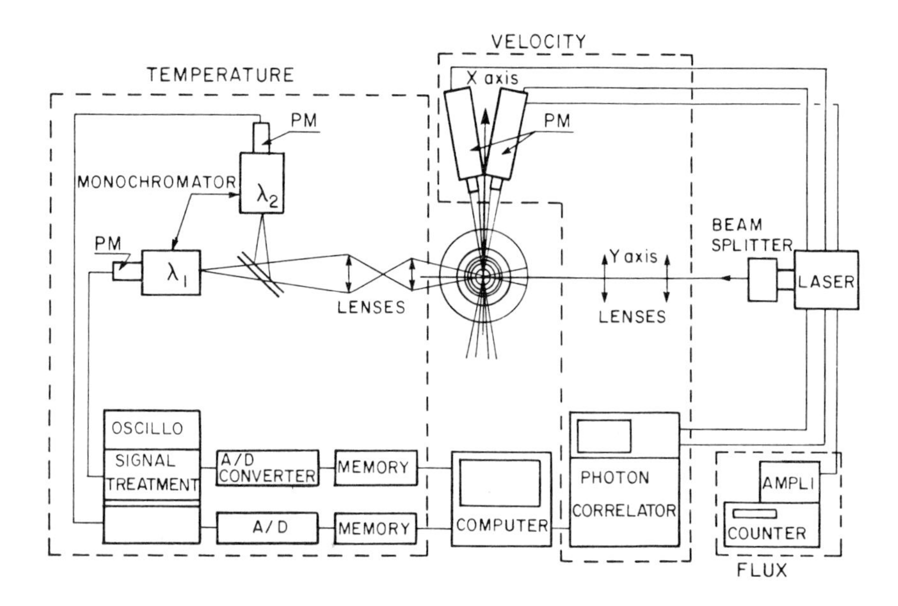

Fig. 8 Experimental set-up for the measurement of velocity, flux, and surface temperature of the particles in flight[33]

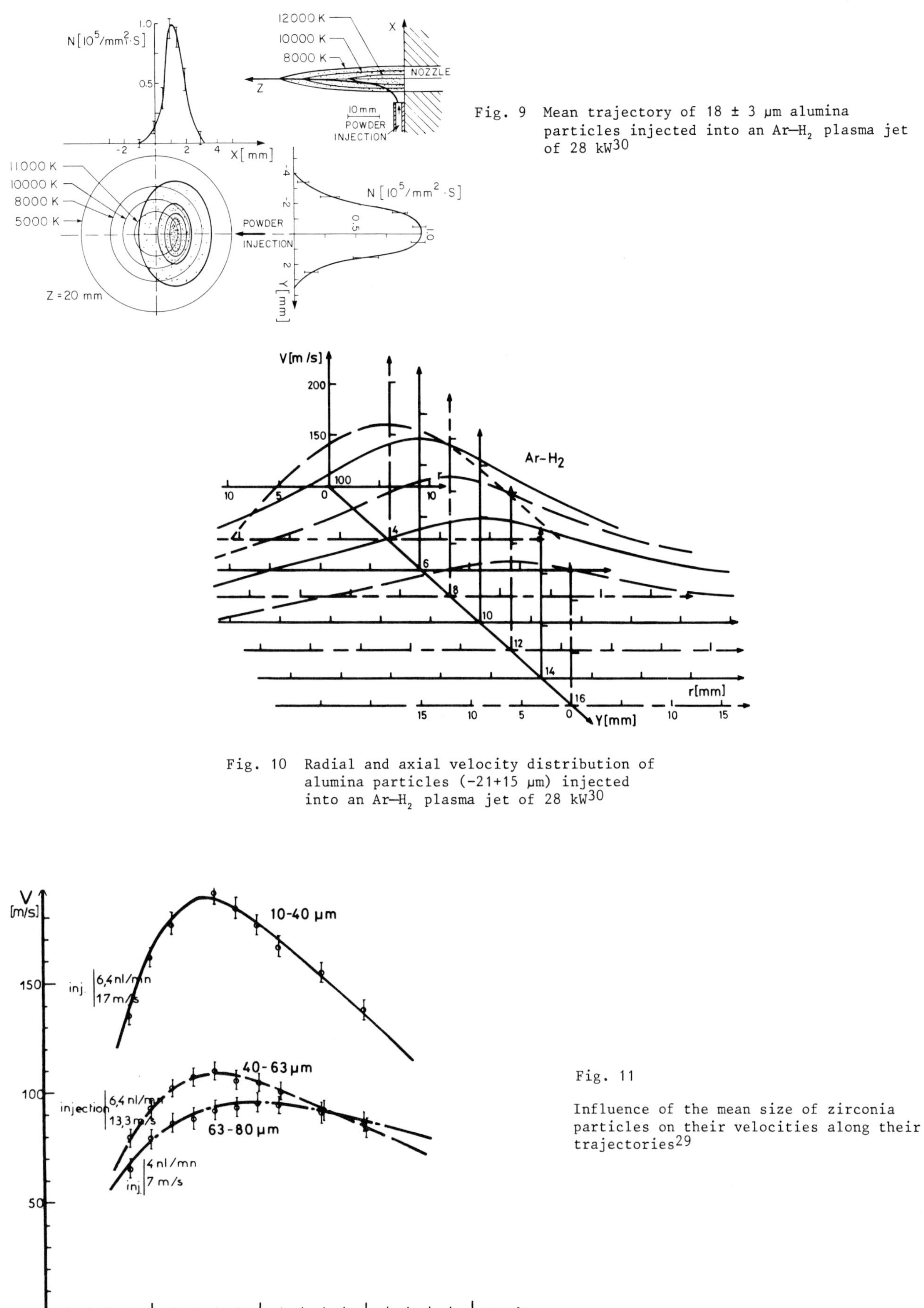

Fig. 9 Mean trajectory of 18 ± 3 µm alumina particles injected into an Ar—H_2 plasma jet of 28 kW[30]

Fig. 10 Radial and axial velocity distribution of alumina particles (-21+15 µm) injected into an Ar—H_2 plasma jet of 28 kW[30]

Fig. 11

Influence of the mean size of zirconia particles on their velocities along their trajectories[29]

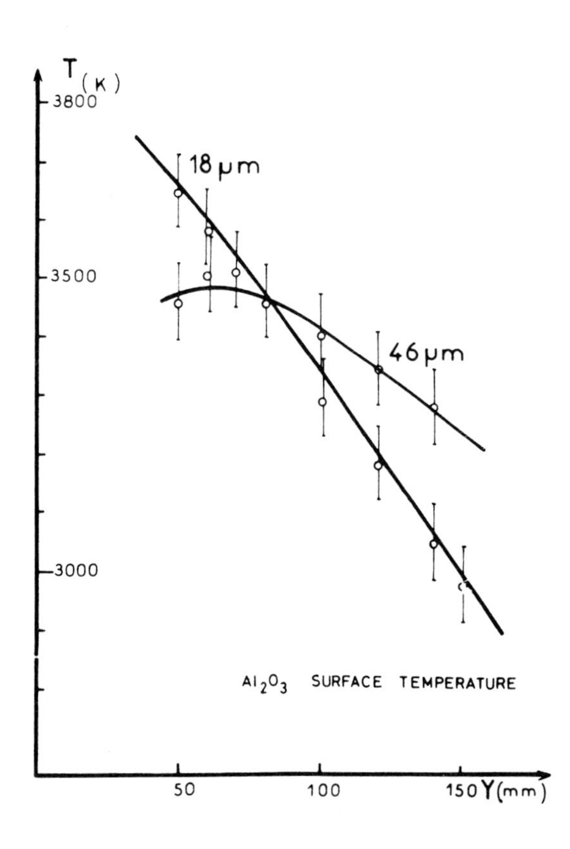

Fig. 12 Influence of the mean size of alumina particles on their surface temperatures along their trajectories[30]

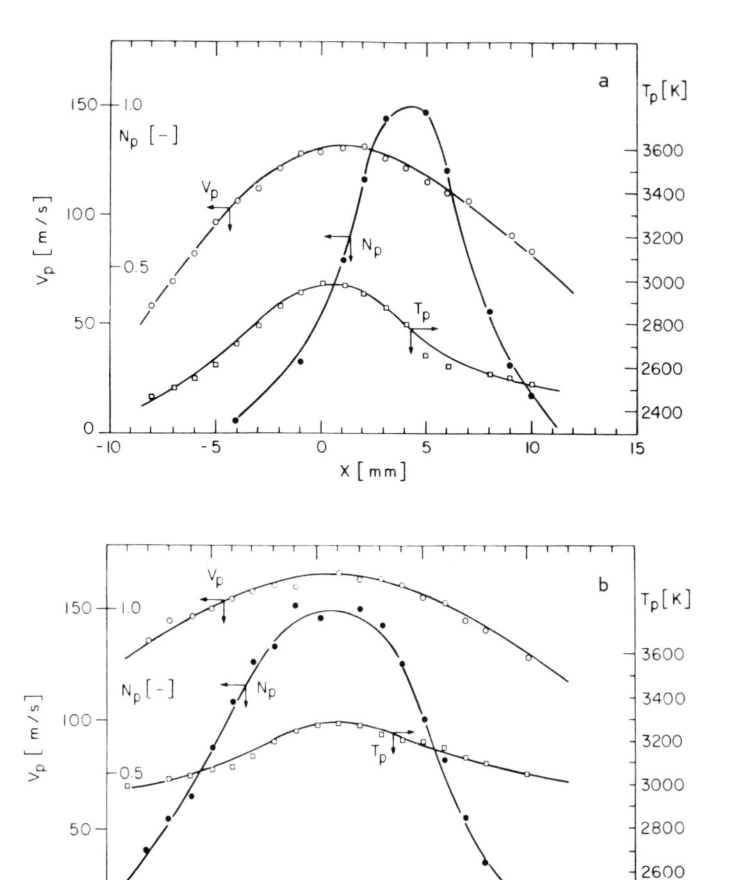

Fig. 13

Alumina particles concentration, velocity, and surface temperature measured 75 mm downstream the nozzle exit in a plasma jet at 29 kW: (a) -75 l/mn Ar; (b) -75 l/mn Ar +15 l/mn H_2[7]

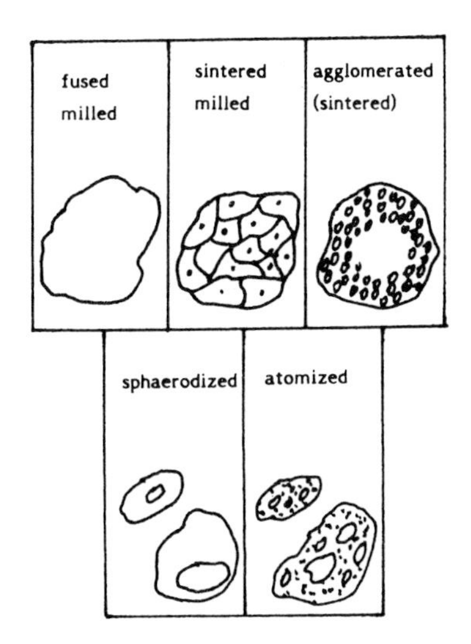

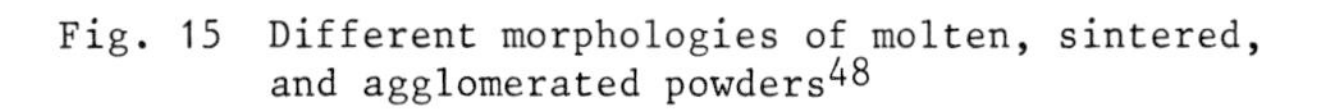

Fig. 15 Different morphologies of molten, sintered, and agglomerated powders[48]

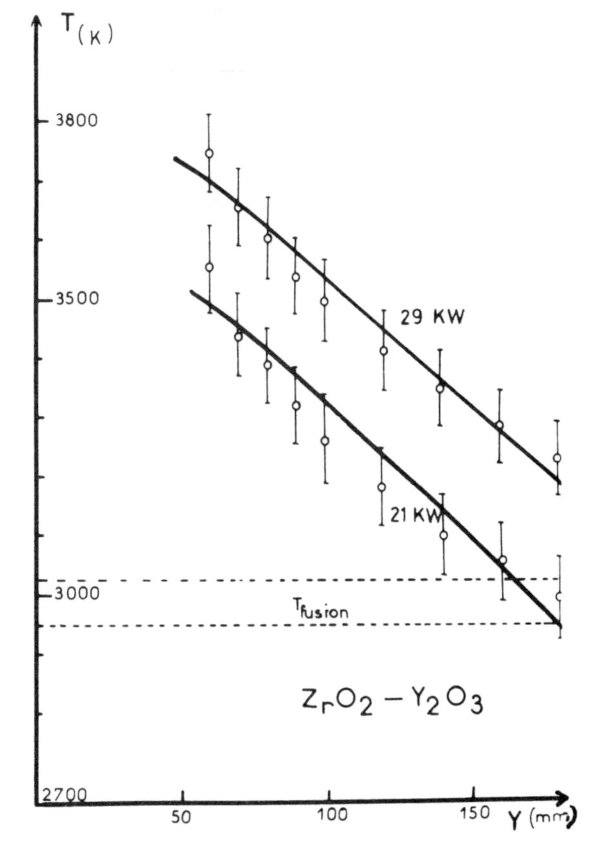

Fig. 14 Evolution of the surface temperature (along the axis of the jet) of ZrO_2 particles (stabilized with 8%(wt) Y_2O_3, -125+35 µm) injected in Ar–H_2 plasma jets with power levels of 21 and 29 kW respectively[29]

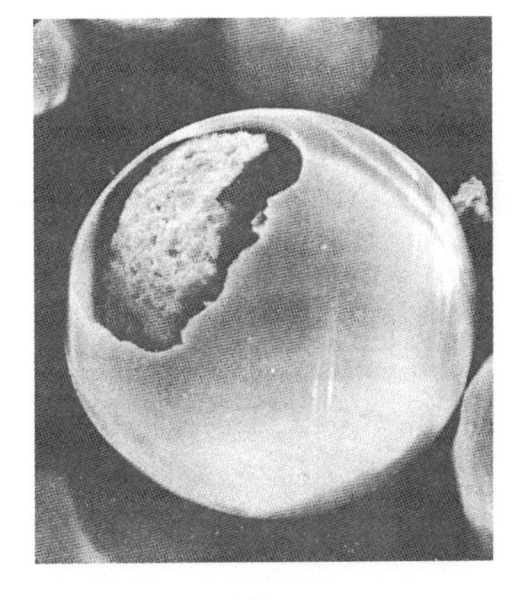

20µm

Fig. 16 Cross-section of a ZrO_2 +8%(wt) Y_2O_3 (-16 +10 µm) particle after plasma spraying and collection in water[8]

□

Effect of nitriding and QPQ surface heat treatments on service life of hot working tools

S O J KIVIVUORI and J J KOHOPÄÄ

SOJK is a supervisor in the Laboratory of Metal Working and Heat Treatment at the Helsinki University of Technology, Finland; JJK is a research scientist in the Laboratory of Production Technology at the Helsinki University of Technology.

SYNOPSIS

In hot working processes the wear due to adhesion or frictional welding has proved to be a very limiting factor on the service life of tools. In the present paper nitriding and QPQ surface heat treatment methods used to reduce the tool wear in lubricated hot backward extrusion are studied. The laboratory wear tests are carried out by a wear testing machine allowing tests in hot working conditions. The results of the laboratory tests are then compared with the experience gained in the industrial processes. On the basis of these results the effects of nitriding and QPQ heat treatment methods on the service life of hot working tools are discussed.

INTRODUCTION

Forging is a metal-forming process in which hot metal, in the form of bar stock or cut pieces, is shaped by forging between impressioned dies. The change in stock shape is usually accomplished gradually by forging in successive die impressions. The process is ideally suited to the repetitive production of large numbers of identical components, such as those required by the mass production industries. During a production run, close shape and weight tolerances must be maintained on forgings and, consequently, only minor changes in the shape of the die cavity can be tolerated.

Forging dies may be taken out of service for a number of reasons. Premature die failures are mostly due to bad tool design, incorrect heat treatment, or improper use of tools /1,2/. To attain acceptable die life, premature failures must be prevented. In these cases the die life depends mainly on the wear resistance of the forging tool surface.

Hot working tools undergo severe thermal and mechanical shocks during each forming blow. Damage can arise due to erosion, plastic deformation, thermal fatigue, and mechanical fatigue /3/. Figure 1 shows the principal modes of tool failure, and also indicates the positions in a tool cavity where each type of failure is most likely to occur.

The service life of hot working tools depends on die material and hardness, work-metal composition, forging temperature, condition of the work metal at forging surfaces, type of equipment used, and workpiece design /4/. Changing one factor almost always changes the effect of another, these effects not being constant throughout the life of the die.

Because of the high cost of the metal-forming tools and the length of time required for their manufacture, the demand for their improvement is urgent. The problem has been approached by increasing the life of forming tools through surface strengthening. Figure 2 lists the possible surface treatments used in increasing the wear resistance of metal-forming tools /5/.

In the present paper the effects of the use of nitriding and QPQ surface treatments on the service life of metal-forming tools will be considered. Nitriding is an old and well-known process which is successfully used in manyny applications where wear resistance is needed. Nitriding increases the hardness and wear resistance of the materials as well as red hardness and fatigue strength.

The QPQ-process (quench-polish-quench), or Nitrotec as it is also called, is based on the established salt bath nitriding process known commercially as Tufftride. The QPQ-process consists of salt bath nitrocarburizing followed by oxidative cooling, mechanical polishing, and brief reimmersion in the oxidizing salt melt. The process sequence is presented in Figure 3. The QPQ-process has been used mainly to increase the corrosion resistance of nitrided parts, but it has also decreased the coefficient of friction, especially in lubricated conditions.

EXPERIMENTAL MATERIALS AND PROCEDURES

The material used in the investigations was chromium hot work tool steel AISI H13, DIN X40CrMoV51, commercially known as UHB Orvar 2 Microdized. Test specimens were austenitized at 1050 °C and tempered twice for 2 h at 600 °C. The room temperature hardness of the specimens was 48-50 HRC. After tempering, one third of the specimens were QPQ-treated, one third were nitrided and the rest were used in the tempered condition.

The laboratory test arrangement is schematically illustrated in Figure 4. Test conditions are designed to simulate hot forming of steels.

The tool specimens to be evaluated are mounted in the flanged disc and the workpiece material is in the form of a 12 mm diameter cylindrical bar. The eccentrically positioned bar has freedom to reciprocate inside an induction heating coil along an axis perpendicular to the die face. In its retracted position, the work bar end region rests inside the heating coil. After heating the bar is forced into contact with the rotating die and after a short period of controlled sliding the bar is drawn back to the heating position. Possible test variables are workpiece reheat temperature, contact pressure, sliding speed, cycle time and lubrication conditions.

In the present investigations two QPQ-treated, two nitrided, and two untreated specimens were mounted in the tool disc simultaneously. The workpiece made of quenched and tempered steel (SIS 2541) 34CrNiMo6 was heated to 1150 °C. The preheating temperature of the tool was approximately 200 °C. The cycle time was adjusted to 5 sec, of which the contact time was 1 sec. Contact load was 1 kN and sliding speed 1.8 m/sec. The tests were carried out both in unlubricated and lubricated conditions. Graphite-water was used as lubricant. 10 % of graphite-based lubricant Berulit 922 was mixed with 90 % water.

The progress of wear was assessed by measuring the profile of the wear groove after 3600 cycles. Wear parameters such as depth and cross-sectional area of the wear track were quantified by averaging from several traces.

Industrial reference tests were carried out in the lubricated hot backward extrusion of C35 carbon steel. The service life of the tools was measured by counting the number of acceptable forgings made by a specific tool.

EXPERIMENTAL RESULTS

The results of the laboratory tests measured with the tool wear test machine are given in Figure 5. Nitriding and QPQ-treatment decreased the wear approximately 30 % compared with untreated specimens when no lubrication was used. In lubricated conditions QPQ-treated specimens had superior wear resistance. The wear of the QPQ-treated specimens was less than 40 % of the wear of the nitrided specimens and only 17 % of the wear of the untreated ones.

The results of the industrial tests are given in Figure 6. The service life of the QPQ-treated tools was approximately 270 % longer than that of the untreated tools. Nitrided tools showed a tool life about 25% inferior to that of untreated tools because of heavy scuffing. The rate of wear was not the limiting factor in the service life of the nitrided tools.

DISCUSSION

During nitriding a nitride layer consisting of the outer ε-iron nitride compound layer and the diffusion layer thereunder is formed. The compound layer is very hard (HV 1000-1500) and it has a particular resistance against wear and seizure. During the tests under dry running conditions nitriding appeared to increase the wear resistance of the tool specimens. The results given in Figure 5 showed that oxidizing used in the QPQ-treating had no effect on the wear resistance when no lubrication was used. Thus the decrease in the wear of QPQ-treated specimens was solely due to the nitriding treatment carried out before oxidizing.

The situation changed drastically when lubricants were used. QPQ-treated specimens had superior wear resistance compared with hardened and nitrided specimens. QPQ-treatment decreased both abrasive and adhesive wear. The decrease in abrasive wear was mainly due to the nitriding treatment and was noticed both on the nitrided and QPQ-treated specimens. The main reason for superior wear resistance of the QPQ-treated specimens was due to the properties of the oxidized surface. A slightly porous oxidized surface has an ability to store lubricants, thus providing well-lubricated contact between work-piece and tool specimen. This can be seen in Figure 7, where the results of the test done under deficient lubrication conditions are given. During testing, lubricants were spread only every fifth cycle thus providing unstable lubrication. The wear of the QPQ-treated specimens was only 14 % of the wear of untreated specimens and 30 % of the wear of nitrided specimens. It is probable that nitrided and hardened specimens were occasionally under dry running conditions but QPQ-treated specimens maintained a lubricated condition. The results of the laboratory tests indicate that by using QPQ-treatment the wear of hot-forming tools can be efficiently reduced in difficult lubrication conditions.

The results of the industrial tests showed good agreement with the results of the laboratory tests. Industrial tests were carried out in a process where lubrication conditions were similar to the conditions in the test presented in Figure 7. The effects of QPQ-treatment on the wear resistance of hot forming tools were very promising both in industrial and laboratory tests.

The heavy scuffing observed in the industrial tests when nitrided tools were used was unexpected, nitrided surfaces having shown excellent scuffing resistance in many engineering applications. The reason for this unexpected scuffing could be the rough surface of the tools nitrided in a salt bath. The scale formed on a workpiece surface during heating, transferred onto the tool surface during the forming blow and stuck to the irregulaties of the rough surface. Because this phenomenon was not observed in the laboratory tests and moreover wear was not the limiting factor of the service life of nitrided backward extrusion tools, no comparisons between wear resistance of nitrided tools in laboratory and industrial tests were made.

CONCLUSIONS

(i) Nitriding has a positive effect on wear resistance of hot forming tools.

(ii) QPQ-treatment decreases both abrasive and adhesive wear of hot forming tools especially when lubricants are used.

(iii) Hot-work tool-wear test machines can simulate the wear of industrial hot work tools, thus giving useful information for predicting tool performance in hot work processes.

REFERENCES

1. Failures of Dies. Metals Handbook, Eighth Edition. Vol. 10, 500-507. ASM 1975. Metals Park, Ohio.

2. Riedel J., Analysis of Tool and Die Failures. Tool and Die Failures. Source Book 11-18. ASM 1982. Metals park, Ohio.

3. Thomas A., Wear of drop forging dies. Tribology in Iron and Steel Works. ISI Publ. 125. 135-141. Iron and Steel Institute, London.

4. Dies and Die Materials for Hammer and Press Forging. Metals Handbook, Eighth Edition. Vol. 5, 19-40. ASM 1975. Metals Park, Ohio.

5. Weist Chr, Westheide H., Application of Chemical and Physical Methods for the Reduction of Tool Wear in Bulk Metal Forming Processes. Annals of the CIRP. 35 (1986)1, 199-204.

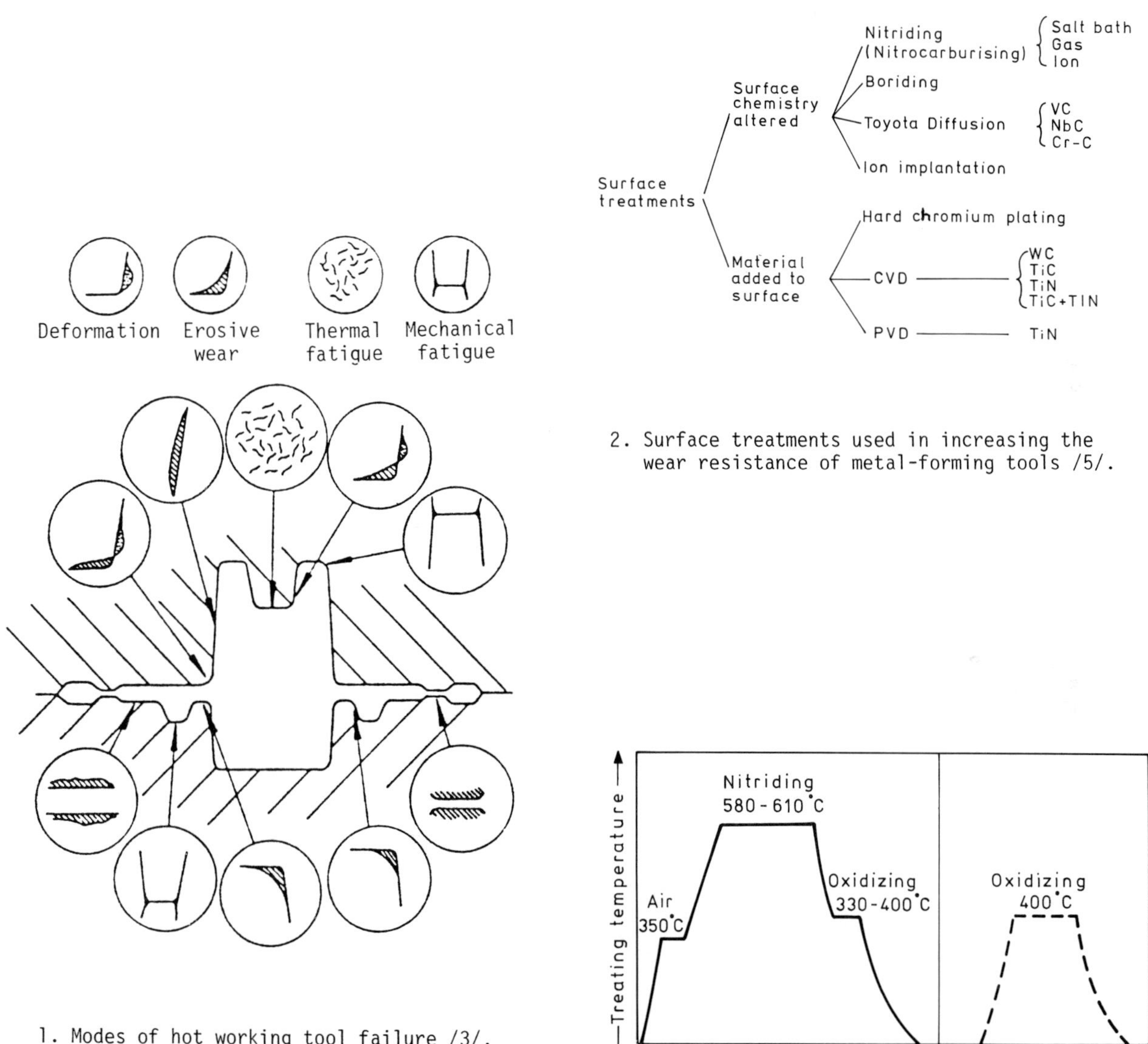

1. Modes of hot working tool failure /3/.

2. Surface treatments used in increasing the wear resistance of metal-forming tools /5/.

3. The process sequence of the QPQ-process.

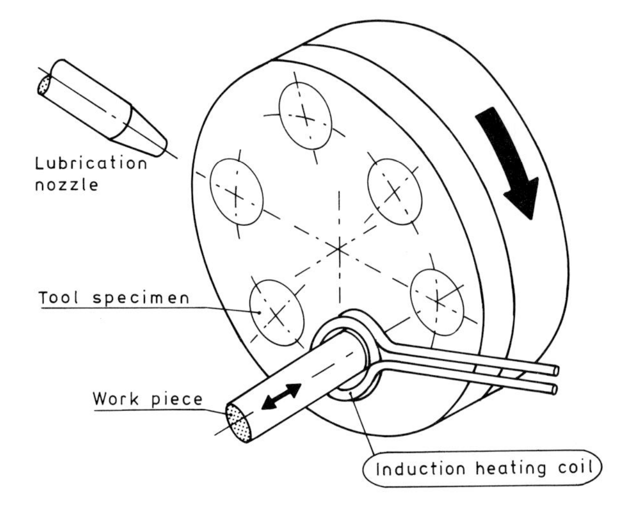

4. Schematic illustration of the laboratory test arrangement used in testing hot working tool-wear.

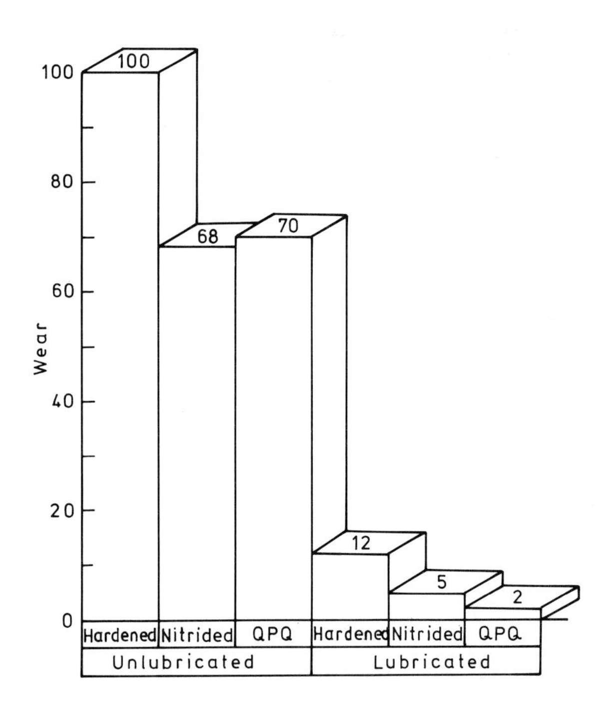

5. The results of the laboratory tests. The effect of nitriding and QPQ-treatment on the wear of hot working tools.

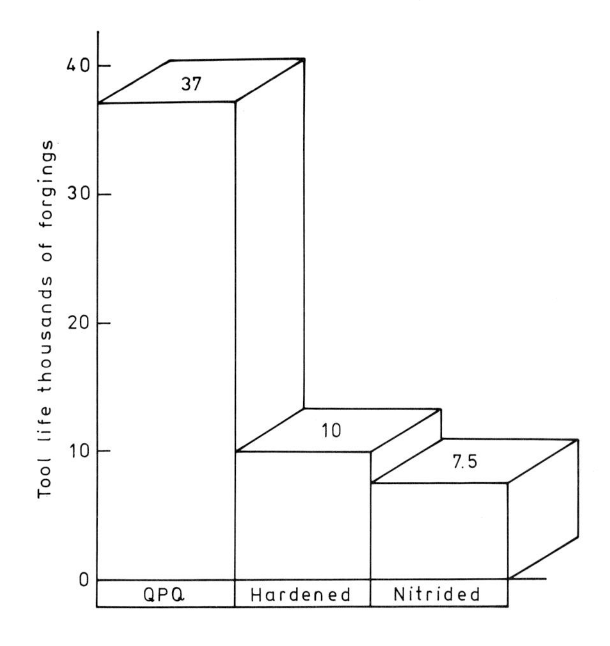

6. The effect of nitriding and QPQ-treatment on the service life of hot backward extrusion tools.

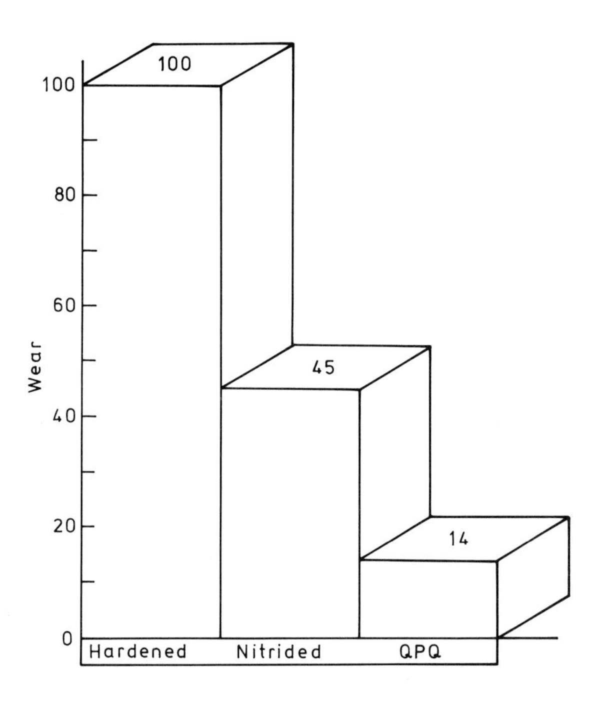

7. The effect of deficient lubrication on the wear of tool specimens tested in laboratory conditions.

Evaluation of adhesion strength of thin hard coatings

T ARAI, H FUJITA and M WATANABE

TA is a Technical Director and Manager of Research Div. VI in Toyota Central Research and Development Labs; HF and MW are Senior Researchers in Toyota Central Research and Development Labs.

SYNOPSIS

Adhesion strength of some thin hard coatings was evaluated by various testing methods. Coatings involve chromium, TiN, Ti(C,N), and carbides (TiC, VC, NbC, Cr_7C_3, and W_3C) made by plating, PVD, CVD, PCVD, and salt bath immersion. The following methods were employed: indentation, scratch, hammering, rolling, and simulation for metal forming. Comparative results obtained provide some information useful for understanding the characteristics of coatings and testing methods.

INTRODUCTION

In recent years new technology has been developed to apply thin coatings of hard materials such as carbides, nitrides, and oxides to metal forming dies and cutting tools. These coatings are extremely high in hardness and generally have excellent frictional and anti-galling properties, providing much profit for industry. Strong adhesion of coatings onto substrates is obviously of prime importance to the reliability of such a coating. The spalling of coatings, as well as other failures (wear, seizure, chipping etc.), has been so frequently encountered in various industries that is is important to evaluate the adhesion strength of coatings. However, there is no optimum method for evaluating high adhesion strength. Although the scratch test [1-4] is the most commonly used technique, an acoustic signal can be generated by not only the spalling but also chippings and crackings and a measurement of the 'adhesion strength' is quite difficult even with the use of a supplementary method, such as microscopic observation and X-ray micro-analysis.

Thus we believe it is important, as for the spalling problem, to accumulate the laboratory data in various testing methods and experience in practical applications, accompanied by careful observation of the origination and development of failures. Such work will provide us with information useful also in preventing the failures of coatings which are encountered during industrial use. Results of research work into the wear problem carried out by us using similar methods have been published [5] and here we will briefly outline a part of our results on the adhesion problem.

EXPERIMENTAL PROCEDURE

the test methods employed involve indentation, scratch, hammering, metal bending, metal coining, metal ironing, and rolling tests, and so on.

In the indentation test, the diamond indenter for the Rockwell C hardness tester was applied to coated specimens under varied loads and the resulting indentations were observed through a microscope.

The scratch test was carried out by means of a Heidon 14 surface testing machine (scratch test I) and the LSRH-Revetest scratch tester (scratch test II). A large load (max. 50 N) was applied through a diamond tip with 0.2 mm radius in the latter and a smaller load (max. 3 N) through a diamond tip with a 25 μm radius in the former. Scratch test II was performed under standard conditions (scratching speed $d\chi/dt$ = 10 mm/min; loading rate dL/dt = 100 N/min). Testing conditions of scratch test I were as follows: scratching speed $d\chi/dt$ = 50 mm/min; loading - constant load. The scratch channels were examined by SEM, X-ray microanalysis, and optical microscope.

The hammering test was carried out using a cemented carbide ball (6.35 mm diameter) to repeatedly strike the surface of a specimen at 380 shots/min (velocity: 36 mm/s) under varied load (max. 343 N) with dry air blowing. Damage occurring at the different numbers of shots was observed by optical microscope and others.

The three methods, bending,[6] coining, and ironing,[7] were employed to simulate loading conditions encountered in the metal forming operation. In the bending test, cold rolled low carbon steel strip 0.8 mm thick was continuously bent into a U-shape by coated dies and the punch corner was examined through microscope, SEM, and X-ray microanalysis. In the coining test, 1.6 mm thick cold rolled carbon steel strip was continuously pressed with V-shaped coated punches to decrease the thickness of the strip by up to 0.8 mm. In the ironing test, cold rolled low carbon steel strip (1.6 mm thick) was continuously ironed by a couple of coated flat dies. Measurement of friction and volume worn was carried out along with microscopic observation, EBS, and X-ray microanalysis. Evaluation of adhesion by metal forming has already been reported.[8]

The rolling test[6] was carried out by an Amsler tester applying the rolling contact stress with 10% slip to a pair of 40 mm diameter steel ring specimens until the coatings were almost completely exfoliated.

As shown in Fig. 1, most test methods employed here caused not only spalling but also other failures. The major stress concerned in each testing method is also shown.

Coatings evaluated in this work involved carbides (TiC, VC, NbC, Cr_7C_3, and W_3C), nitrides (TiN), carbonitrides (Ti(C,N)), and others (chromium, nickel-phosphor, and amorphous carbon). These are produced by means of electrolytic plating, electroless plating, ion plating, sputtering, plasma assisted CVD (PCVD),[9] CVD at low,[10] medium,[11] and ordinary temperatures or salt bath immersion method.[6,12] Sputtering, PCVD, and salt immersion methods were performed by us. Most other coatings were carried out by commercial treaters or process developers under the conditions they recommended. Two specimens were cut away with commercially available gear cutters TiN coated by ion plating. preparation of specimens and coating procedures were carried out in more than one batch in most coatings (as shown in Table 1) in order to achieve various combinations of substrate materials and coating thicknesses, and to confirm stability in quality of coatings.

Specimens for low temperature coating (plating, PVD, PCVD, and low temperature CVD) were preliminarily hardened under the standard conditions for substrate steel so that they received a substrate hardness of HRC61-63 and then, after coating, tested without any subsequent heat treatment. In the case of the salt bath immersion method, reaustenitizing hardening by use of a vacuum furnace was subsequently applied only for high speed steel substrate. It was also applied to all specimens by CVD at medium and ordinary temperature. The specimens prepared by PVD, PCVD, and CVD at low temperature were all tested without any additional finishing. Those prepared by CVD at medium and ordinary temperatures and salt bath immersion were tested after lapping. The chromium plated specimens were ground before bending, coining, rolling tests, lapping for indentation, scratch, hammering, and ironing tests. Surface finishing of the specimens on test was maintained, with some exceptions, at Rmax. 0.03~0.1 μm for indentation, scratch, and hammering, while Rmax. 0.1~0.8 μm was achieved for bending, coining, ironing, and rolling. However, CVD-TiC and TiC+TiN at ordinary temperature and salt bath VC had a Rmax. of 1~3 μm.

RESULTS AND DISCUSSION

Table 1 lists the results of all kinds of tests. The rating of A, B, C, and D in the table is determined based on the relation between the degree of spalling and the load applied, the number of shots or the number of rotations at which spalling occurred. Other failures as shown in Fig. 1 are not taken into account for the rating. The typical examples of such failures are chipping and the resulting exposure of the substrate surface observed in the hammering test and cracking in chromium coatings by plating, which produced a clear acoustic signal accompanied by no spalling in the scratch test. However, in some tests (bending, coining, and rolling) there is no denying the possibility that some spalling was induced by the chipping which occurred previously. It is obvious that large deformation and the resulting cracking in the indentation test is a determining factor in the extent and shape of spalling.

In the indentation test, only cracks or small spallings surrounded by cracks were observed on most specimens at smaller load (600 N) and larger load (1500 N) respectively. The shapes of cracks and spallings induced by cracking varied from coating to coating. This means that physical properties of coatings influence the evaluation results for adhesion strength. As an extreme case, some TiN coatings by PVD spalled in a large annular area surrounding the indentation and only this coating can clearly be concluded to be the worst in this testing method.

Spalling was observed only on some specimens coated by PVD and PCVD in the scratch test under smaller loading (scratch test I). Large loading by LSRH testing machine (scratch test II) produced spalling on all except chromium plated specimens. Although clear acoustic signals were detected on chromium coated specimens at very small load, only cracking was recognized through microscopic observation.

Coatings came off in two different ways in the hammering test. Spalling was detected at a small number of shots only on TiN coating by ion plating and W_3C coating by low temperature CVD. It is concluded through microscopic observation and examination by X-ray microanalyser and an instrument for measuring the surface roughness that the damage observed on other coatings was caused by wear and chipping during repetition of hammering.

Remarkable wear, scoring, and resulting increase in friction were detected in the metal forming tests shown in Fig. 2 for the ironing test and Fig. 3 for the bending test. These methods are also useful for evaluating wear and scoring resistance. Spalling was clearly observed only on W_3C coating by low temperature CVD and some TiN coatings by ion plating and by PVD in the ironing test. Bending test and coining test under proper conditions also caused spalling on the die radius of specimens in the bending test (Fig. 4) and on the edge of specimens in the coining test (Fig. 5).

In the rolling test, spalling occurred on almost all specimens at different numbers of rotations (as shown in Fig. 6) and a clear difference is achieved between coatings. None of the other failures were found.

The following are observations made while compiling the information contained in Table 1. It is quite important to select the proper testing condition to clarify the difference in spalling behaviour between each coating. For example, in the coining test only a few testing conditions out of seven are wholly useful for a wide variety of coatings as far as this research is concerned.

Comparative results between coatings were reversed in some cases by the testing procedure. Blowing away of wear debris was effective in increasing the number of shots to spalling in the hammering test. However, moisture in the compressed air used for blowing away remarkably accelerated the spalling for PCVD-TiN coatings but negligibly affected that for IP-TiN coatings. In the case of PCVD, chlorine possibly included in the coatings[13] may cause the problem. The results in Table 1 were thus obtained by blowing with dry air.

Surface finishing of specimens also considerably affected the results. In the coining test finishing by oilstone and the resulting rough surface on the edge (Rmax. 1 μm) (Table 1, coining III) caused spalling of TiN coating by PVD and PCVD at a smaller number of shots compared with the finishing by a rubber stone (Rmax. 0.3 μm) (Table 1, coining IV). Thus the comparison between coatings was made clearly by oilstone finishing. The effect of surface finishing was recognized in a similar way on TiC coatings by CVD and TiN coatings by CVD in the bending test and on VC coating by the immersion method in the rolling test. The phenomenon, also recognized in the bending test, should be kept in mind during practical application of these coatings onto dies for metal forming.

Spalling behaviour was considerably affected by substrate hardness. A typical example can be seen in the bending test of chromium plated specimens. Chromium plating on grey cast iron sustrate was spalled far more quickly than that on hardened D2. As further evidence, W_3C coating by low temperature CVD, which has a nickel layer between substrate and coating, was rated lowest in all testing methods in which specimens are subjected to high compression loads. Although nickel-phosphor plating prior to CVD coating is effective in improving adhesion strength as assessed by means of the bend tests,[10] lower hardness of the plating seems to have a counter effect in such a high loading. The harmful effect of soft substrate is often discussed.[1]

The significant effect of coating thickness on the critical load as detected by acoustic signal has been reported.[2-4] However, to date it has not been clearly recognized within the present research because of poor experimental data.

As shown in Table 1, results of PVD coatings vary quite considerably lot by lot. Since the specimens were prepared in a lot by us (except those cut off by gear cutters) and then subjected to the coating operation, no consideration is required of difference in grinding and polishing conditions which could be a causal factor for the scattering of adhesion strength of PVD coatings.[14] Therefore, some factors involved in coating and pretreatment procedures carried out by the treaters may cause this large scattering. On the other hand, little scattering was found in the lots of VC coating by immersion method even in the rolling test, which makes a clear distinction between coatings, as shown in Fig. 7.

Salt bath immersion and CVD at the medium and ordinary temperatures produced coatings found to be in the foremost rank in almost all tests. Diffusion of coating elements and carbon in substrate which occurs during the coating operation may cause these good results. Adhesive superiority of these coatings to the low temperature coatings (PVD and plating) is often recognized also in practical applications to dies for metal forming, etc. The information obtained especially in these three test methods simulating metal forming accords well with experiences in industrial applications.

It is noticeable that under appropriate conditions the PCVD coating achieved better evaluations in comparison with PVD coatings in some testing methods, especially in the hammering and rolling tests. The PCVD method has been attracting the attention of researchers throughout the world recently.[13,15]

The tests employed here not only caused spalling but also other failures (wear, galling, chipping and cracking) in most cases irrespective of the kind of coating. Unless an adhesive with a greater adhesion strength than that of the coatings tested could be employed spalling would occur. The force which induces spalling is exerted by the loading tools which penetrate into the substrate through the coating or are brought into contact with the surface of the coating. In the former, the depth of penetration of the loading tools (indentors) may be considerably affected by thickness, mechanical properties, etc. of coatings and the loading tools produce other failures, such as cracking and chipping, whose extent may also be affected by the abovementioned factors. In the latter, energy applied to the interface may also be determined by frictional behaviour between the coating surface and the loading tools, which is highly influenced by surface finishing, scoring behaviour, and mechanical properties of coatings. Wear and chipping also have an effect on the energy brought to the interface.

It is observed in many tests that other failures such as cracking and chipping often induced the spalling and coatings were sometimes removed from substrate by wear prior to the spalling. Thus it is quite difficult to evaluate the 'adhesion strength' whichever method is employed.

CONCLUSION

The adhesion strength of hard coatings produced by various methods (CVD, PVD, PCVD, salt bath immersion, and conventional plating) have been evaluated using seven methods. The following conclusions have been reached.

In most cases spalling is accompanied by other failures (cracking, chipping, wear, and scoring) which make it difficult to evaluate the adhesion strength except in connection with the occurrence of these other failures.

Among these testing methods rolling with strip induced almost only spalling, providing a clear difference in spalling durability between the coatings.

Edge finishing and surface finishing have a considerable effect on the results, as well as substrate hardness. This should be borne in mind during industrial application of such coatings to dies for metal forming, etc.

Coating by the high temperature process (CVD at ordinary and medium temperatures, and salt bath immersion) showed high adhesive strength in all testing methods. Among the low temperature coatings, PCVD seem to be the most promising from the point of view of adhesion according to the experimental results so far.

ACKNOWLEDGEMENTS

The authors are grateful for the assistance of Y Ohta, Y Tsuchiya, Y Takada, and Dr T Konaga, who contributed experimental skill, sustained effort and a grasp of the objectives for the accomplishment of the experimental program.

Table 1 Summary of the results

A ⟶ D : Rating based only on occurrence of spalling
Superior Inferior
CR: Accompanied with cracking
W: Accompanied with wear
CH: Accompanied with chipping
-: No any failures

	Process	Plating							PVD														PCVD						
		Chemi-cal	Electrolytic						Ion plating										Sputtering								
Specimens	Coating	Ni-P	Chromium						TiN									i-c	TiN			TiN				TiC		
	Lot No.	a	a	b	c	d	e	f	a	b	c	d	e	f	g	h	i	a	a	b	c	a	b	c	d	a	b	c
	Coating temp. °C (Nominal)		35															90 ~ 150				550						
	Thickness μm	18	25	25	25	3	6	16	1	2	4	1	3	1	3	0.3	0.5	3	2	2	2	2	2	3	14	2	2	3
	Substate material	D2	D2	Cast Iron	M2				M2							Co-Hss	Co-Hss	M2	M2			M2				M2		
	Substrate hardness HRC	63	62	12	62	63	64	63	63	62	63	63	62	63	62	67	67	65	62	62	61	63	61	62	63	62	64	64
Evalution methods	Indentation	—	—	—	—	A^{CR}	A^{CR}	A^{CR}	B	D	D	B or C	B or C	—	—	—	—	B or C	D	B	D	B or C	—	—	—	B or C	—	—
	Scratch I	—	—	—	—	A^{CH}	A^{CR}	A^{CR}	A^{CH}	A^{CH}	C	—	—	—	—	—	—	—	D	C	C	C	—	—	—	C	—	—
	Scratch II	—	—	—	—	A^{-}	A^{CR}	A^{CR}	—	—	—	—	B	—	—	—	—	D	—	—	—	—	—	—	—	—	—	—
	Hammering	—	—	—	—	—	—	A^{CH}	—	—	—	C	C	—	—	D	D	—	—	—	—	A^{CH}	—	—	—	A^{CH}	—	—
	Bending I	D	C^{W}	D	—	—	—	—	—	—	—	—	—	—	—	—	—	—	—	—	—	—	—	—	—	—	—	—
	Bending II	—	—	—	D	—	—	A^{W}	D	—	D	A^{-}	A^{-}	—	—	—	—	—	D	A^{-}	—	A^{-}	—	—	—	A^{-}	—	—
	Bending III	—	—	—	—	—	—	D^{W}	D	—	—	—	—	—	—	—	—	—	—	—	—	—	—	—	—	—	—	—
	Bending IV	—	—	—	—	—	—	D^{W}	D	—	D	—	—	—	—	—	—	—	—	—	—	—	—	—	—	—	—	—
	Coining I	—	—	—	—	—	—	A^{W}	—	—	—	—	—	—	C	—	—	—	—	—	—	—	—	—	—	—	—	—
	Coining II	—	—	—	—	—	—	—	—	—	—	C	C	—	—	—	—	—	—	—	D	—	A^{CH}	C	—	—	A^{CH}	D
	Coining III	—	—	—	—	—	—	A^{W}	—	—	—	C	C	—	—	—	—	—	D	A^{CH}	—	C	—	—	—	C	—	—
	Coining IV	—	—	—	—	—	—	—	—	—	—	A^{CH}	A^{CH}	—	—	—	—	—	D	A^{CH}	—	A^{CH}	—	—	—	A^{CH}	—	—
	Coining V	—	—	—	—	—	—	A^{W}	—	—	—	D	D	—	—	—	—	—	—	—	—	—	—	—	—	—	—	—
	Coining VI	—	—	—	—	—	—	A^{W}	—	—	—	—	—	D	D	—	—	—	—	—	—	—	—	—	—	—	—	—
	Coining VII	—	—	—	—	—	—	—	—	—	—	D	—	—	—	—	—	—	—	—	—	—	—	—	—	—	—	—
	Ironing	—	—	—	—	—	—	A^{W}	—	C	—	—	—	—	—	—	—	A^{W}	—	—	—	—	—	A^{W}	D	—	—	—
	Rolling with 10% slip	—	—	—	—	—	—	D	—	D	D	—	D	—	—	—	—	—	—	—	—	B	—	—	—	B	B	—

Remark Bending (I) and (II) : U shape bending of cold rolled mild steel strip 0.8 mm thick, Difference in number of shots
Bending (III) and (IV): U shape bending of stainless steel strip 1.0 mm thick, Difference in number of shots
Coining (I)-(IV) : Coining to reduce thickness of 1.6 mm thick mild steel strip, Diffrence in number of shots and die finishing
Coining (V) and (VI) : Coining to reduce thickness of 1.0 mm thick W1 steel strip, Diffrence in die finishing
Coining (VII) : Coining to reduce thickness of 1.0 mm thick 304 steel strip
Scratch (I) : By Heidon 14 surface testing machine under smaller load (less than 3N)
Scratch (II) : By LSRH Revetest scratch tester under larger load (max 50N)
Surface roughness of specimens: Rmax 0.03-0.1 for indentation, scratch and hammering with some exceptions
Rmax 0.3-0.8 for metal bending, coining and rolling with some exceptions

	Process	CVD								Salt bath immersion														
Specimens	Coating	W_3C (Ni-P)		TiCN	TiN	TiC		TiC+TiN		VC												NbC		Cr_7C_3
	Lot No.	a	b	a	a	a	b	a	b	a	b	c	d	e	f	g	h	i	j	k	l	a	b	a
	Coating temp. °C (Nominal)			850	900	1020		1020 900		1025												1025		900
	Thickness μm	7 (4)	5 (8)	17	6	6	6	7	6	7	2	4	7	5	5	7	5	7	7	6	5	8	5	5
	Substate material	M2	D2	M2		D2	M2	D2	M2	D2	M2											D2	M2	O1
	Substrate hardness HRC	61	58	63	63	61	62	62		62	64	64	63	65	63							61	64	63
Evalution methods	Indentation	B or C	B or C	A^{CR}	C	—	C	—	A^{CR}	—	B	B	C	—	B or C	B or C	B or C	B or C	B or C	B or C	B or C	—	—	—
	Scratch I	A^{CR}	—	B	A^{CH}	—	A^{CR}	—	A^{CH}	—	A^{CH}	A^{CH}	A^{CH}	—	—	—	—	—	—	—	—	—	—	—
	Scratch II	D	—	A	B	—	C	—	B	—	—	—	C	—	—	—	—	—	—	—	—	—	—	—
	Hammering	D	—	A^{CH}	A^{CH}	—	A^{CH}	—	A^{CH}	—	—	—	A^{CH}	—	—	—	—	—	—	—	—	—	—	—
	Bending I	—	—	—	—	A^{CH}	—	A^{CH}	—	A^{-}	—	—	—	—	—	—	—	—	—	—	—	A^{-}	—	—
	Bending II	D	—	—	A^{-}	—	A^{-}	—	—	—	—	—	A^{-}	—	—	—	—	—	—	—	—	—	—	—
	Bending III	—	—	—	A^{-}	—	A^{-}	—	—	—	—	—	—	—	—	—	—	—	—	—	—	—	—	—
	Bending IV	—	—	—	—	—	—	—	B	—	—	—	A^{-}	—	—	—	—	—	—	—	—	—	—	—
	Coining I	D	—	—	—	—	A^{CH}	—	—	—	—	—	A^{CH}	—	—	—	—	—	—	—	—	—	—	—
	Coining II	—	—	—	—	—	—	—	—	—	—	—	A^{CH}	—	—	—	—	—	—	—	—	—	—	—
	Coining III	—	—	—	—	—	—	—	—	—	—	—	A^{CH}	—	—	—	—	—	—	—	—	—	—	—
	Coining IV	D	—	—	—	—	A^{CH}	—	A^{CH}	—	—	—	A^{CH}	—	—	—	—	—	—	—	—	—	—	—
	Coining V	—	—	—	—	—	—	—	—	—	—	—	—	—	—	—	—	—	—	—	—	—	—	—
	Coining VI	—	—	—	—	—	—	—	—	—	—	—	A^{CH}	—	—	—	—	—	—	—	—	—	—	—
	Coining VII	—	—	—	—	—	—	—	—	—	—	—	A^{CH}	—	—	—	—	—	—	—	—	—	—	—
	Ironing	D	—	A^{-}	A^{-}	—	A^{-}	—	A^{-}	—	—	—	—	A^{-}	—	—	—	—	—	—	—	—	A^{-}	A^{W}
	Rolling with 10% slip	D	—	B	C	—	A	—	C	—	—	—	A	—	A	A	A	A	A	A	A	—	—	—

	Indentation	Scratch	Hammering	Bending	Coining	Ironing	Rolling with slip
Methods							
Main stress concerned							
Failures observed	Spalling Cracking	Spalling Cracking Chipping	Spalling Cracking Chipping Wear	Spalling Cracking Chipping Wear Scoring	Spalling Chipping Wear Scoring	Spalling Galling Wear	Spalling

Fig. 1 Method, main stress concerned, and resulting failures in the tests

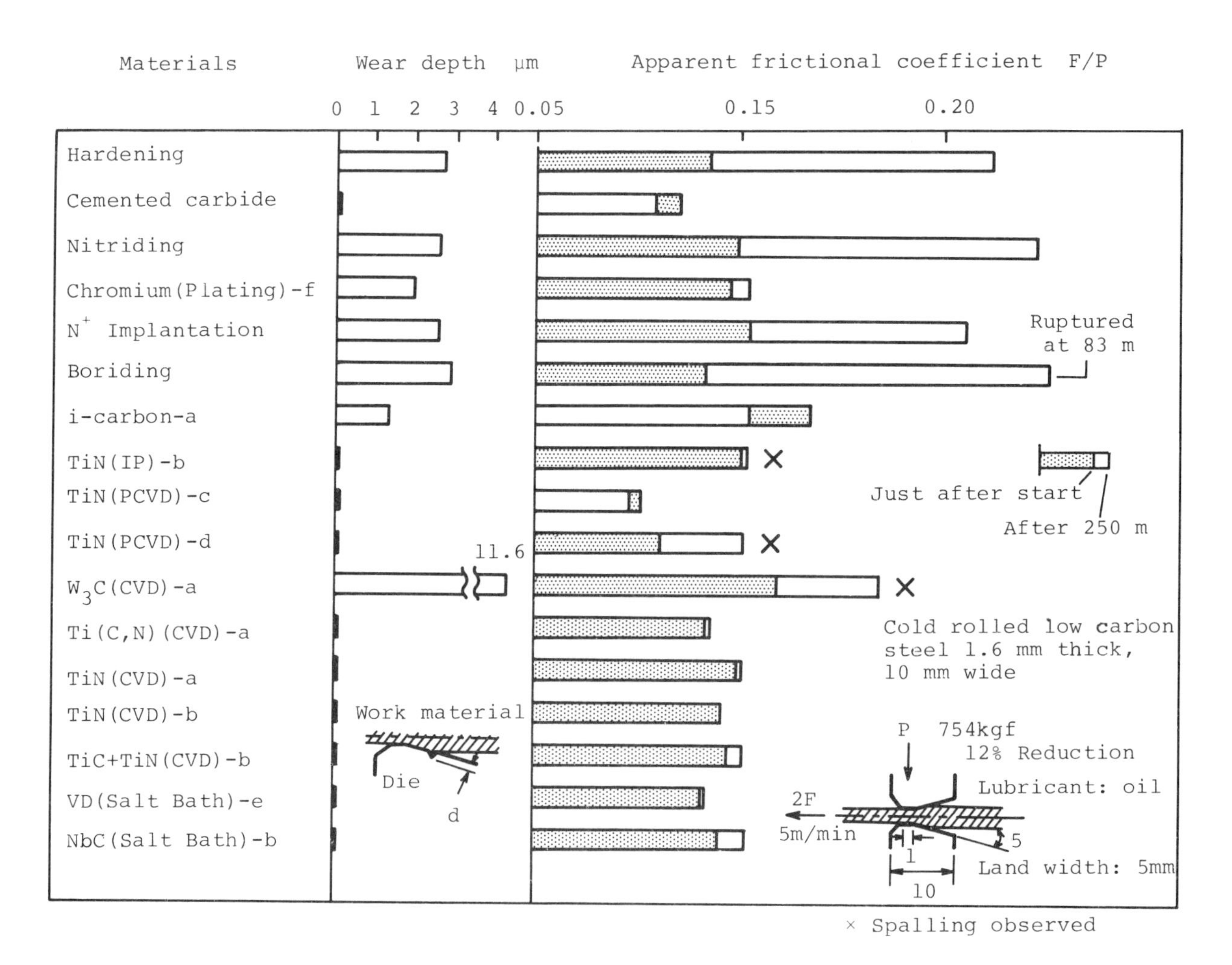

Fig. 2 Comparative friction coefficient, depth of wear, and scoring observed on dies in the ironing test

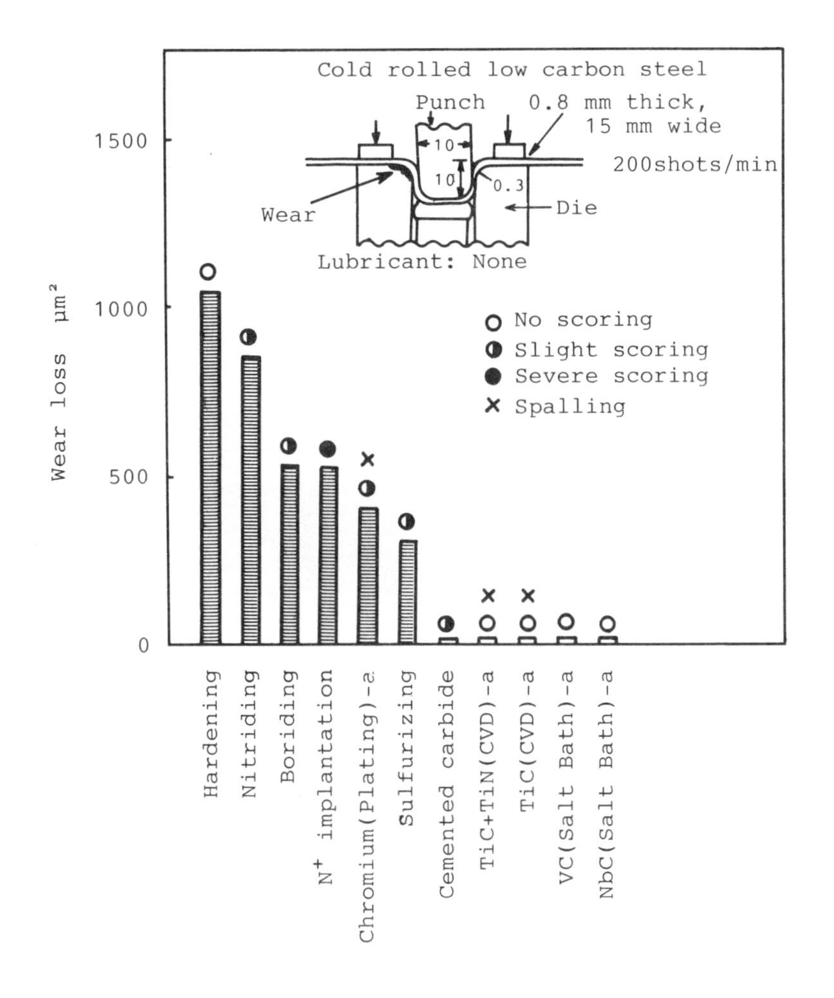

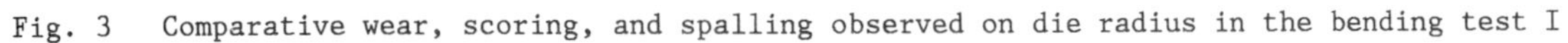

Fig. 3 Comparative wear, scoring, and spalling observed on die radius in the bending test I

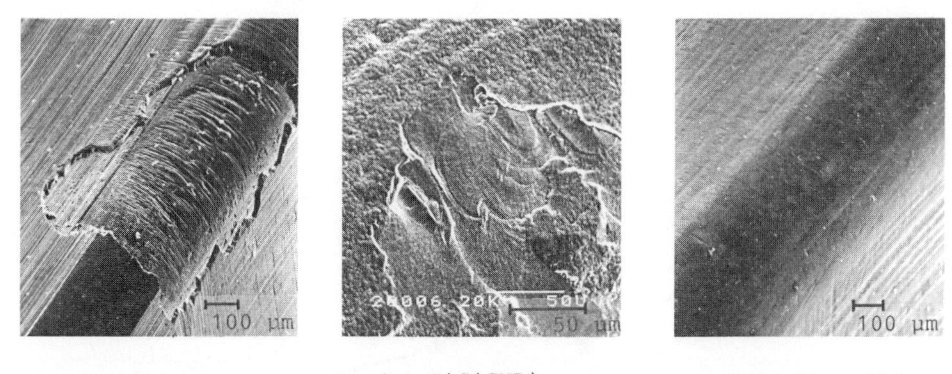

(a) Chromium(Plating)-a 5,000shots

(b) TiC(CVD)-a 30,000shots

(c) VC(Salt Bath)-a 30,000shots

Fig. 4 Spalling and chipping observed in the bending test I: (a) spalling; (b) chipping; (c) no failure

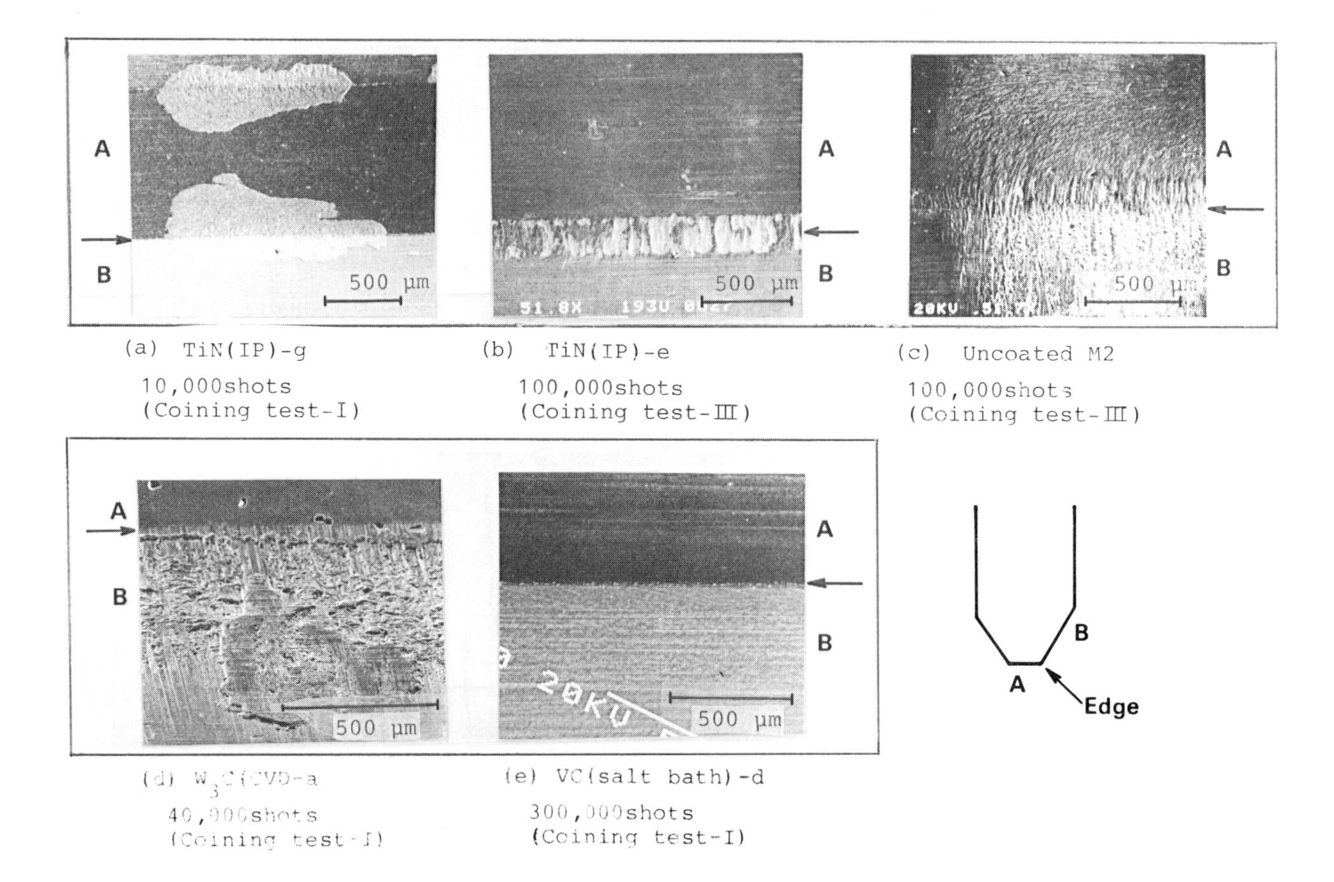

Fig. 5 Spalling and wear observed in the coining test I and II: (a), (b), and (c) spalling; (d) wear; (e) no failure

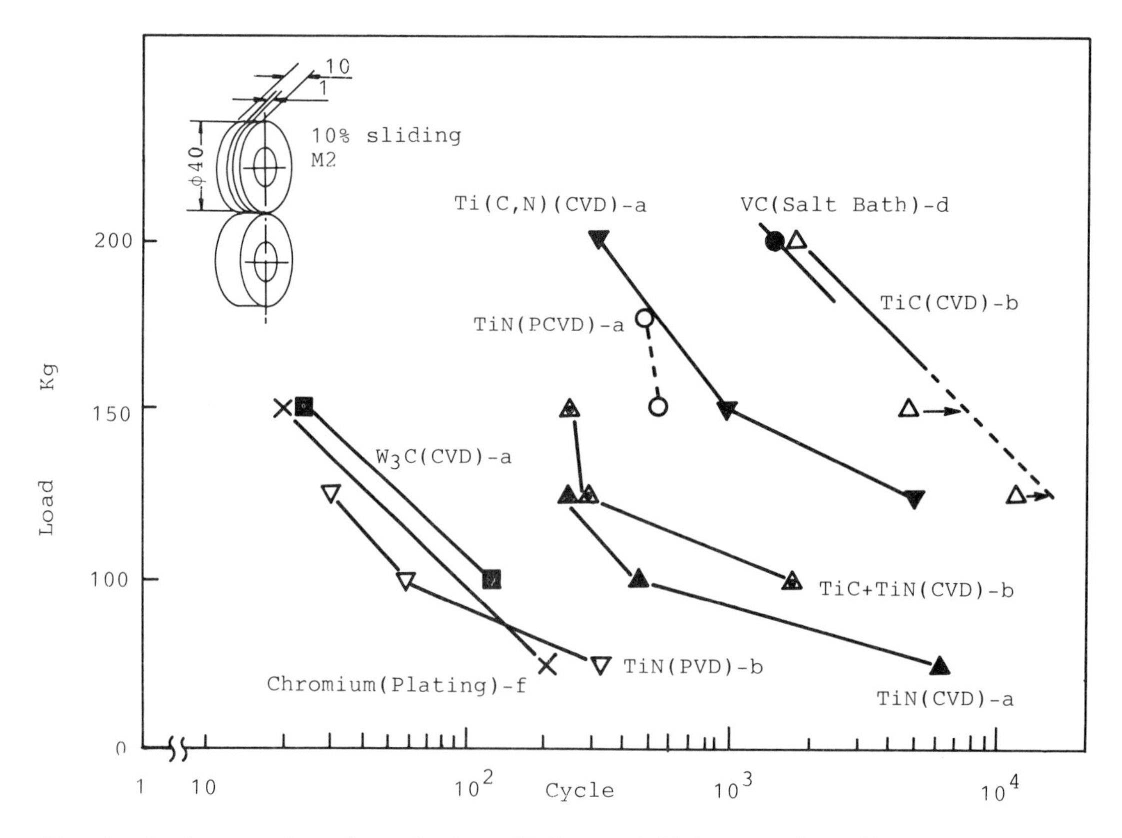

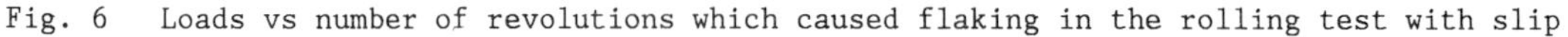

Fig. 6 Loads vs number of revolutions which caused flaking in the rolling test with slip

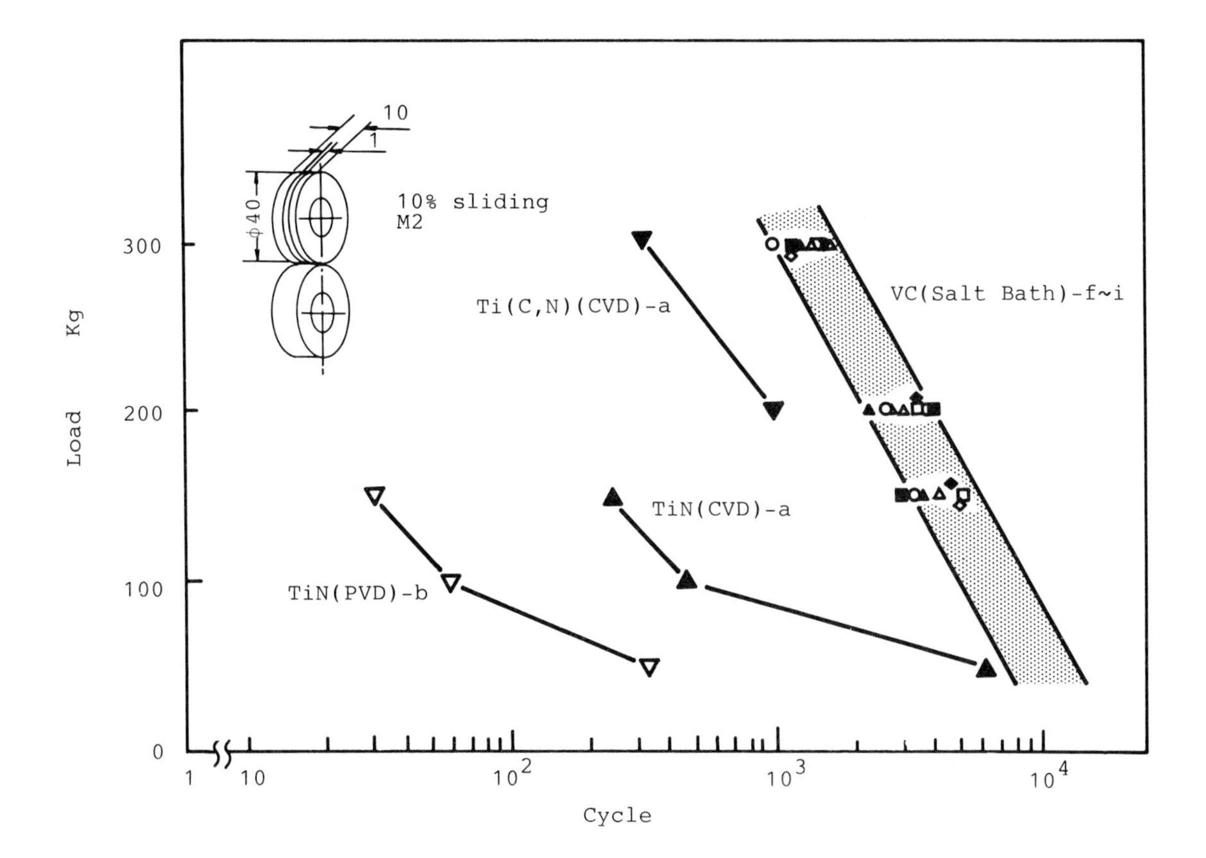

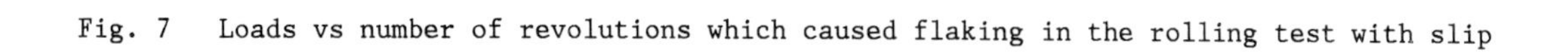
Fig. 7 Loads vs number of revolutions which caused flaking in the rolling test with slip

Stability, wear resistance and microstructure of Fe–Cr–C and Fe–Cr–Si–C hardfacing alloys

S ATAMERT and H K D H BHADESHIA

The authors are with the University of Cambridge.

ABSTRACT

Fe-34Cr-4.5C wt-% hardfacing alloys deposited by manual metal arc welding have a metastable microstructure consisting of large, primary M_7C_3 carbides in a matrix which is a eutectic mixture of austenite and more M_7C_3 carbides. Given thermal activation and time, the austenite in the microstructure tends to decompose into a Cr depleted ferrite and M_7C_3 carbides.

In an effort to improve the high-temperature stability, oxidation, and corrosion resistance of such hardfacing alloys, a systematic study has been carried out on the role of Si in the microstructure, phase chemistry, and abrasive wear resistance of Fe-Cr-C base alloys. The investigations have been carried out on model alloys cast by an argon arc melting technique, which is in general found to simulate adequately the corresponding structures obtained by manual metal arc welding.

In all cases, Si is found to partition strongly into the austenite during solidification; Si concentrations of up to 18 at-% have been found in the matrix austenite, even though the average Si concentration used was far less.

Si is found to influence significantly the morphology of M_7C_3 carbides, possibly through reducing the orientation dependence of interface energy. This is believed to be beneficial in enhancing the toughness and hence impact wear resistance of the alloys. A further effect of Si is to reduce the Cr concentration of the austenite in the as-cast alloys. This is advantageous because the Cr is used better in the formation of M_7C_3 carbides. The carbon concentration of the austenite is found to increase with Si level, although the reasons for this await detailed thermodynamic analysis. Preliminary work indicates that the high-Si alloys have a better abrasion wear resistance.

INTRODUCTION

Conventional hardfacing alloys can be classified into five main classes: hard steels, hard cast irons, highly alloyed iron base alloys, cobalt base alloys, and nickel base alloys.[1] These alloys are in general very effective and have been established after many years of steady development work. They do however contain relatively large amounts of expensive and sometimes strategically important alloying additions. It is still the case that the detailed role of some of these additions is not understood. This work is part of a systematic re-examination of iron, nickel, and cobalt base hardfacing alloys[2] with a view to establishing a quantitative alloy design procedure to enable the optimisation properties (corrosion, oxidation, and wear resistance), alloy content, cost, and microstructural stability. The alloy design methods involve the use of thermodynamic calculations and will be reported elsewhere.

For reasons which will become apparent later, this work deals specifically with a microstructural characterisation of the influence of silicon additions to iron base hardfacing alloys of nominal base composition Fe-4.5C-34Cr wt-%. These alloys, when deposited by arc welding techniques, have a microstructure consisting of very hard (≈ 1400 HV, volume fraction ≈ 0.6) primary M_7C_3 carbides in a matrix which is a eutectic mixture of austenite (γ) and M_7C_3 carbides, when the alloy is not diluted by the metal on which it is deposited.[*3] Although after solidification the ambient temperature microstructure is a mixture of austenite and M_7C_3, the equilibrium microstructure should consist of ferrite and M_7C_3, because at sufficiently slow cooling rates the austenite should undergo a diffusional transformation to ferrite and more M_7C_3.[4] However, during welding the structure becomes configurationally frozen[†3] at a temperature of around 1150°C, because such diffusional transformations require considerable time to accomplish both the redistribution of substitu-

* The term M_7C_3 refers to the fact that the carbide is not a pure chromium carbide, but also contains iron and other substitutional alloying additions. 'M' therefore stands for some combination of metal atoms.

† A configurationally frozen microstructure is one which does not change during cooling from the temperature at which it becomes configurationally frozen.[5] The term frozen is not to be confused with the freezing of liquid. In the present context, the alloy solidifies and then becomes configurationally frozen at some lower temperature where the mobility of atoms becomes inadequate, for the given cooling rate, to support diffusional transformation. It should be noted that a higher cooling rate should increase the temperature at which a microstructure becomes configurationally frozen.

tional alloying additions and to rearrange the atoms as required for the lattice changes accompanying transformation (the diffusion necessary for the latter is called reconstructive diffusion[6]). It follows that the microstructures obtained are thermodynamically unstable and the γ should tend to decompose if the hardfacing deposit is in service at high temperatures (≈ 700°C). The diffusional transformation may not influence significantly the wear resistance, but it might alter the corrosion and oxidation resistance of the alloy; the metastable austenite obtained after deposition has a relatively high Cr concentration (≈ 16 wt-%)[3] but that of the ferrite is much lower. Of course, as far as the M_7C_3 is concerned, its Cr concentration (≈ 40 wt-%) is more than adequate for corrosion and oxidation resistance.[3] A major aim of the present work was to design a Fe-Cr-C base alloy which has a better matrix corrosion and oxidation resistance at all service temperatures even following inevitable diffusional transformations. It was proposed to achieve this by adding high levels of silicon, which is known to have a very low solubility in carbides[7] (and specifically also in M_7C_3[3]). Hence, most of the silicon in the alloy should during solidification be partitioned into the liquid phase, and subsequently (during the eutectic transformation) into the austenite, giving a much enhanced level of Si. There are several advantages in this; Si is known to enhance the oxidation and corrosion resistance of iron[8] and it is a very cheap alloy addition. Furthermore, if the austenite subsequently transforms to ferrite and M_7C_3, then the Si concentration of the ferrite should be even higher, so that oxidation resistance should be enhanced. It was also anticipated that Si may refine the carbides in the microstructure; such an effect is well established for high strength steels.[9] At the same time, any influence of further alloy additions on the partitioning of Cr between the M_7C_3 and matrix also needed to be investigated; any change may influence the volume fraction of hard phase and therefore wear properties.

The relationship between microstructure and wear is known to be complex and ill understood.[2,10,11] For example, the behaviour of an alloy under abrasive conditions may be quite different under impact conditions.[12] A working hypothesis is nevertheless necessary during alloy development and in the present work the rather simple and probably crude hypothesis is adopted that good abrasion resistance is obtained when the volume fraction of the hard phase is high, and when the hard phase is well supported by a relatively tough matrix. For impact wear resistance, the hard phase is prone to cracking, so that a more equiaxed hard phase should give better wear resistance in these circumstances.

EXPERIMENTAL PROCEDURE

The experiments utilised both welds deposited by a manual metal arc welding technique and experimental casts (weight ≈ 65 g) made from high purity elements in an argon arc furnace with a water cooled copper mould.

The welds were deposited in three layers so that the top layer could be examined in an essentially undiluted condition. Electrodes 4 mm in diameter were used, the welding conditions being 160 A, 23 V a.c., with a welding speed of about 0.004 m/s and an interpass temperature of about 350°C. The electrical energy input is therefore ≈ 920 J/mm. These conditions are the same as used by Svensson *et al.*[3]

Thin-foil specimens for transmission electron microscopy were prepared from 0.25 mm thick discs spark machined from the weld deposits. The discs were subsequently thinned and electropolished in a twin-jet polishing unit using a 5% perchloric acid, 25% glycerol, and 70% ethanol mixture at ambient temperature and 55 V. The foils were examined in a Philips EM400T transmission electron microscope operated at 120 kV. Microanalysis experiments were also carried out on this microscope, using an energy dispersive X-ray analysis facility. The specimens, which were about 100 nm thick, were held in a beryllium holder tilted 35° from the normal which is equal to the take-off angle. The X-ray count rate was optimised at about 200 counts/s, over a count period of 100 s, giving a typical statistical accuracy of ≈ 1%. The data were analysed using the LINK RTS 2 FLS program for thin foil microanalysis; this corrects the data for atomic number and absorption and accounts for overlapping peaks by fitting standard profiles. Even though the probe diameter used was about 3 nm, beam spreading due to scattering of electrons within the thin foil gave an estimated broadened beam diameter of ≈ 20 nm. The elements analysed were iron, silicon, chromium, and manganese; at typical levels found in the present alloys, it has been established that none of these elements cause significant fluorescence effects in thin foils.[3]

The microanalysis results reported for the primary carbides and for the regions near primary carbides were obtained using an EDAX system on a scanning electron microscope, and the data are fully corrected for atomic number, absorption, and fluorescence.

The beryllium window on the detector used in the microanalysis system absorbs X-rays from light elements; the concentration of carbon in the M_7C_3 is therefore determined by assuming stoichiometry, and of austenite from lattice parameter measurements. However, the raw data obtained prior to any corrections are also presented later in the text. The raw data excludes carbon; the term y_i refers to the concentration of element i in atomic percent when the presence of carbon is ignored. The corresponding true concentration (at-%, obtained after correcting for carbon) is denoted x_i.

The carbon concentration of the austenite was determined from lattice parameter measurements carried out on an X-ray diffractometer with step-scanning at 0.02° 2θ intervals for the range 2θ = 10-120°.

Abrasion wear tests were performed on 5 mm diameter cylindrical specimens under a 0.43 kg load (P), the flat face of the specimen being in contact with a rotating (sliding velocity 0.63 m/s) 75 mesh SiC coated disc. The disc was renewed after ever 10 min interval. At least three experiments were carried out for each alloy, the wear rate being deduced from the weight loss (w) as a function of time. The specific wear resistance (R) is given by $R = w/(\rho PL)$, where ρ is the specimen density and L is the travel distance in metres for each 10 min interval. The chemical compositions of the alloys used are presented in Table 1.

RESULTS AND DISCUSSION

Comparison between weld and experimental alloy

Much of the initial work has been carried out on experimental 65 g melts discussed earlier, because the design of suitable welding electrodes is a more complex problem. It was therefore felt necessary to check that the essential microstructures of experimental welds compared well with those obtained by manual metal arc welding.

Alloy S1 is essentially of the same composition as alloy M1; the slightly lower carbon and chromium concentrations of S1 should not significantly influence the nature of the γ or M_7C_3, although the volume fraction of M_7C_3 should be lower in S1. Optical microscopy (Fig. 1) demonstrated that the alloys have identical microstructures; microanalysis results (Fig. 2, Table 2) show that the austenite in S1 has a higher Cr content, and this is consistent with the fact that the casting technique involves a somewhat higher cooling rate than that associated with arc welding. This would mean that the alloy becomes configurationally frozen at a higher temperature, where the Cr concentration in γ in equilibrium with M_7C_3 is expected to be higher. With this exception, the results show that argon arc melts can be used in simulating manual metal arc weld deposits, especially for revealing trends in microstructure and phase composition. A comparison of the compositions of M1 and S1 is presented in Table 2.

Higher silicon alloys

Optical micrographs of alloys S2 and S3 are presented in Fig. 3. A major effect of silicon is to change the morphology of the primary carbides. At very low silicon concentrations, the primary carbides are elongated (Fig. 1) and tend to become more equiaxed with increasing silicon concentration (considerable microscopy confirms that this is not just a sectioning effect). The changes imply that the orientation dependence of the liquid/M_7C_3 interface energy becomes less orientation dependent with increasing Si. A similar effect is also evident for the eutectic M_7C_3 carbides which also appear more globular in alloy S3. These morphological changes should be beneficial to the toughness of the alloy, perhaps imparting better impact abrasion wear resistance, and experiments to confirm this are in progress.

The microanalysis data (Fig. 4, Table 2) show that the silicon partitions strongly to the matrix austenite. This should considerably enhance the oxidation and corrosion resistance of the alloys.

The effect of Si is also to reduce the level of Cr in the austenite during the eutectic decomposition of liquid. This may at first sight seem detrimental, but the austenite should in any case eventually decompose to low-chromium ferrite. The decrease in the Cr level of γ formed during solidification should in fact be beneficial since the Cr is best used in forming the hard M_7C_3 carbides during solidification.

The carbon concentration of the austenite seems to increase significantly as the silicon concentration increases. The reason for this is not clear and will have to await detailed thermodynamic analysis.

Results from the abrasion wear tests are presented in Table 3. They show that a generally improved wear resistance is obtained with the addition of Si, consistent with the interpretation of the microstructural and microanalytical data.

CONCLUSIONS

An attempt has been made to rationalise the influence of silicon on the detailed microstructure and phase composition of Fe-34Cr-4.5C alloys. The investigation has been carried out on model alloys cast by an argon arc melting technique, which is in general found to simulate adequately the corresponding structures obtained by deposition during manual metal arc welding.

In all cases, Si is found to partition strongly into the austenite during solidification; Si concentrations of up to 18 at-% have been found in the matrix austenite, even though the average Si concentration used was much lower.

It is found that Si causes significant changes in the morphology of M_7C_3 carbides. Both primary and eutectic M_7C_3 carbides tend to adopt more equiaxed morphologies and this has been attributed tentatively to Si causing a decrease in the orientation dependence of the liquid/M_7C_3 interfacial energy. In any case, the effect may be advantageous in enhancing the toughness and hence impact wear resistance of the high-Si alloys.

A further effect of Si is to reduce the Cr concentration of the austenite in the as-cast alloys. This is regarded as beneficial since the Cr is used better in the formation of M_7C_3 carbides whose volume fraction should therefore increase with Si level. It is pointed out that if the austenite eventually decomposes to low-Cr ferrite, as required by equilibrium, then there is little point in having a high Cr content in the γ in any case.

The carbon concentration of austenite significantly increases with Si; the reasons for this are not clear and an explanation must await detailed thermodynamic analysis of the equilibrium between austenite and M_7C_3 carbide. This effect can be compensated for by increasing the general carbon level of the alloys.

Preliminary abrasion wear tests indicate that the high silicon alloys have a better wear resistance, and welding electrodes are currently being prepared for further investigations in which the effect of dilution on the first two layers of the usual manual metal arc deposits will also need to be studied.

ACKNOWLEDGEMENTS

The authors are grateful to Professor D. Hull for the provision of laboratory facilities, to B. Gretoft, L.-E. Svensson, and B. Ulander for helpful discussions, and to ESAB AB (Sweden) for the provision of samples. Financial support from the Turkish Government is also gratefully acknowledged.

REFERENCES

1 Metals handbook, 9th edn, vol. 6, 1983, p.771, ASM, Ohio, USA
2 S. Atamert, CPGS Thesis, University of Cambridge, 1985
3 L.-E. Svensson, B. Gretoft, B. Ulander, and H.K.D.H. Bhadeshia, J.Mat.Sci., 21, 1986, 1015
4 Metals handbook, 7th edn, vol. 8, 1973, p.402, ASM, Ohio, USA
5 D. Turnbull, Metall. Trans. 12A, 1981, 695
6 H.K.D.H. Bhadeshia, Prog. Mat. Sci., 29, 1985, 321
7 W.S. Owen, Trans. ASM, 46, 1954, 812
8 N. Birks and G.H. Meier, 'Introduction to high temperature oxidation of metals', 1983, p.85, London, Arnold
9 J. Gordine and I. Codd, J.I.S.I., 207, 1969, 461
10 K. Gahr, 'Wear of materials', p.266
11 K. Gahr and V. Doanne, Metall. Trans. 11A, 1980, 613
12 B. Ulander, Private communication, 1987

Table 1 Chemical compositions of Fe-C-Cr-Si alloys, in wt-%. Alloy numbers beginning with the letter 'S' are experimental casts while the others have been deposited using manual metal arc welding

No.	C	Cr	Mn	Si
M1	4.50	37.9	1.41	0.86
S1	3.90	33.6	0.00	0.20
S2	4.48	34.4	0.00	3.60
S3	3.60	31.2	0.00	6.90

Table 2 Mean concentrations (at-%) of phases found in various alloys. The concentrations reported for regions near primary M_7C_3 are probably unreliable due to some overlap of information from both primary M_7C_3 and γ

	Primary M_7C_3	Near Primary M_7C_3	γ	Eutectic M_7C_3
M1				
Cr	44.9	14.7	15.0	42.8
Mn	0.9	1.9	1.7	0.1
Si	0.1	1.9	1.7	0.1
C	30	5.8	5.8	30
S1				
Cr	50.8	22.9	20.9	
Si	0.1	0.2	0.2	
C	30	2.4	2.4	
S2				
Cr	48.9	11.5	11.5	
Si	0.2	12.2	12.4	
C	30	3.7	3.7	
S3				
Cr	50.5	10.9	14.3	
Si	0.1	19.2	17.2	
C	30	6.3	6.3	

Table 3 Specific wear resistance data

Alloy	R, $m^2/(kg\ 10^{-11})$
S1	0.860
S2	0.014
S3	0.004

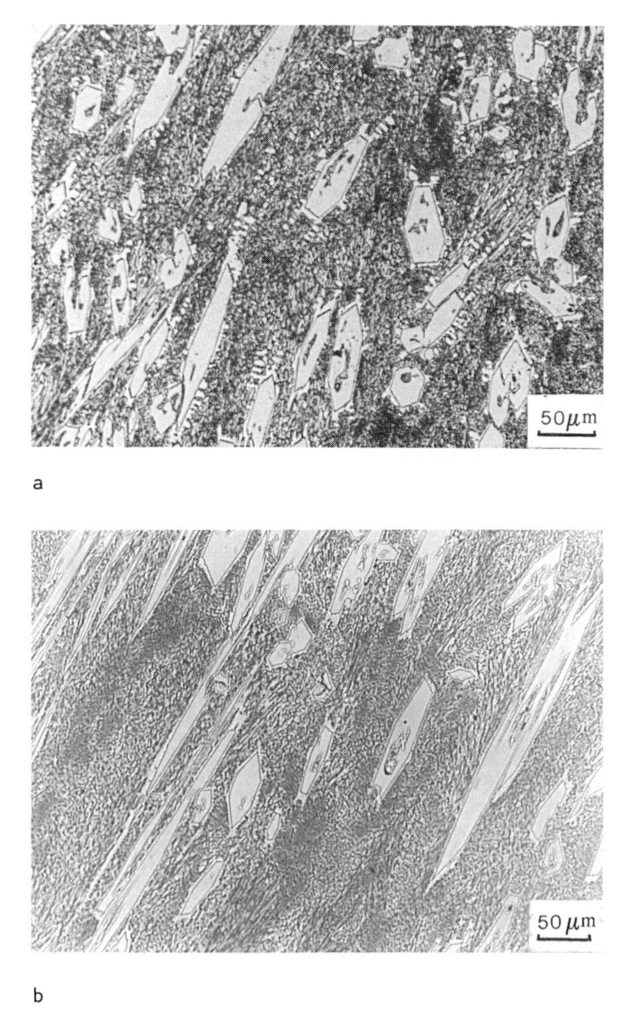

Fig. 1 Optical micrographs showing large primary M_7C_3 carbides in a eutectic mixture of γ + M_7C_3: (a) top, undiluted layer of alloy M1 deposited by manual metal arc welding, (b) alloy S1 argon arc melted in a water-cooled copper mould

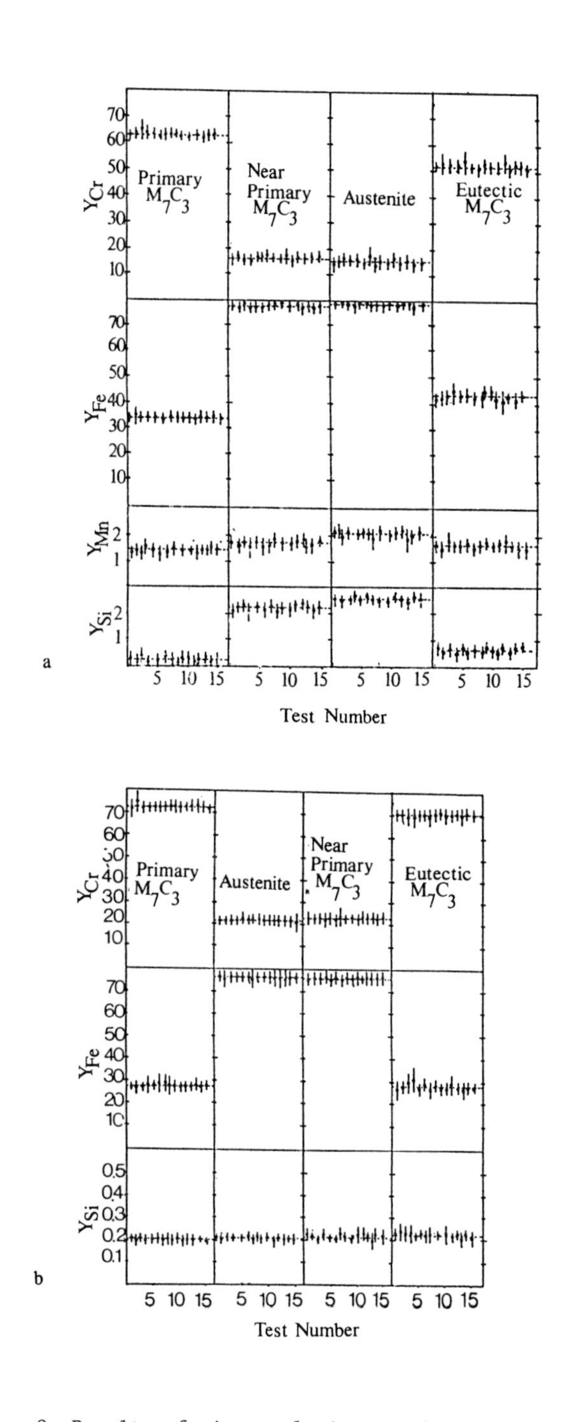

Fig. 2 Results of microanalysis experiments carried out in a transmission electron microscope. Continuous lines refer to average composition; concentrations y_i are in at-% and ignore the presence of carbon: (a) alloy M1, (b) alloy S1

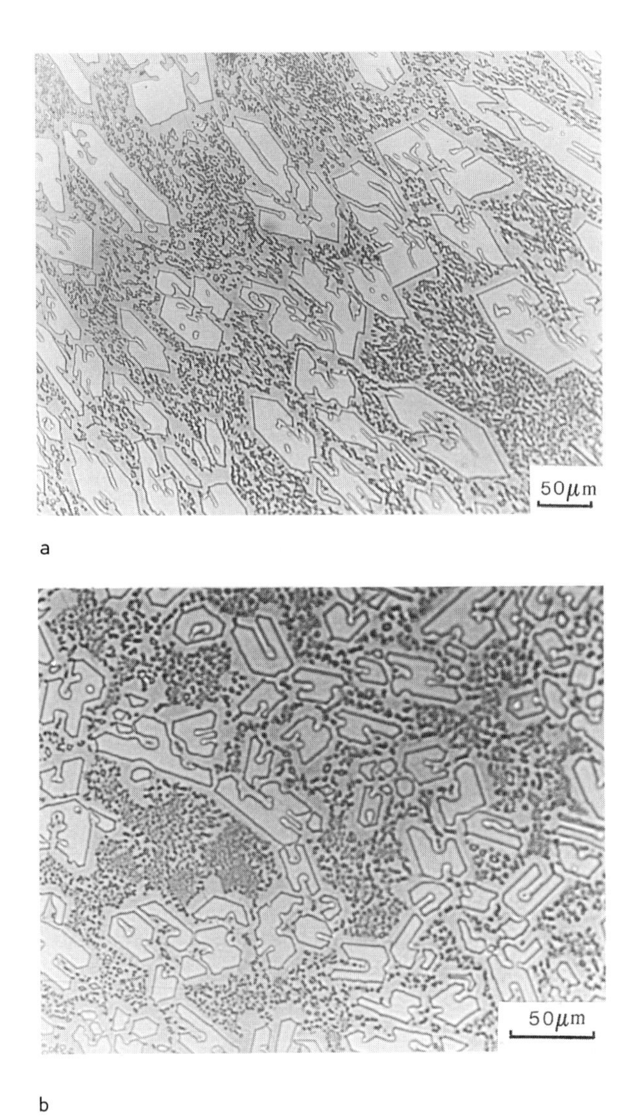

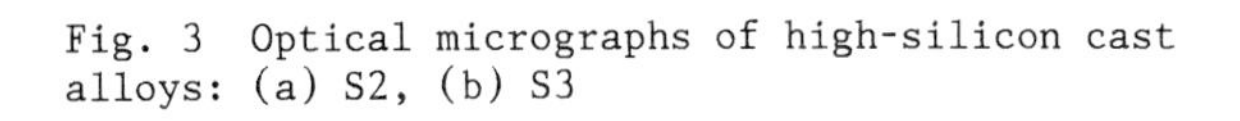

Fig. 3 Optical micrographs of high-silicon cast alloys: (a) S2, (b) S3

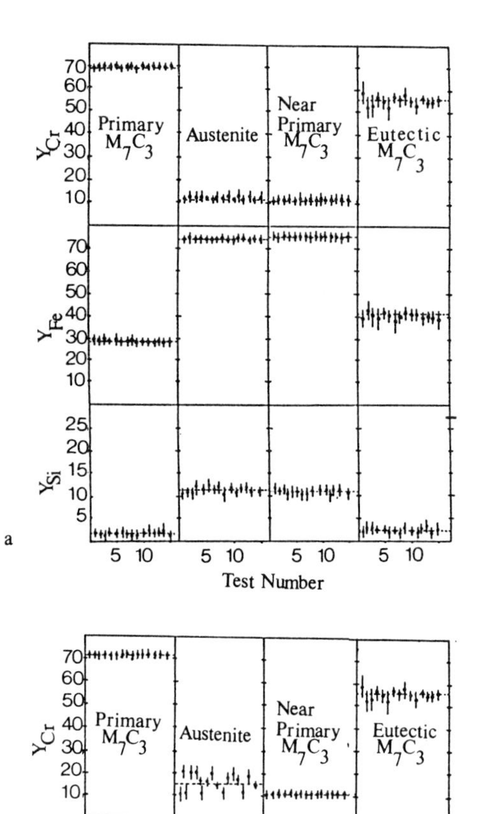

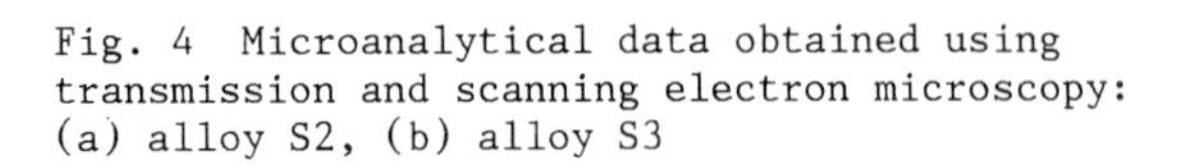

Fig. 4 Microanalytical data obtained using transmission and scanning electron microscopy: (a) alloy S2, (b) alloy S3

Ion-nitriding techniques for titanium and titanium alloys

B TESI, T BACCI, C BADINI and C GIANOGLIO

BT and TB are in the Dipartimento di Meccanica e Tecnologie Industriali, University of Florence, Italy; CB and CG are in the Dipartimento di Scienza dei Materiali e Ingegneria Chimica, Polytechnic of Turin, Italy.

SYNOPSIS

This study aimed to identify the influence of the working conditions and of the alloy chemical composition on the characteristics of surface layers in samples of titanium and some α-ß titanium alloys when subjected to ion-nitriding. It is pointed out that, for each alloy, the maximum depth of the nitrided layer is obtained using well defined pressure values.
Moreover, the nitrided layer phases appear to depend both on the alloy chemical composition and on the working conditions employed; in particular, the presence of ε nitride (Ti_2N) is correlated with specific values of the treatment temperature, pressure and gas composition.

INTRODUCTION

During the last decades, titanium and its alloys have been increasingly employed for special mechanical components used in mechanical and aeronautic constructions and in the bio-engineering field. These mechanical components must **often be** characterized also by a high wear resistance and by a surface hardness far higher than that of titanium and its alloys. To achieve the required surface hardness, it is necessary to apply appropriate thermochemical treatments.
Therefore, titanium surface nitriding has been studied in order to define industrial techniques capable of building up surface layers characterized both by high hardness and wear resistance. While the first studies /1-5/ analyzed the reaction of titanium alloys with gas mixtures containing nitrogen in furnaces at high temperature with an atmospheric or higher pressure, recent studies have investigated the possibility **of carrying** out the thermochemical treatment by means of more **sophisticated** techniques, using, for example, both radio frequency plasma and ionic glow-discharge units /6-10/.
In the reaction of titanium with nitrogen, a nitrided surface layer consisting of δ nitride TiN_x (with x in the range of 0.6-1 /11/), ε nitride Ti_2N and an interstitial solid solution of nitrogen in α titanium (maximum solubility 25% /3/) is produced; the presence and the quantity of these phases depend both on the adopted specific technique and on the selected working conditions /2,5,6/. In fact, it has been observed that, when using a traditional nitriding process, the diffusion layer (interstitial solid solution of nitrogen in α titanium) becomes thinner as the temperature is increased, contrary to what occurs in the compound layer (δ+ε nitrides) /3-5/. It has also been demonstrated that it is possible to realize compound layers consisting of only one phase (δ nitride) by using particular working conditions in the ion-nitriding /7/.
Further, the nitrided layer, in its heterogeneity, presents different hardness values according to the prevalent constituent phase: they have been measured between 2000 and 1500 Kg/mm^2 in the compound layer and between 1500 and 300 Kg/mm^2 in the diffusion layer, taking into consideration that these values depend also on different adopted gauge modalities /6,8,10/.
Thus, the specific goal of the present work aimed to identify how both the working conditions and the alloy chemical composition affect the characteristics of the ion-nitrided layers.

EXPERIMENTAL PROCEDURE

The ion-nitriding process for the treatment of titanium and titanium alloy samples was carried out in a laboratory plant similar, in its general scheme, to industrial scale units used for ion-nitriding of steel. The principal parts of the plant consist of (fig. 1):

- High vacuum cylindrical chamber in carbon steel, chromium plated, inside which a double screen in stainless steel AISI 304 forms the real treatment chamber (Φi=150 mm, H=150 mm). The equipment frame and the screens are earthed and form the anode of the plant. The sample support - cathode of the plant - is placed in the treatment chamber centre; it is made in titanium c.p. grade 4, insulated from the other metallic parts and connected to the negative pole of the power supplies.
- High vacuum pumps with appropriate measuring instruments to gauge the pressure in the treatment chamber during the different working phases.
- Two glow-discharge power supplies, connected in parallel to the anode and cathode of the treatment unit; each supply is capable of providing adjustable stabilized current from 0 to 500 mA and with a maximum output voltage of 5 KV.
- Gas feed system: high purity nitrogen and hydrogen are fed separately to the treatment unit through two gas purifiers (O_2 and H_2O less than 1 ppm); the flows are controlled by means of flow-meters.

For the experimental tests, prismatic samples 12x12x40 mm, with ground lateral surfaces (Ra=0.3 μm), were used. A hole (Φi=5 mm, H=30 mm), along the sample axis, allows insertion of an insulated thermocouple for the treatment temperature control.

Titanium c.p. grade 4 and α-β IMI 318, IMI 550 and OT4 alloys were used for the samples; the chemical composition of these materials is reported in **Table I**.

The working conditions were selected on the basis of literature data, including those which refer to titanium and titanium alloy nitriding techniques different from that of ion-nitriding.

Thus, particularly on the basis of data reported in /6,7,10/, for the different tests, pressure was fixed in the range of 0.5 - 32 Torr and temperature in the range of 800 - 1000 °C. The temperature values were selected in order to point out the influence on the process development both of the temperature and of the lattice structures of the examined alloys. The treatment gas composition (N_2 + H_2) was varied in a large range (from 10% to 80% of H_2) in order to evaluate the influence of H_2 percentage in the gas both on the nitrided case depth and on the phase nature. Finally, the time of the glow-discharge was selected, for the different tests, equal to 4, 8, 16 and 32 h.

For the ion-nitriding process the following procedure was adopted:

- Cleaning of the sample surface by cathodic sputtering for a period of about 30 min at a pressure of 0.1 Torr with a tension equal to 800 V. At the end of this phase the temperature was equal to about 350°C.
- Raising of the gas pressure to the prefixed values with a gradual current increase to raise the sample temperature to the selected value in about 30 min.
- Ion-nitriding phase at the prefixed values of pressure, temperature and gas composition and for the selected treatment period.

The values of temperature and pressure define the values of current and voltage; the required currents were between 250 and 450 mA and the voltage of about 450 V, while the current densities, depending on the different working conditions, were in the range of 15-25 mA/cm^2.

The nitrided samples were systematically subjected to the following analyses:

- Measurements of hardness and thickness of the nitrided layers through microdurometer with Knoop indentor.
- Morphology analysis and control of the nitrided layer thickness by means of optic microscope, scanning electron microscope and microanalysis through energy dispersion probe.
- Analysis and identification of present phases by means of X-ray diffractometer (Cu Kα λ=154.18 pm).

The analyses of the samples were carried out both on the external surfaces and on their sections perpendicular to the sample axis.

RESULTS AND DISCUSSION

Firstly, it must be observed that all the samples, subjected to the ion-nitriding at the selected working conditions, show a more or less extended nitrided surface layer characterized by high hardness values over 1500 HK; in some cases, on the external part of the layer, they reach 2000 HK. However, while there is a high level of hardness on the first part of the nitrided layer, subsequently, the hardness level decreases with a more or less steep gradient to the matrix values. For example, fig. 2 shows the microhardness profiles of samples ion-nitrided at 8 Torr and 1000°C for 8 h. It can be seen that, in general, the nitrided layer depth is higher in IMI 550 and OT4 alloys than in the other examined materials.

Considering the influence of the different working parameters, these results can be set out.

Both titanium and titanium alloy samples, nitrided at the lowest test temperature (800°C), show very thin nitrided layers (in general, minor than 15 μm). Thus, it is clearly evident that sufficiently thick surface layers can be obtained by ion-nitriding only at a working temperature over 900°C. This signifies that the thermochemical treatment must be carried out on alloys which are in the existence field of β phase /10,12/.

Considering the influence of the gas mixture pressure, the experimental results demonstrated that an increase in the value of this parameter produces an increase in the surface layer depth. Fig. 3 shows, for the examined materials, the depths of the nitrided layers which are reported as a function of the gas pressure, while the other working parameters are fixed. The distance, measured on a normal sample section, from the surface to a plane where the hardness is 550 HK has been assumed as the depth of the nitrided layer. These data, as reported in fig. 3, clearly show that even a treatment at 0.5 Torr produces an appreciable nitrided depth, and that, for the tested materials, the maximum depth of the nitrided layer is reached through treatments at a pressure of between 8 and 16 Torr.

Finally, the tests have confirmed that the nitrided layer depth is proportional to the square root of the treatment time /10/, as reported in fig. 4.

To analyze the morphology of the the nitrided layers, optic and scanning electron microscopes were employed. After appropriate chemical etching, all the examined samples presented a well defined interface between the surface layer and the metallic matrix. The morphology of the surface layer is typically characteristic of each examined alloy; the working parameters determine only the layer depth variations. Thus, samples in titanium c.p. grade 4 and in OT4 alloy present a sharp, rectilinear interface between the matrix and the nitrided layer, as clearly shown in figs. 5 and 6. On the other hand, samples in IMI 318 and IMI 550 alloys present long crystals that initiate at the interface and move towards the matrix, as shown in figs. 7 and 8. It is of particular importance to note that while in the first alloy these crystals are isolated and very far apart, in the second alloy these crystals are very numerous and close together.

In addition, the surface phases, produced by the ion-nitriding, were examined by means of X-ray diffraction. From these analyses, it can be stated that the nitrided layers present different structures depending on the alloy nature and the adopted working conditions. However, in all the titanium samples, independently of the adopted working conditions, the resulting ion-nitrided layers, beginning at the sample surface, consist of δ nitride (TiN), ϵ nitride (Ti_2N) and a solid solution of nitrogen in α titanium. This fact is in agreement with the phase diagram for Ti-N system, which shows an ϵ phase field extending up to 1065°C. On the contrary, in the other alloy samples the ion-nitrided layers were found to have the ϵ nitride only under particular working conditions: after ion-nitriding at 1000°C and for 8 h, the ϵ nitride was absent in OT4 and IMI 318 alloy samples treated at a pressure above 2 Torr, and in IMI 550 samples treated at any tested pressure level, as reported in Table II. Moreover, considering the influence of the gas composition on the nitrided layer structures, the ϵ nitride was observed not only in titanium samples but also in OT4 alloy samples ion-nitrided in gas mixture consisting of N_2 60% and H_2 40% (Table III).

Furthermore, the ϵ nitride was found to be present in all samples ion-nitrided at temperatures equal to or less than 900°C (Table IV).

The presence of ϵ phase in the nitrided layers may be correlated with the differentiated distribution of the alloy elements in the matrix and in the nitrides. In fact, the microanalysis through energy dispersion probe revealed that there is a higher aluminium concentration and a lower metal transition concentration in the nitrides than in the matrix.

Finally, it should be noted that ϵ nitride crystals present a preferred orientation in the nitrided layers: normally the c axis of the elementary cell is perpendicular to the sample surface, as can be seen, for example, in the diffractogram reported in fig.9.

CONCLUSIONS

The conducted research on the ion-nitriding of titanium and some α-β titanium alloys enables us to formulate these final considerations.

Using the ion-nitriding treatment at temperature above 900°C, hardened surface layers of sufficient thickness can be obtained also in short treatment time (8 h).

The maximum depth of the nitrided layers is reached, for the different tested materials, at treatment pressures in the range of 8-16 Torr.

The hardness in the nitrided surface layers depends on the microstructural morphology, on the present nitrides and on the extension of the diffusion layer (solid solution of nitrogen in α titanium). The compound layer structures (δ+ϵ nitrides) are determined by the alloy chemical composition and the adopted working conditions: while the δ nitride is always present, the ϵ nitride is present only for specific testing conditions. This fact may be correlated with the differentiated distribution of the alloy elements in the structure.

Among the examined alloys, the IMI 550 samples present a maximum depth of the nitrided layer in accordance with the very extended diffusion layer built up by ion-nitriding of this material.

ACKNOWLEDGEMENTS

The authors wish to thank Dr. Maltagliati, of the "Officine Galileo" in Florence, for having put at their disposal the nitriding equipment and the "Elettrochimica Marco Ginatta" firm in Turin which furnished the titanium alloys.

REFERENCES

1) E.Mitchell, P.J.Brotherton: J. Inst. Met.,1964-65, vol 93, 381-386.
2) K.N.Strafford, J.M.Towell: Oxidation of Metals, 1976, vol. 10, 69-84.
3) J.P.Bars, E.Etchessahar, J.Debuigne: J. of Less Common Metals, 1977, vol. 52, 51-76.
4) J.P.Bars, D.David, E.Etchessahar, J.Debuigne: Metall. Trans. A, 1983, vol. 14 A, 1537-1543.
5) E.Etchessahar, J.P.Bars, J.Debuigne, A.P.Lamaze, P.Champin: Titanium Sci. Technol.; Proc. of 5th Int. Conf. Titanium, 1985, vol. 3, 1423-1430.
6) H.Michel, M.Gantois: Mem. Sci. Rev. Met., 1972, vol. 69, 739-749.
7) Ming - Biann Liu, D.M. Gruen, A.R.Krauss, A.H.Reis, S.W.Peterson: High Temp. Sci., 1978, vol.10, 53-65.
8) T.Bell, H.W.Bergmann, J.Lanagan, P.H.Morton, A.M.Steines: Proc. of 4th Int.Cong. on Heat Treat. of Materials 1985, vol. 2, 1008-1032.
9) M.Komuna, O.Matsumoto: J. of Less Common Metals, 1977, vol.52, 145-152.
10) K.T.Rie, T.Lampe: Mat. Sci. and Eng., 1985, vol. 69, 473-481.
11) B.Holmberg: Acta Chem. Scand., 1962, vol. 16, 1255-1261.
12) R.J.Wasilewski, J.L.Kehl: J. Inst. Met., 1954-1955, vol. 83, 94-104.

TAB. I - Chemical composition of examined materials (weight percent).

	Ti c.p.	IMI 318	IMI 550	OT4
Al	-	5.98	4.09	4.50
Fe	0.03	0.09	0.04	0.05
V	-	3.25	-	-
Mo	-	-	4.10	-
Sn	-	-	2.09	-
Mn	-	-	-	1.96
Si	-	-	0.48	-
O	0.11	0.16	0.16	0.07
N	0.04	0.03	0.01	-

TAB. II – Influence of gas pressure on the nitrided layer phases (treatment at 1000°C for 8 h)

Materials	Treatment conditions %N_2	%H_2	Pressure (Torr)	Phases in the nitrided layer
Ti c.p.	90	10	1	TiN, Ti_2N, Ti
"	"	"	10	TiN, Ti_2N, Ti
"	"	"	30	TiN, Ti_2N, Ti
"	80	20	2	TiN, Ti_2N, Ti
"	"	"	4	TiN, Ti_2N, Ti
"	"	"	8	TiN, Ti_2N, Ti
"	"	"	16	TiN, Ti_2N, Ti
"	"	"	32	TiN, Ti_2N, Ti
OT4	80	20	2	TiN, Ti_2N Ti
"	"	"	4	TiN, Ti
"	"	"	8	TiN, Ti
"	"	"	10	TiN, Ti
"	"	"	16	TiN, Ti
"	"	"	32	TiN, Ti
IMI 318	80	20	2	TiN, Ti_2N, Ti
"	"	"	4	TiN, Ti
"	"	"	8	TiN, Ti
"	"	"	10	TiN, Ti
"	"	"	16	TiN, Ti
"	"	"	32	TiN, Ti
IMI 550	80	20	2	TiN, Ti
"	"	"	4	TiN, Ti
"	"	"	8	TiN, Ti
"	"	"	16	TiN, Ti
"	"	"	32	TiN, Ti

TAB. III – Influence of gas composition on the nitrided layer phases (treatment at 1000°C and 10 Torr for 8 h)

Materials	Gas composition %N_2	%H_2	Phases in the nitrided layer
Ti c.p.	20	80	TiN, Ti_2N, Ti
"	60	40	TiN, Ti_2N, Ti
"	80	20	TiN, Ti_2N, Ti
"	90	10	TiN, Ti_2N, Ti
OT4	60	40	TiN, Ti_2N, Ti
"	80	20	TiN, Ti
IMI 318	20	80	TiN, Ti
"	40	60	TiN, Ti
"	60	40	TiN, Ti
"	80	20	TiN, Ti
IMI 550	60	40	TiN, Ti
"	80	20	TiN, Ti

TAB. IV – Influence of temperature on the nitrided layer phases (treatment at 10 Torr for 8 h)

Materials	Treatment conditions			Phases in the nitrided layer		
	$\%N_2$	$\%H_2$	Temperature (°C)			
OT4	80	20	1000	TiN,		Ti
"	"	"	850	TiN,	Ti_2N,	Ti
IMI 318	80	20	1000	TiN,		Ti
"	"	"	800	TiN,	Ti_2N,	Ti
"	60	40	1000	TiN,		Ti
"	"	"	900	TiN,	Ti_2N,	Ti
IMI 550	60	40	1000	TiN,		Ti
"	"	"	900	TiN,	Ti_2N,	Ti

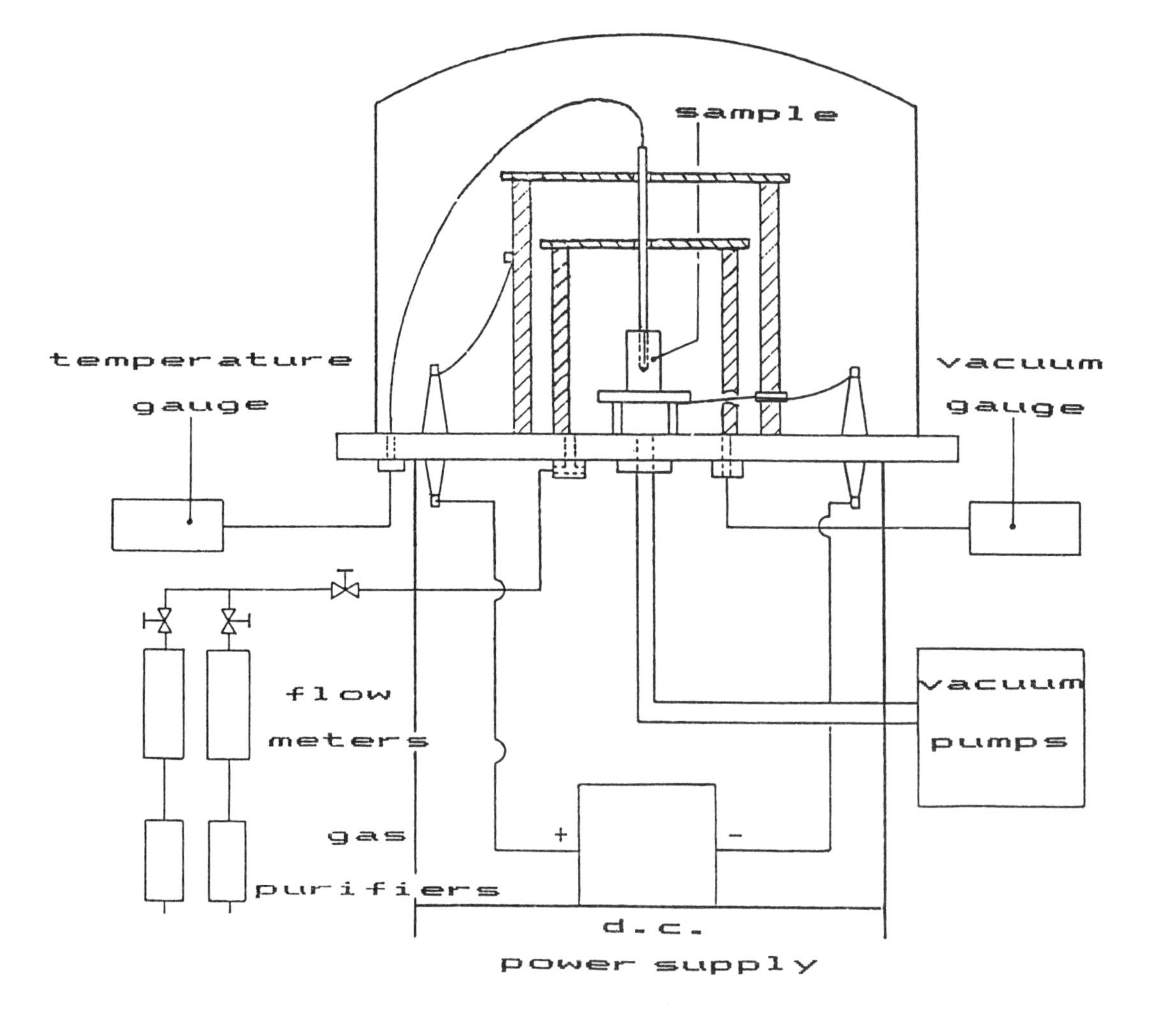

Fig.1 – Scheme of the ion-nitriding unit.

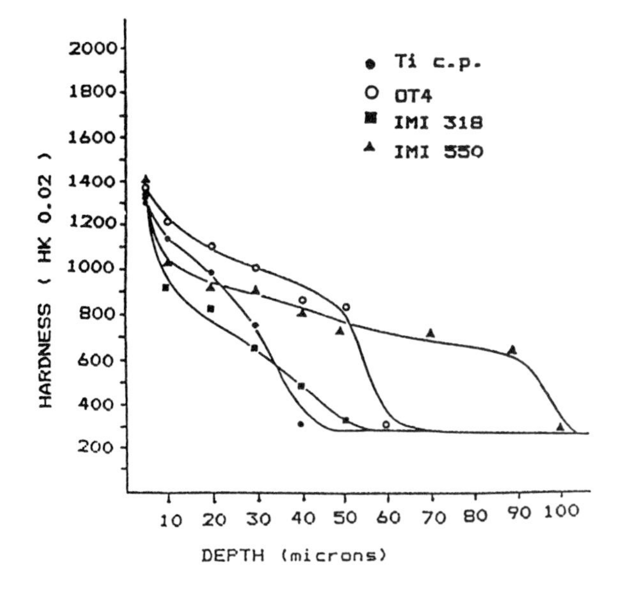

Fig.2 - Microhardness profiles in the ion -nitrided layers (1000°C, N_2 80% H_2 20%, 8 Torr, for 8 h).

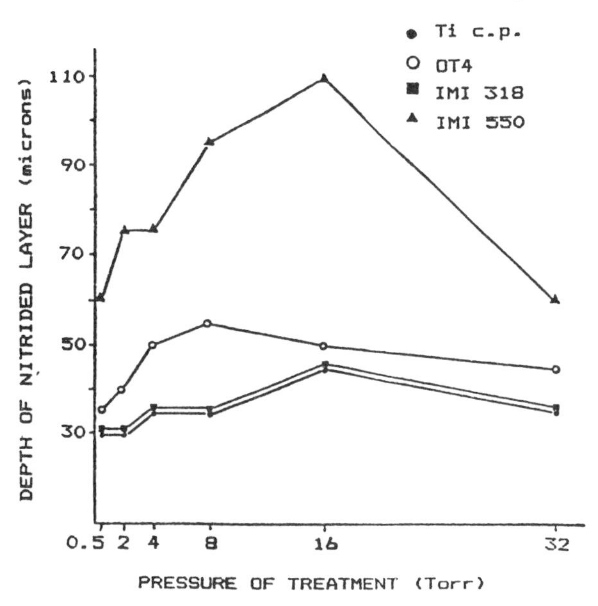

Fig.3 - Nitrided layer depth vs.treatment pressure (1000°C, N_2 80% H_2 20% for 8 h).

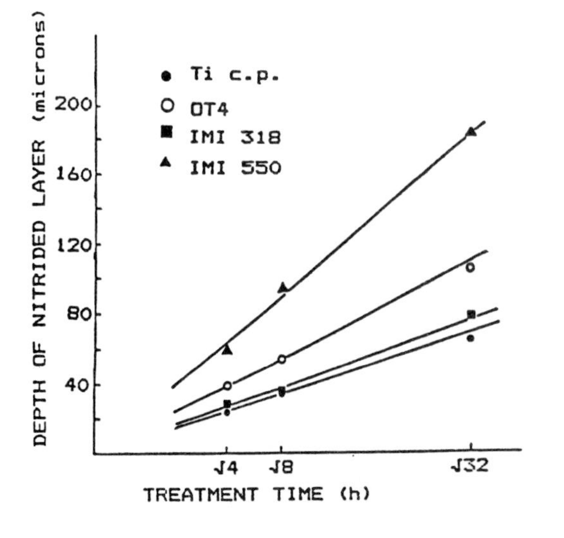

Fig.4 -

Nitrided layer depth vs.treatment time (1000°C, N_2 80% H_2 20%, 8 Torr).

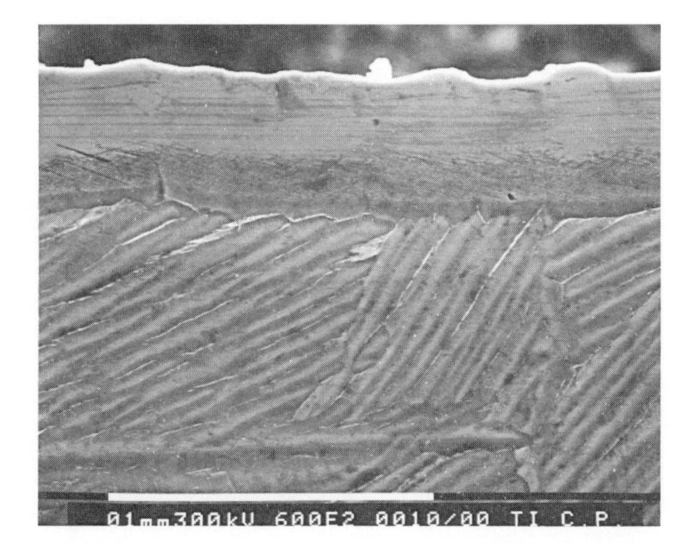

Fig.5 - Microstructure of a Ti c.p. grade 4 sample (1000°C, N_2 80% H_2 20%, 8 Torr, for 8 h).

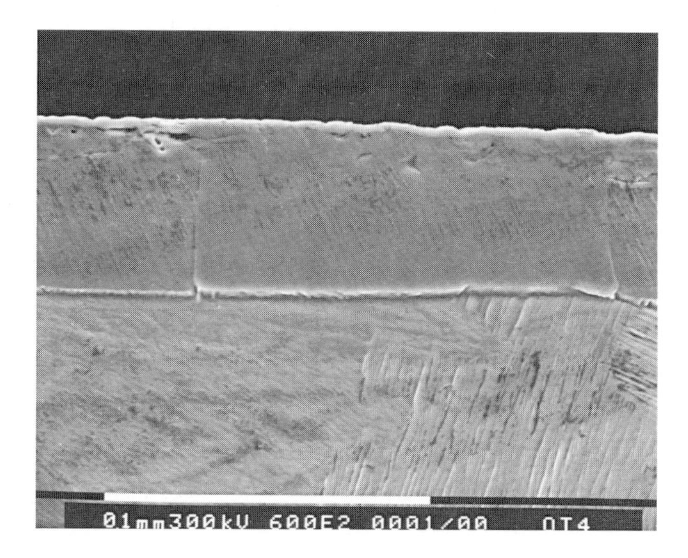

Fig.6 - Microstructure of an OT4 alloy sample (1000°C, N_2 80% H_2 20%, 8 Torr, for 8 h).

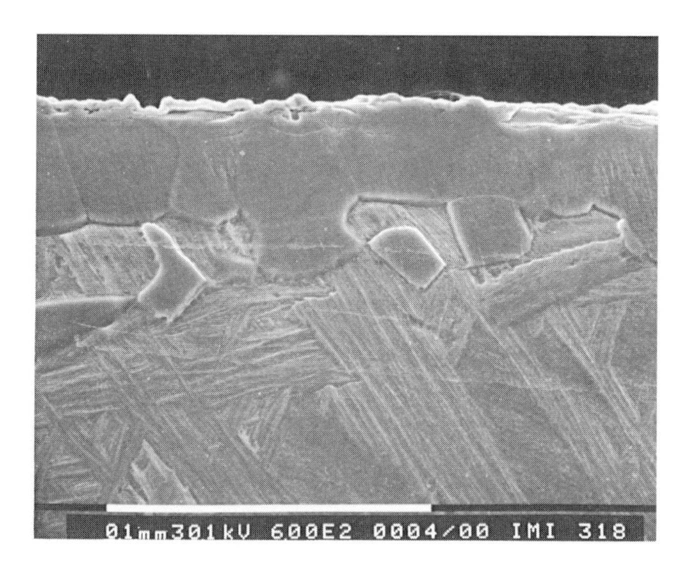

Fig.7 - Microstructure of an IMI 318 alloy sample (1000°C, N_2 80% H_2 20%, 8 Torr, for 8 h).

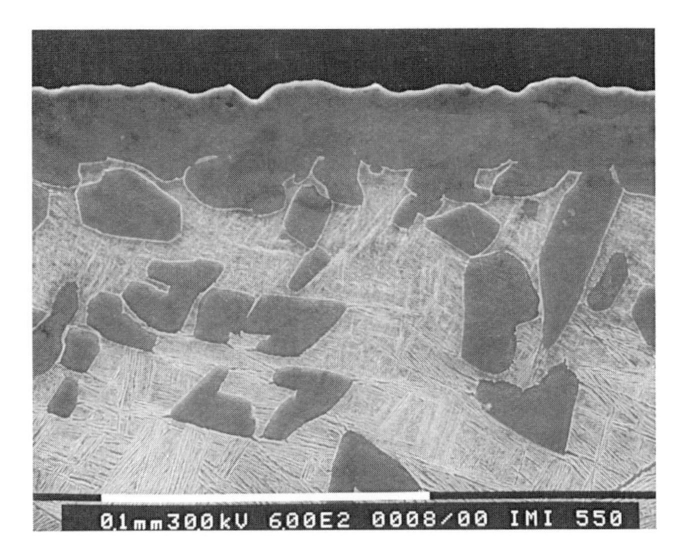

Fig.8 - Microstructure of an IMI 550 alloy sample (1000°C, N_2 80% H_2 20%, 8 Torr, for 8 h).

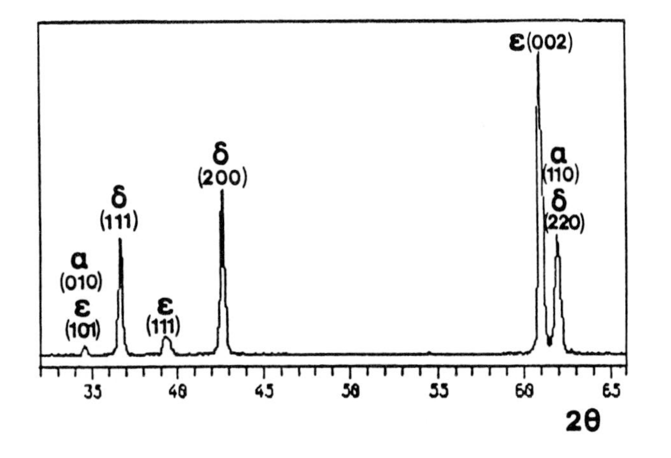

Fig.9 - X-ray diffractogram of a nitrided Ti c.p. grade 4 sample (1000°C, N_2 60% H_2 40%, 10 Torr, for 16 h).

Study of wear resistance of electroless co-deposited Ni–P–SiC coating

DENG ZONGANG and HUANG XINMIN

The authors are in the Department of Materials and Science at the Hefei Polytechnic University, PRC.

SYNOPSIS

This paper shows that there is no linear dependence between load and wear volume of electroless co-deposited coating, as there is an abrupt change from lubricating into dry wear when the load comes up to a critical volume. Among the electroless deposited Ni–P and Ni–P–SiC coatings, the wear resistance of composite coating with high phosphorus content is the best. This is because the SiC particle plays an important part in resisting wear, and a good matrix supports the SiC particle properly. If the coatings are heated at a temperature range from the as-plated temperature to 600°C, the wear resistance is increased with the rise in heating temperature. The optimum heating time is one hour.

1. INTRODUCTION

Since 1966 the co-deposited coatings have been used extensively in industry as they offer improved wear and low friction. Many researchers are working towards application of the coating,[1-4] so that many studies concentrate on the co-deposited process and simple comparisons of wear resistance.[5-7]

Few works deal with the effect of test load coating structure and properties on wear resistance. The paper looks into the influence of test load, coating composition, and heat treatment process on wear resistance which has been studied by means of advanced instruments and equipment such as SEM and AES.

2. EXPERIMENTAL PROCEDURES

The electroless deposition process and composition of coating is presented in Table 1. Because an abrupt change in the structure and properties of electroless nickel occur at a phosphorus content of about 7%,[8] the Ni–8.73P–SiC and Ni–4.5P–SiC coatings were chosen to study the effect of a different matrix on the wear resistance of composite coating. The Ni–8.9P coating was chosen for comparison with composite coating, which shows the effect of SiC grain on wear resistance.

The wear test was carried out on the MM-200 wear testing machine (rotating ring and block type) and lubricated by machine oil. A high carbon steel was used as the substrate and an annealed medium carbon steel as the rotating ring. The sliding speed was set at 50 m min^{-1} and the total sliding distance was 3000 m. All these were chosen to imitate the working condition of die.

The wear volume was calculated by means of measuring the width of wear scar. The wear scars were observed in S-570 SEM and some were analysed by LAS-600 AES. The friction temperature near the wear scar was measured with a thermocouple and the friction coefficient was read straight off the wear testing machine.

3. EXPERIMENTAL RESULTS

3.1. Wear results of coatings under different loads

All the coatings are heated at 400°C for one hour. The coating structure and the microhardness are listed in Table 2.

The variation of wear volume with load is plotted in Fig. 1, which shows that no curve has any linear dependence. There is a turning point in every wear curve, which indicates that there is a critical load for every coating during lubricating wear. When the critical load is exceeded, the wear volume of coating increases severely.

Compared with other coatings, the Ni–8.73P–SiC composite coating has the highest critical load and the smallest wear volume in the whole load range. In the low load range (below critical load) the wear volume of Ni–4.5P–SiC composite coating is less than that of Ni–8.9P coating, whereas in the high load range (above the critical load) the wear volume of Ni–4.5P–SiC composite coating is much greater than that of Ni–8.9P coating.

Fig. 2 is the surface morphology of wear scar tested at 60 kg. Besides plowing grooves, there are many peelings on the wear scar of composite coating. The peelings on the wear scar of Ni–4.5P–SiC coating are more numerous and deeper than that of Ni–8.73P–SiC coating. There is no peeling on the wear scar of Ni–8.9P coating, but the microcut and plowing grooves are sharp.

3.2. Wear results of Ni–8.73P–SiC composite coating treated by various heat treatment processes

Heating temperature versus wear volume of Ni–8.73P–SiC coating is shown in Fig. 3. The general trend of the curve is that the higher the temperature in the range from as-plated to 600°C, the smaller the wear volume. The variation of hardness with heating temperature is also illustrated in Fig. 4. By studying the two curves, we can see that there is a turning point located at 400°C on both curves. This point divides the curves into two ranges: low temperature range (as-plated to 400°C) and the high temperature range (400°C to 600°C).

In the low temperature range, with the increase of hardness the wear volume decreases, but in the high temperature range, the variation of wear volume with heating temperature is the same as that of hardness.

Fig. 4 shows the wear scars of coatings heated in the two temperature ranges. It can be seen that there are many peelings and plowing grooves on the wear scar of the coating heated in the low temperature range, on the one heated in the high temperature range, however, there is only the plowing groove.

The transfer of elements of the wear scar surface is investigated by AES (see Fig. 5). From plot (b) it is seen that the element variation on the wear scar of the coating heated at 600°C is very small, with only a slight increase of phosphorus. However, on the wear scar of the coating heated at 400°C, the transfer of elements is severe, not only does iron transfer to the coating surface from medium carbon steel, but Ni and P also transfer to the steel ring surface.

The relation between wear volume and heat treating time is illustrated in Fig. 6. The coating is heated at 600°C and the heating time is from 30 min to 240 min. As the heating time increases, the wear volume falls at first and then at 60 min starts to rise. The hardness dependence of heating time is also shown in Fig. 6. In the time range from 30 min to 240 min, the hardness gradually decreases.

4. DISCUSSION

4.1. Effect of load on wear resistance of coatings

Load is one of many effects on wear and, in general, with the increase of load the friction heat will increase. Fig. 7 shows the variation of friction heat measured near the wear scar with load. This curve is the same as that shown in Fig. 1, they all have an alike critical load, which indicated that there must exist a relation between wear and friction heat. Because our testing is lubricating wear and, as is well known, the lubricant oil has a critical temperature, if the temperature of the surroundings is beyond the critical temperature (as may be caused by friction heat), the oil will be separated from the sample and the lubrication fail. Therefore, it is the load that changes the lubricating state from lubricated into dry wear by suddenly increasing friction heat, which leads to the abrupt change in wear volume (Fig. 1). Thus, if we control the friction heat or use a lubricant oil with a high critical temperature, the wear volume of coating decreases.

4.2. Effect of coating composition on wear resistance

The wear resistance of Ni–8.73P–SiC coating is the best of the three coatings. Comparing Ni–8.73P–SiC with Ni–8.9P coating, it can be seen that the hardness of composite coating is higher than that of Ni–P alloy coating. Generally, the harder the coating, the better the wear resistance. In addition, there is a difference between the structure of two coatings, i.e. the composite coating content with some SiC grain dispersed in the Ni–P alloy matrix.

Because the SiC grain is a compound with high hardness and melting points, it prevents the matrix grain from coarsening and growing at high temperature,[9] increases the stability of composite coating, and resists microcut (see Fig. 8), so that the Ni–8.73P–SiC coating displays a greater wear resistance.

Comparing the Ni–8.73P–SiC with the Ni–4.5P–SiC coating, the difference between the two coatings is the matrix structure and hardness. The matrix of the Ni–8.73P–SiC coating contains more Ni_3P phase and less Ni solid solution than that of the Ni–4.5P–SiC coating. As the Ni_3P is hard phase and its melting point is high, the Ni–8.73P alloy matrix is harder and has higher heat stability. The difference in hardness between matrix and SiC is not great, so that the matrix of Ni–8.73P alloy can support SiC particles very well. In the matrix of Ni–4.5P alloy, Ni solid solution is more than 50%. This is a ductile phase and, although it can be strengthened by Ni_3P phase, it will still soften considerably when the friction heat causes the Ni_3P grain to coarsen and grow. Thus the matrix of the Ni–4.5P alloy is soft and cannot support SiC particles properly and the SiC particles lose their advantages in this coating.

From the above analysis, it can be clearly seen that the SiC particles make a great contribution to the wear resistance of coatings. However, a good matrix is necessary if the SiC is to give full play to wear resistance.

4.3. Effect of heating temperature on wear resistance of composite coating

Fig. 3 shows that the heating temperature exerts a different effect on wear resistance and hardness.

When the coating is heated within the low temperature range, as the temperature increases the hardness gradually rises to maximum and the wear volume decreases. Therefore, the wear resistance is controlled by hardness.

When the coating is heated in the high temperature range, the hardness and wear volume decrease while the temperature increases. The relation between hardness and wear volume in the high temperature range is opposite to that in the low temperature range.

Analysing Fig. 3, if we heat two coatings respectively in two temperature ranges, the hardness of coatings may be alike, but the wear volume is different. Associating Fig. 3 with Fig. 4 and Fig. 5, it can be seen that many peelings and severe transfer of elements occur at wear scars of coatings heated within the low temperature range, which results in serious wear. Because of no peeling and slight transfer of elements on the wear scar, the wear volume is very small if the coating is heated within the high temperature range.

4.4. Effect of heat treating time on wear resistance

In Fig. 6, the hardness falls throughout, whereas the wear volume has a minimum which is located at 60 min. This indicates that the wear resistance does not depend only on the hardness. As we know, when the coating is heated at 600°C, with the increase in the heat treating time the toughness rises and the hardness falls owing to the grain growth. Therefore, when the hardness and toughness are under optimum coordination, the coating shows a maximum wear resistance, i.e. the wear resistance depends on both hardness and toughness. This conclusion is also put forward by recent researchers.[10-12]

5. CONCLUSION

(1) There is a critical load for every coating studied in this paper. If the coating is applied to production, the real load should be below the critical load.
(2) The composite coating is more wear resistant. This comes from the coordination of hard particles with good matrix.
(3) Under identical test conditions, the wear resistance of the coating is dependent on its structure and properties.
(4) For the co-deposited Ni–8.73P–SiC coating there is an optimum heat treatment process, i.e. heating at 600°C for 60 min. The heat treatable property of electroless deposited nickel coating has a great advantage over other coatings such as electroplated chromium.

REFERENCES

1. F. N. Hubbell, Plat.Surf.Finish.,1978,Dec, 58-62
2. J. K. Dennis, S. T. Sheikh, and E. C. Silverstone, Trans.Inst.Met.Finish.,1981,59,118-122
3. E. C. Kedward, C. A. Addison, and A. A. B. Tennett, Trans.Inst.Met.Finish.,1976, 54,8-15
4. D. Buchkov and G. Garrilov, Metall.,1980,34,(3)
5. W. Metzger and Th. Florian, Trans.Inst.Met. Finish.,1976,54,174-177
6. N. Feldstein, T. Lancsek, D. Lindsay, and L. Salerno, Met.Finish.,1983,Aug,35-41
7. J. Lukschandel, Trans,Inst,Met,Finish.,1978, 56, 118-120
8. A. H. Graham, R. W. Lindsay, and H. J. Read, J.Electrochem.Soc.,1965,112,401-413
9. H. Xinmin, The Structure and Properties of Electroless Co-deposited Ni–P–SiC Coating, to be published
10. K. H. Zum Gahr, Z.Metallkd,1978,69,643-650
11. E. Hornbogen, Wear,1975,33,251-259
12. N. Saka, Fundamentals of Tribology, 1978

TABLE 1 Bath composition and coating phosphorus content

	1	2	3
bath composition (g l^{-1})			
$NiSO_4 \cdot 7H_2O$	10	10	30
$NaH_2PO_2 \cdot H_2O$	15	15	8
$CH_3COONa \cdot 3H_2O$	15	15	15
H_3BO_3	5	5	5
SiC (<3 µm)	15		15
temperature (°C)	84	84	84
ph volume	4.8-5	4.8-5	5-5.4
coating phosphorus content (wt%)	8.73	8.9	4.5

TABLE 2 Phase component and hardness of coatings

coating	Ni–8.73P–SiC	Ni–8.9P	Ni–4.5P–SiC
phase component (wt%)			
Ni	36.4	36.5	65.8
Ni P	60.4	63.5	31
SiC	3.2		3.2
hardness (HV)	1257	1033	1021

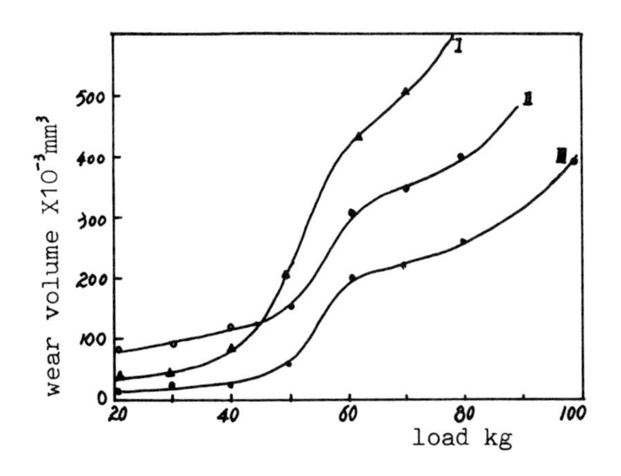

Fig. 1 The dependence of wear volume on load
- ● - Ni–8.73P–SiC
- ○ - Ni–8.9P
- ▲ - Ni–4.5P–SiC

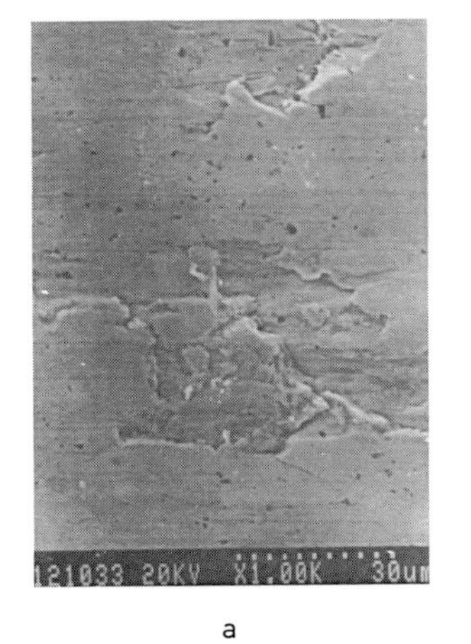

a

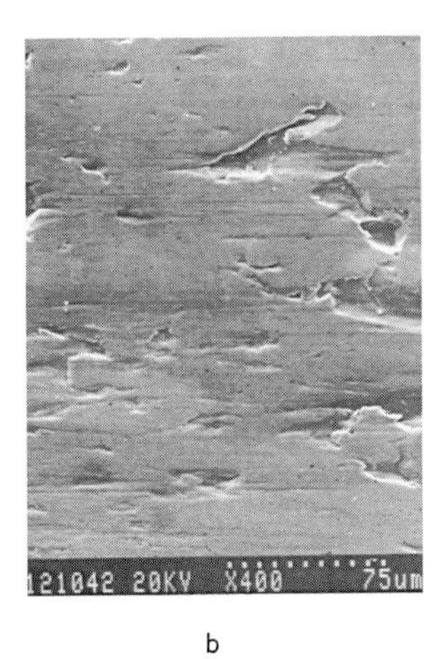

b

c

Fig. 2 The wear scar of coating: (a) Ni–8.73P–SiC, (b) Ni–4.5P–SiC, (c) Ni–8.9P

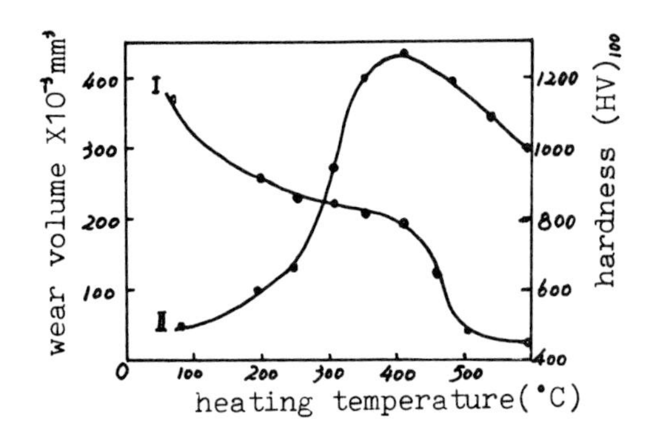

Fig. 3 Heating temperature versus hardness and wear volume
II hardness curve
I wear volume curve

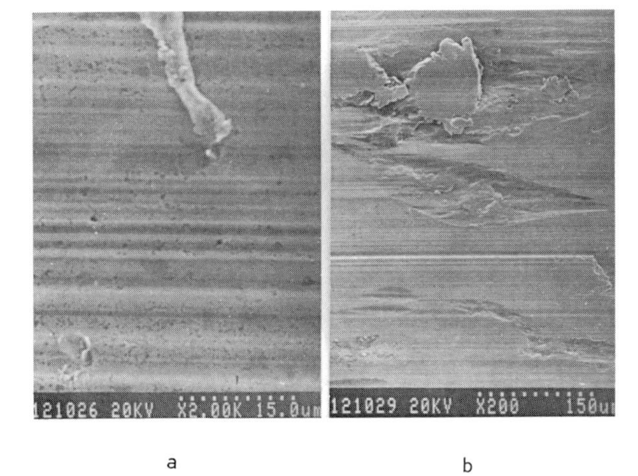

a b

Fig. 4 The wear scar of coating heated at (a) 290°C and (b) 600°C

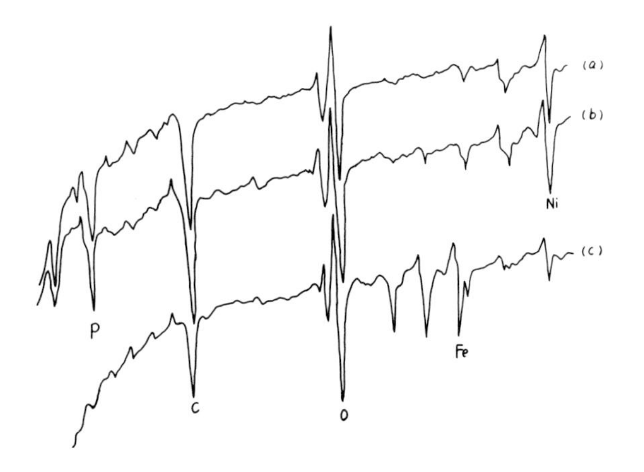

Fig. 5 AES curve of (a) coating surface, wear scar of coating heated at (b) 600°C and (c) 400°C

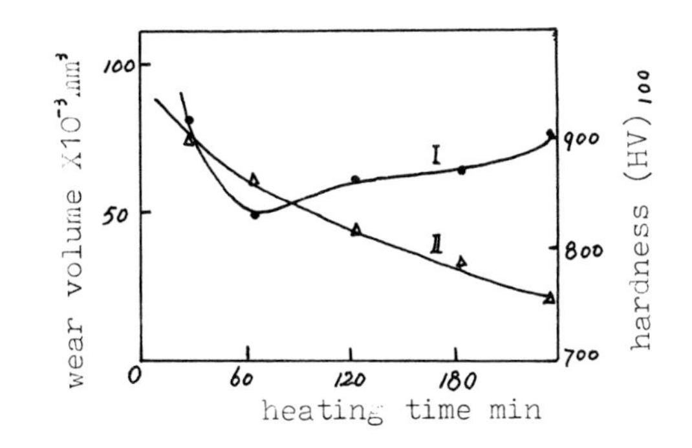

Fig. 6 The heating time versus wear volume (I) and hardness (II)

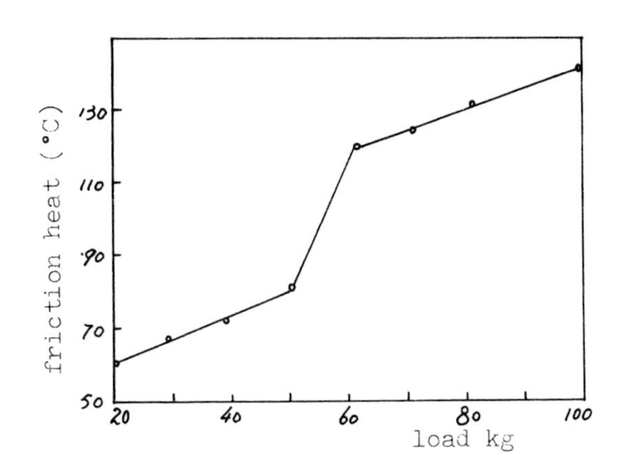

Fig. 7 The dependence of friction heat on load

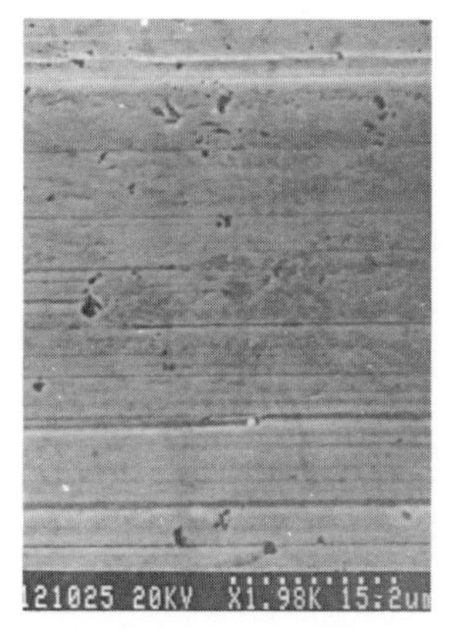

Fig. 8 The trace of SiC particles resisting microcut

Influence of their thermal history on mechanical properties of zirconia ceramics

M A HEPWORTH and D T PINDAR

The authors are in the Ceramics Section at T&N Materials Research Ltd., Rochdale, Lancs.

SYNOPSIS

Zirconia powders, containing yttria and rare earths as alloying constituents, prepared by electrorefining/comminution exhibit high sinter activity comparable with similar compositions produced by coprecipitation from aqueous solutions. Careful choice of the composition and sintering conditions allows ceria doped powders to be converted into ceramics having many of the mechanical characteristics of low ductile metals such as cast iron.

INTRODUCTION

During the last decade two main types of toughened zirconia ceramics have been developed. Both employ the principle of transformation toughening in which a significant quantity of the high temperature tetragonal zirconia phase, which is metastable at room temperature, is produced in the finished ceramic. As a result of its ability to absorb the energy of an advancing crack tip during its transformation to the monoclinic phase, the stable phase at ambient temperature in the chosen alloy, cracks are effectively blunted and the ceramic exhibits improvements in fracture toughness which are claimed to be up to 10 fold, effectively from about 1-2 to as high as 15 MPa $m^{\frac{1}{2}}$.

Both systems use compositions in which the zirconia is only partially stabilised by adding a carefully chosen amount of the alloying oxide which is less than that required for its complete conversion to the stable cubic phase. In the case of magnesia partially stabilised zirconia, Mg PSZ, this is accomplished by adding 3-3.5 w/o i.e. 8-10 mol% magnesia as the alloying oxide, sintering at a temperature in the range 1700-1800°C, and then heat treating the dense ceramic by holding it at a temperature in the range 1400-1450°C for several hours in order to form a precipitate of crystals of the tetragonal phase in the remaining cubic matrix. On cooling to room temperature the tetragonal crystals do not revert to the stable monoclinic modification but are retained in the desired metastable condition.

An alternative processing route is used to achieve a similar end in the manufacture of yttria partially stabilised zirconia, YPSZ, ceramics. Aqueous solutions of zirconium salts containing 5-6 w/o i.e. about 3 mol% of yttrium salts are coprecipitated as hydroxides and, after careful dehydration, are heated to a temperature of about 800°C to produce the oxide 'alloy'. Under these conditions the alloy is found to consist of submicron crystallites, largely in the metastable tetragonal modification which, after conventional ceramic forming, can be sintered to almost theoretical density at temperatures in the range 1400-1600°C. The resulting ceramics continue to retain a high proportion of the tetragonal phase on cooling to room temperature.

This study has carried these developments forward in two directions. In the first of these a new processing route has been developed in which PSZ containing one or more alloying oxides is produced by electrorefining of the components followed by comminution to produce powders which are capable of being converted into ceramics of near theoretical density at sintering temperatures as low as 1350°C. Secondly, in addition to yttria, several of the rare earth oxides have been employed as alloying oxides. In this latter context, the use of ceria has been found to be of particular interest because of it being able to form tetragonal phases which are metastable at room temperature.

EXPERIMENTAL

Powder Processing

Materials containing 6 w/o (3.2 mol%) Y_2O_3, 11.4, 12.1 and 14.3 w/o (8.6, 9.1 and 10.8 mol% respectively) ceria CeO_2, 12 w/o (9 mol%) neodymia Nd_2O_3 and 14 w/o (10 mol%) praeseodymia Pr_6O_{11} were prepared by an electrofusing and refining technique which ensures that the alloying oxide is homogeneously incorporated into the zirconia host. The melts were solidified under closely controlled conditions to give polycrystalline intermediates with minimal segregation of the constituents. A specialised comminution route based on proprietary fine milling and classification techniques resulted in their reduction to, predominantly, sub micron powders.

The composition of the powders, determined using XRF, are shown in Table 1. Particle size distributions were determined using either a Sedigraph or a Coulter Counter; mean particle diameters for each powder are shown in Table 1 and a typical sedimentation curve is shown in Fig. 1. Powder morphology was examined using an SEM. This revealed the presence of dense, roughly equiaxed particles typified by those shown in Fig. 2.

Consolidation and Sintering

The poor flow characteristics of the powders were improved by blending with small quantities of an organic binder lubricant, such as polyethylene glycol wax, added as a solution in either water or IPA, the solvent removed by heating and the mixture rubbed through a 500 μm sieve. Compacts 30 mm diameter were prepared by die pressing the granulated powder under a total load of 1000 kg followed by wet bag isostatic pressing at a peak pressure of 200 MPa. Green densities were generally in the range 53-55% of theoretical density.

Sintering was carried out in air using a programmed electrically heated furnace capable of attaining a maximum temperature of 1600°C. Typical overall linear shrinkage was about 20%. Densities of sintered specimens were determined by mercury displacement.

Microstructure Evaluation

Estimates of the proportions of the tetragonal and monoclinic phases were obtained from the relative intensities of the neighbouring (111) and (11$\bar{1}$) peaks measured using an X-ray diffractometer. Polished compacts were thermally etched by heating for 15 minutes at 1500°C and their microstructure and those of fracture surfaces examined using either scanning or transmission electron microscopy. Mean grain sizes were estimated using a line intercept method.

Mechanical Properties

Bars, 4 x 4 x 30 mm were cut from sintered blocks using a diamond saw and their surfaces ground with a 400 grit diamond wheel. Flexural strength was measured on a 3 point loading fixture with an outer span of 20 mm and a crosshead speed of 0.5 mm/min. Sintered discs were diamond ground and highly polished and their hardness determined from the impression of a Vickers indenter under a load of 1 kg.

Similar indentations were made with loads of up to 30 kg and the cracks produced used to calculate the fracture indentation toughness, K_{1c}. Alternatively, work to fracture measurements were carried out on 4 x 4 x 30 mm bars notched using a 220 μm diamond wheel.

RESULTS

Sintering Behaviour

Despite having been processed via electrorefining the YPSZ powders exhibited excellent sinter activity comparable to those produced by chemical coprecipitation. Thus, sintering at 1500°C with a hold time of 2 hours resulted in 98% of theoretical density (taken as 6.06 g/cm^3 for this composition) being attained (Table 2); one experiment demonstrated that similar results could be achieved by sintering in a nitrogen atmosphere although in this instance the colour of the sinters changed from yellow to a deep grey-black. Phase analysis of the as sintered specimens revealed the presence of <5% of the monoclinic phase, the remainder consisting of the tetragonal/cubic modification. Varying the sintering temperature had little effect on the texture of the specimens which SEM examination revealed to consist of equiaxed grains with a narrow range of sizes averaging 0.6 μm.

Neodymia and praeseodymia powders exhibited contrasting behaviour. Although both were highly sinter active producing dense ceramics at 1500°C the sintered materials were found to contain fine cracks. Analysis using XRD revealed that they consisted primarily of the monoclinic phase.

The behaviour of ceria doped powders was found to be more complex. Again, their sinter activity was comparable to that of powders prepared by chemical coprecipitation;[1] for example the 11.4 w/o CePSZ material could be sintered to >95% theoretical density (taken as 6.25 g/cm^3 for this composition) at temperatures as low as 1400°C. When sintered at 1500°C however, the specimen was found to be cracked on emerging from the furnace and examination using XRD revealed the presence of >90% of the monoclinic phase in the surface. Sintering powders with higher ceria contents, 12 and 14 wt%, resulted in the production of crack-free specimens over the temperature range 1350-1500°C with the tetragonal phase predominating at all temperatures (Table 2).

Examination of the microstructure of the as sintered materials using SEM and TEM (Figs 3 and 4) revealed dense assemblies of equiaxed grains whose average size was larger and showed a greater variation with sintering temperature than that exhibited by YPSZ materials. For example, when sintered at 1400 and 1500°C the 12 wt% CePSZ compositions had average grain sizes of 1.1 and 1.7 μm respectively. TEM micrographs at the highest magnification revealed the presence of a minute amount of liquid phase at triple points.

Two observations indicate that CePSZ materials are more sensitive than YPSZ to sintering conditions. Compacts sintered at 1500°C in contact with an alumina support stained the latter blue; phase analysis and EDAX examination revealed that the CePSZ contact face had >90% monoclinic phase present coinciding with the Ce content falling from 9 to 7.5 mol%. Secondly, compacts sintered at 1500°C in nitrogen were found to be cracked on removal from the furnace.

Mechanical Properties

Strength The variation of flexural strength with sintering temperature for the YPSZ and CePSZ compositions is shown in Table 3 and Fig 5. Generally, the values for YPSZ, about 800 MPa, are slightly lower than those determined for comparable compositions prepared from chemically coprecipitated powders;[2] this can be understood in terms of the latter materials having smaller grain sizes in the range 0.1-0.3 μm, depending upon the sintering temperature employed.

Strengths of the CePSZ materials are illustrated in Fig 5 and are found to vary markedly with both the Ce content and the sintering temperature. At the lowest sintering temperature, 1350°C, 14 w/o CePSZ has an average flexural strength of 700 MPa comparable to that of YPSZ; indeed individual values exceeding 1 GPa have been observed. As the sintering temperature is raised the strength falls; this is most marked with the 12 wt% CePSZ composition where the value is halved as the sintering temperature is raised from 1400 to 1500°C and coincides with a significant increase in grain size over this temperature range. A marked feature is the very high values of m, the Weibull modulus, recorded for all the CePSZ materials.

Hardness H_v values for the YPSZ ceramics lie in the range 1350-1400 kg/mm^2 which again coincides with data obtained for materials derived from powders prepared via the chemical route.

The hardness of the CePSZ ceramics is lower, the maximum value observed being about 1000 and falling to below 900 kg/mm^2 as the sintering temperature is raised (see Table 3). It may be noted that even the latter value is comparable to those observed for very hard steels.

Fracture toughness Measurements were derived from the size of the cracks generated from the corners of a Vickers hardness indentation using a load of 10 Kg. YPSZ material gave K_{1c} values of up to 7 MPa m$^{\frac{1}{2}}$ more or less as expected for a material with the phase composition, strength and grain sizes already noted.

Attempts to apply this technique to the CePSZ materials did not produce the expected cracks even when the load was increased to 30 kg. Fig 6 illustrates the observed effect of the material undergoing a quasi-ductile deformation around the impression created by the indenter. Clearly, although the critical stress intensity factor is high no quantitative interpretation is possible.

Further insight was obtained from work to fracture measurements carried out on notched beams which gave results illustrated in Fig 7. Of particular interest is the 12 w/o CePSZ composition sintered at 1500°C which undergoes progressive deformation very much in the manner of a fibre reinforced composite yielding an exceptionally high work to fracture of 860 J m^{-2}. CePSZ sintered at lower temperatures has a lower work to fracture of about 400 J m^{-2} which, although halved, should be compared to the typical value of the work to fracture of an alumina ceramic of 20 J m^{-2}.

DISCUSSION

Clearly electrorefining followed by sophisticated crystallisation and comminution is capable of producing highly sinter active powders which have improved packing characteristics and can be processed to dense ceramics at comparatively low temperatures. YPSZ ceramics containing 6 w/o Y_2O_3 produced by this route have properties comparable to those derived from chemically coprecipitated powders with a similar composition.

The properties of CePSZ ceramics show a marked dependence on their composition and the sintering temperature. Early phase diagrams of the ZrO_2 end of the ZrO_2/CeO_2 system[3] which showed the tetragonal/monoclinic phase boundary extending towards ambient temperatures for 10 mol% CeO_2 have now been revised with the conclusion that the tetragonal phase is metastable at temperatures <1000°C for all compositions.[4] Hence the principles of transformation toughening are applicable in this system in an analogous manner to that of the now well documented Y_2O_3/ZrO_2 system. The reasons for the differing behaviour of lanthana, praesodymia and neodymia are perhaps related to the relative stability of the Ce^{4+} state; some evidence for this is provided by the cracked specimens produced when CePSZ powders are sintered in nitrogen under reducing conditions instead of in air.

The main features of transformation toughening have been well summarised by Gupta and Andersson.[5] They point out that at a temperature, M_s, where the free energies of the two phases are equal the metastable tetragonal phase can undergo a diffusionless martensitic-type transformation to the stable monoclinic form under the action of an applied stress. The transformation is enhanced by the proximity of the M_s to ambient temperature which may be arranged by controlling the composition and grain size; M_s decreases with decreasing grain size and increasing alloying constituents. It follows that there is a critical grain size for a given composition at which the material will show maximum fracture toughness.

The behaviour of the CePSZ composition can be interpreted against this background. Those compositions about 12 w/o CeO_2 are particularly interesting as they exhibit some remarkable properties. After being sintered at 1400°C the composition containing 12.1 w/o CeO_2 has an M_s below ambient temperature resulting in a material which exhibits a comparatively high flexural strength (700 MPa) and some transformation toughening, as measured by work to fracture, equivalent to a K_{1c} value of 10 MPa m$^{\frac{1}{2}}$.

Increasing the sintering temperature to 1500°C produces an increase in grain size which raises the M_s so that it becomes very near to room temperature. Although the flexural strength of this material is now halved it exhibits exceptional fracture toughness. This is well illustrated in Fig 7 which shows its quasi-ductile behaviour under the action of an increasing load; indeed microcracking of the ceramic is audible long before failure finally occurs during the measurement of flexural strength. Its mechanical properties are similar to those of a good quality cast iron; the high work to fracture is equivalent to a fracture indentation toughness K_{1c} of 20 MPa $m^{\frac{1}{2}}$ and a critical flaw size of 2-3 mm leading to values of the Weibull modulus, m, >20. The results of a detailed study of the failure mechanism and microstructure of this material will be published elsewhere.(6)

The effect of composition may be seen by reference to the behaviour of the material containing 11.4 w/o CeO_2. After sintering at 1400°C the M_s is at or slightly above ambient temperature and it has partly autotransformed, resulting in it consisting primarily of the monoclinic phase; sintering at 1500°C results in more grain growth so that the M_s is raised further above ambient temperature resulting in massive transformation leading to cracking of the material as it cools in the furnace.

These properties however decrease as the testing temperature is raised and future work will need to focus on combinations of alloying constituents in order to improve their retention of properties in more severe environments.

CONCLUSIONS

1. Electrorefined zirconia powders can be produced with sinter activities comparable to those prepared by chemical coprecipitation routes.

2. Certain CePSZ compositions are probably the toughest monolithic ceramics so far developed. Their mechanical properties are functions of their composition and sintering temperature and close control of these parameters allows materials to be produced which at or near ambient temperature are either very strong or very tough.

ACKNOWLEDGEMENTS

The authors gratefully acknowledge the extensive help of Unitec Ceramics Division of Universal Abrasives Ltd in the provision of electrorefined zirconia powders, Dr R Stevens of the University of Leeds for work to fracture and TEM studies, and the directors of Turner and Newall plc for permission to publish this work.

REFERENCES

1. K Tsukuma and M Shimada J. Mats Science 20 1178-1184 (1985)

2. V Gross and M V Swain J. Australian Ceram Soc 22 (1) p1-12 (1986)

3. P Duwez and F Odell J.Am.Cer.Soc. 33(9) 274-83 (1950)

4. E Tani, M Yoshimura and S Samiya J.Am.Cer.Soc. 66(7) 506-10 (1983)

5. T K Gupta and C A Andersson Ceramic Engineering and Science Proceedings, 13th Automotive Materials Conference Sept-Oct 1986 pp.1150-1157

6. R Stevens and D T Pindar To be published.

Table 1

Typical Powder Characteristics

Wt w/o	UY2	UC1	UC3	UC6/7
ZrO_2	90.7	85.1	85.8	83.1
HfO_2	1.56	1.55	1.58	1.48
Y_2O_3	5.97	-	-	
CeO_2	-	12.1	11.4	14.3
SiO_2	0.19	0.11	0.22	0.14
TiO_2	0.17	0.15	0.13	0.15
Al_2O_3	0.15	0.06	<0.02	<0.02
Fe_2O_3	0.03	0.07	0.06	0.09
CaO	0.05	0.07	0.05	0.02
MgO	<0.05	<0.05	<0.05	<0.05
Na_2O	<0.1	<0.1	0.1	<0.1
LoI	1.14	0.14	0.14	0.18
Mean particle size d_{50} µm	0.62	0.55	0.57	0.46

Table 2

Sintering behaviour		Sintered density		% Phase composition	
Material	Sintering temperature °C	g/cc	% theoretical	Tetragonal	Monoclinic
UY2 6 w/o YPSZ	1500	5.98	98.6	>95	< 5
UC1 12 w/o CePSZ	1350	5.65	90	-	
	1400	6.11	97.7	>90	<10
	1450	6.13	98.1	>90	<10
	1500	6.15	98.4	>90	<10
UC3 11 w/o CePSZ	1400	6.08	97.3	>20	80
	1500	cracked			>95
UC6 14 w/o CePSZ	1350	6.08	97.3	90	10
	1400	6.15	98.4	>95	<5
	1450	6.17	98.7	-	-
	1500	6.14	98.2	>85	<15

Table 3

Mechanical Properties

Material	Sintering temp. °C	Flexural Strength MPa	Weibull Modulus m	Vickers Hardness H_v kg/mm²
UY2 6 w/o YPSZ	1500	793	6.5	1392
UC1 12 w/o CePSZ	1400	695	11	984
	1450	426	13	832
	1500	326	21.4	916
UC6 14 w/o CePSZ	1350	701	12.7	1003
	1400	608	11.3	983
	1500	574	7.5	884

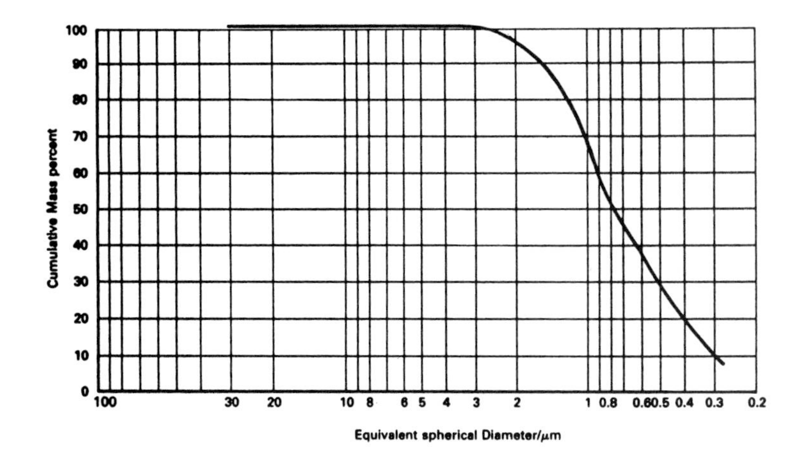

FIGURE 1

Particle size distribution of YPSZ powders

FIGURE 2 Typical powder morphology

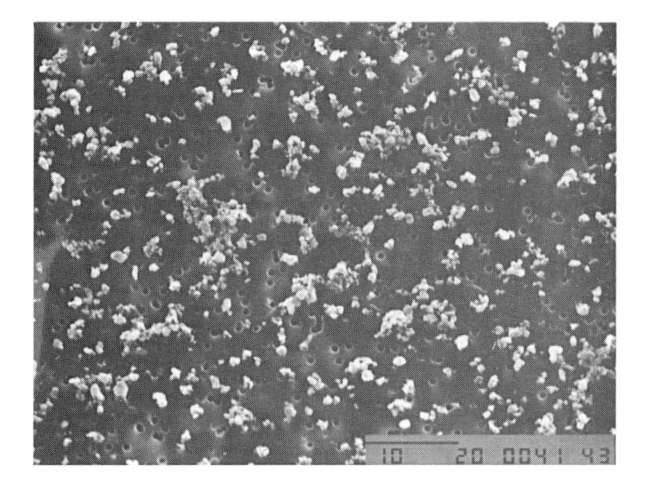

FIGURE 3 Fracture surface of 12 w/o CePSZ (SEM)

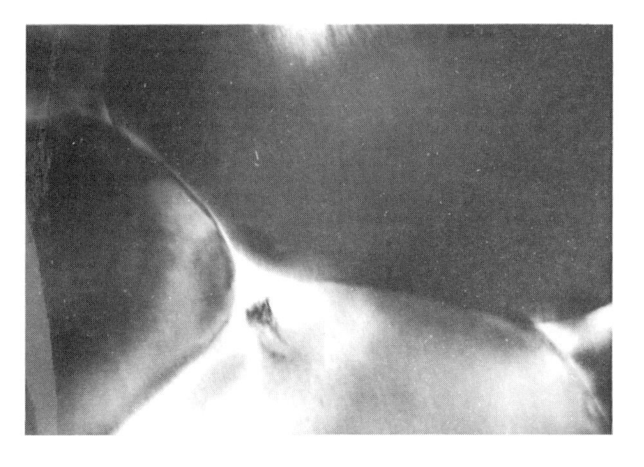

FIGURE 4 Detail of 12 w/o CePSZ showing presence of liquid at triple point (TEM; 22,500x)

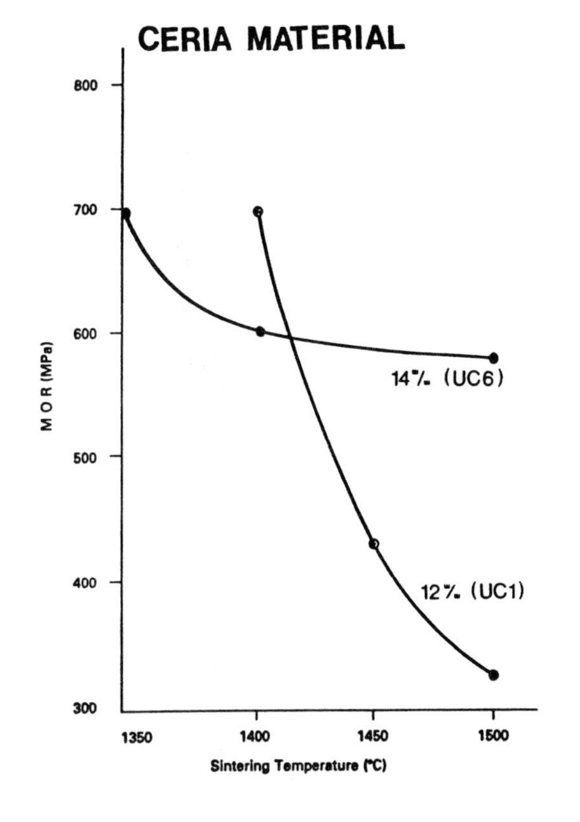

FIGURE 5 Variation of flexural strength of 12 w/o CePSZ and 14 w/o CePSZ with sintering temperature

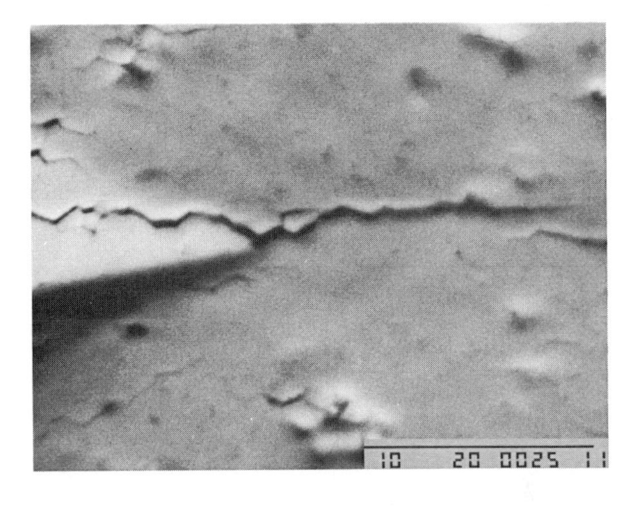

FIGURE 6 Deformation and cracking round the impression made by a Vickers indenter in 12 w/o CePSZ

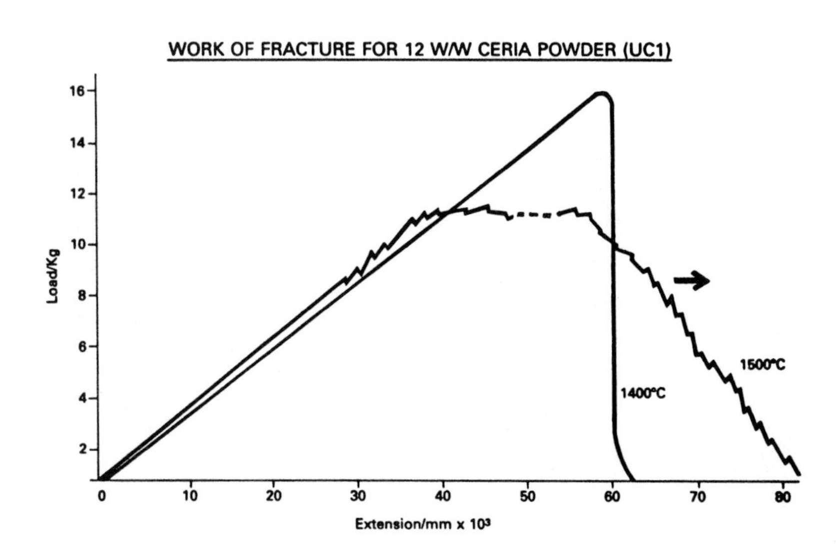

FIGURE 7

Quasi-ductile behaviour of CePSZ

Effect of high temperature on phase stability of plasma-sprayed zirconia ceramics

J R BRANDON and R TAYLOR

The authors are with the Manchester Materials Science Centre, University of Manchester/UMIST.

Synopsis

Thermal barrier coatings of ZrO_2 + 8.7 mole% $YO_{1.5}$ conventionally applied by plasma spraying are currently used in critical areas in the turbine section of modern jet engines. In the as-sprayed condition the coatings exist as a metastable, non-transformable tetragonal phase T'. This has been shown to be stable after long thermal cycling tests up to 1200°C.

However of fundamental interest is the long term stabilty at temperatures in excess of 1200°C. In this investigation coatings were annealed at temperatures of 1200-1600°C for times of 1-100 hours and the change in the phase structure during ageing studied by X-ray diffractometry. A complex pattern of decomposition is observed in which the original T' decomposes to form either two different T' phases of high and low yttria, or low yttria T' and cubic phases. These results are discussed in terms of the equilibrium diagram.

Introduction

Thermal barrier coatings are currently being developed to operate in the high temperature regions of gas turbines. Such materials need to possess low thermal conductivity, good thermal shock resistance, and impact toughness and chemical stability. However the thermal expansion coefficient needs to match the superalloy substrate. The latter criteria focuses attention on zirconia. Pure zirconia exists as three polymorphs. Below 1200°C the structure is monoclinic, from 1200-2370°C the structure is primitive tetragonal T and from 2370°C to the melting point ZrO_2 exists in the cubic (fluorite) form. To mitigate problems associated with these transformations zirconia is alloyed with other oxides such as MgO, CaO or $YO_{1.5}$, usually to stabilise the cubic phase. Recent studies have centred on alloys of zirconia partially stabilised, with $YO_{1.5}$ concentrations less than that needed to fully stabilise the cubic phase (YPSZ). Work carried out at UMIST has studied the thermal and mechanical properties of plasma sprayed coatings with compositions in the range 7-12.5 mole% $YO_{1.5}$. A composition of 8.7 mole% $YO_{1.5}$ possesses the optimum mechanical and thermal properties and exhibits the best performance during cycling tests up to 1200°C.

Three versions of the ZrO_2-$YO_{1.5}$ phase diagram exist[1-3] which differ in detail at the high yttria end. The most widely accepted is due to Scott[1] (figure 1) who first reported the existence of a non-transformable face centred tetragonal phase T' which he and other workers[4,5] suggested formed from the cubic ZrO_2 solid solution via a displacive phase transformation. Its non-transformability relates to its reluctance to undergo the stress assisted martensitic transformation from T' ZrO_2 to M ZrO_2 found in alloys of lower $YO_{1.5}$ content. This form of non-transformable T'-ZrO_2 has been widely encountered in plasma ZrO_2 and is believed to form only under non-equilibrium cooling conditions. Andersson, Greggi and Gupta[6] have studied transformations in the ZrO_2-$YO_{1.5}$ system and concluded that diffusional or equilibrium transformations would only occur at temperatures above 1227°C. Below this temperature diffusionally controlled reactions are too sluggish and only martensitic transformations occur. They produced a phase diagram of the quench products (figure 2) predicting the existence of the T' phase via boundaries representing the start of the diffusionless martensitic transformation $M_s^{c-T'}$ and $M_s^{T'-M}$. In our tests at UMIST[7] X-ray diffraction studies show that all compositions in the range 7-12.5 mole% $YO_{1.5}$ form 100% T' when plasma sprayed, in accordance with the findings of Andersson and Gupta.

From the phase diagram it can be predicted that the ageing in the two phase region should cause the T' to decompose. Miller et al[4] aged several samples of ZrO_2 + 9.5 mole% $YO_{1.5}$ for times up to 100 hours at temperatures up to 1600°C. They concluded that T' decomposes to cubic ZrO_2 + transformable tetragonal ZrO_2, the latter transforming to monoclinic zirconia on cooling. However there is doubt over the quality and interpretation of this data. Lanteri et al[8] carried out a TEM analysis of annealed ZrO_2 + 8.7 mole% $YO_{1.5}$ and showed that on annealing at 1600°C T' decomposes into the equilibrium phases: cubic ZrO_2 and colonies of low yttria tetragonal ZrO_2. Interestingly the latter did not transform to monoclinic on cooling.

Since the gas temperatures of a modern jet engine can exceed 1600°C it is of considerable interest to carry out a detailed study of the stability of the T' phase. in this investigation specimens of ZrO_2 + 8.7 mole% $YO_{1.5}$ were aged at temperatures of 1200, 1300, 1400, 1500 and 1600°C for times up to 500 hours, and the phase structure studied by X-ray diffraction.

Experimental Details

Powders of ZrO_2 + 8.7 mole% $YO_{1.5}$ were plasma sprayed by Rolls Royce Ltd. using standardised spraying parameters onto nickel based superalloy plates 5cm square. The superalloy substrate was subsequently dissolved using nitric acid. Fragments of these coatings (about $1cm^2$) were annealed in air inside a platinum crucible at temperatures of 1200, 1300, 1400, 1500 and 1600°C for times of 1, 3, 10, 30 and 100 hours. At the end of the ageing period specimens were removed from the furnace as quickly as possible in order to maximise cooling rates thereby minimising diffusional processes during cooling.

X-ray diffraction analysis of all coatings was carried out using a Philips diffractometer and Philips PW1710 programmable control unit. The radiation source was nickel filtered copper Kα (λ $K\alpha_1$ = 1.54056Å λ $K\alpha_2$ = 1.54183Å). A program was constructed to perform the following scans.

i) Continuous scan of the region 2θ=10° to 2θ=120° at $0.1°s^{-1}$. This scan was primarily used to detect the presence of free monoclinic zirconia and any other oxide present. This would also reveal the presence of any primitive tetragonal phase if mixed indices reflections were observed.

ii) A very slow scan of the region 2θ=72-76° at $0.005°s^{-1}$ to examine the (400) cubic and (400) (004) tetragonal T' type reflections.

iii) Slow scan of the region 2θ=27-32.2° at $0.1°s^{-1}$ to examine the (111) type reflections from the monoclinic and cubic tetragonal.

Most information was obtained from the slow scans covering the (400) region for which the traces are a composite of several components ie. (400)T', (004)T', (400)F etc. These were deconvoluted by hand. The individual peaks from the kα1 and kα2 radiation were also resolved during the deconvolution process. The two separate wavelengths were taken as; Cu $K\alpha_1$ (λ=1.54056) and Cu $K\alpha_2$ (λ=1.54183). The net effect is a component peak kα which has an offset Gaussian distribution in some cases and in other cases two peaks relating to each wavelength can be identified. Since the $K\alpha_1$ constituent is approximately twice the intensity of the $K\alpha_2$ it was used as the reference point for the calculation of d spacings. The peaks produced by the low angle (111) type reflections do not have a significant offset making their measurement much more straightforward. The areas of the peaks were measured as accurately as possible by counting the squares under the peaks on the trace.

To estimate the phase percentages the integrated intensities of individual reflections were determined and molar ratios of the monoclinic and cubic/tetragonal calculated from

$$\frac{M_M}{M_{F,T'}} = 0.82 \frac{I_M(111) + I_M(11\bar{1})}{I_{F,T'}(111)} \qquad (1)$$

This equation relates the I_M, IT'_F, intensities of the monoclinic and cubic/tetragonal (111) type peaks to their relative mole fractions. This is based on a relationship by Porter and Heuer[9] and has been used by a number of workers including Miller[4] and Boch et al [10]. Porter and Heuer calculated a volume fraction ratio using structure factors and a small Lorentz polarisation correction. The only modification made by Miller is the conversion from volume ratio to mole ratio. In a similar manner Miller modified a relationship by Teufer[11] to produce the relation for the mole fraction of the cubic C_F and face centered tetragonal T' phases based on (400) peak intensities

$$\frac{M_F}{M_{T'}} = 0.88 \frac{I_F(400)}{I_{T'}(400)+I_{T'}(004)} \qquad (2)$$

Equation (1) was used first to calculate any monoclinic fraction. The remaining fraction was then divided accordingly between the cubic and tetragonal phases using equation (2). The ratio of intensities for the (400) and (004) reflections of the T' can be calculated as being twice the appropriate structure factors squared times a small Lorentz polarisation correction[11]. The value obtained is 3.15. This ratio is used as a guideline when deconvoluting cubic and tetragonal composite peaks. Values of 2θ were measured to the nearest 0.02°. The d values for each phase were calculated using the 2θ values for the resolved $K\alpha_1$ peaks using the Bragg equation. The c/a ratio measured was used to estimate the $YO_{1.5}$ composition of the tetragonal phase, using the following expression derived by Miller[4] using data given by Scott[1].

$$\%YO_{1.5} = (1.0223 - c/a)/0.001309 \qquad (3)$$

Relatively large errors may be present when using this equation because of the sensitivity to c/a ratio. However it provides a useful method of comparing the relative yttria concentrations of different T' phases.

Results and Discussion

The as sprayed coatings consisted entirely of non-transformable tetragonal T' phase as noted by Morrell and Taylor[7] and by Noma[12]. No mixed indices peaks were noted confirming the absence of any equilibrium transformable tetragonal phase. This observation is at variance with the observations of Miller et al[4] and Suhr et al[13] who noted significant quantities of cubic and monoclinic in the as sprayed coatings. One possible reason for this discrepancy could be inhomogeneity of the starting powder and incomplete melting and homogenisation in the plasma arc. Figure 3 shows a typical peak from the (400) region showing deconvolution into the (400)T' and (004)T' peaks. The lattice parameters of the as sprayed T' material are

c = 5.1616
a = 5.1055
c/a = 1.0110

corresponding to the starting composition of 8.7% $YO_{1.5}$.

When samples are annealed in the temperature range 1200-1600°C the original T' phase initially decomposes into a low yttria non-transformable phase we shall call T'_1 and another phase of higher yttria content (cubic or T'_2). In principle the lower yttria content T'_1 phase should transform martensitically to monoclinic. The high yttria phase at elevated temperatures should exist as cubic zirconia according to the equilibrium diagram but may transform on rapid cooling to a high yttria tetragonal phase depending on the annealing time and temperature. This tetragonal high yttria phase we shall call T'_2. In principle therefore deconvolution of the (400) peaks could, and in some cases does involve resolution of 5 peaks (400)T'_1, (004)T'_1, (400)F, (400)T'_2 and (004)T'_2. A typical example of peak resolution for the 30 hour anneal at 1400°C is shown in figure 4 and deconvoluted peaks for specimens aged for 100 hours are given in figure 5.

Data for all annealed specimens is summarised in Table 1. Transformation is initially fairly rapid at 1600°C with a clear separation into high and low yttria peaks being observed after 1 hour. The separation into high and low ytrria phases appears to continue for the 3 and 30 hour anneals. However after the 100 hour anneal no low yttria T'_1 phase was present (figure 5a). Instead 27% of the monoclinic phase was detected. The composition of the non-transformable T'_2 phase is estimated using equation 3 at 12.4% $YO_{1.5}$ which is similar to that measured from the phase diagram for the equilibrium cubic phase at 1600°C. Also the monoclinic fraction of 27% calculated using equation 1 is in good agreement with the equilibrium diagram prediction of 30% using the Lever rule. If we consider data for the samples aged at 1500°C a clear pattern emerges. The original T' decomposes into a cubic phase of 13.75% $YO_{1.5}$ and a low yttria phase in which the yttria composition decreases with annealing time to 4.2% $YO_{1.5}$ after 100 hours. The high temperature cubic phase transforms on cooling to a high yttria T'_2 phase with a low (1.0043) c/a ratio and the low yttria phase being retained as T'_1 (see figure 5b). The sequence of 400 peaks as a function of annealing time are given in figure 6. A detailed compilation of phase analysis is shown in Table 2a. The results for the specimens aged at 1400°C and 1300°C are perhaps the most intriguing. The sequence of (400) peaks after ageing are shown in figure 7 and a more detailed compilation of the phase analysis for 1400°C anneals is shown in Table 2b. The feature of all the coatings aged at this temperature is the gradual decomposition of the original T' to a mixture of a low yttria T'_1 and a high yttria cubic phase seemingly via an intermediate high yttria T'_2 phase. After 1 hour at 1400°C it can be seen that the material is still T' but that there is a significant broadening of the (400) T' peak around the base. This is caused by the gradual diffusion of the yttria, forming high and low yttria regions with correspondingly smaller and larger c/a ratios than the parent T'. No separation of the peak can be seen, but it is likely that a small amount of cubic phase is contributing to the shape. Three hours of ageing produced a similar trace but with a more noticeable shoulder on the (400) peak. The shoulder gives the peak an offset towards the low yttria t' direction suggesting that the peak is in fact a composite of low and high yttria T'. However no definite separation can be seen. The cubic phase cannot be clearly resolved from the background. The specimen aged for ten hours continues this trend, only now it is possible to recognise two different composition T' phases. Using equation 3 the concentration of yttria ($YO_{1.5}$) in the low yttria T'_1 has fallen to 5.58% whereas that of the high yttria T'_2 has increased to 10.24%. The cubic phase is now indicated by a distinct peak. The deconvolution of an XRD trace such as this is made easier after examination of the traces obtained for the coatings aged for 30 and 100 hours. The observed features become more distinctive, with a steady increase of the low yttria T'_1 phase and a decrease in the high yttria T'_2 in favour of the cubic phase. The yttria content of these phases has changed to 5.04% and 10.78% respectively for the 30 hour coating. After 100 hours it can be seen in figure 5c that only the cubic and low yttria t'_1 phases are present, with the T'_1 having an yttria content of only 4.5%. Similar behaviour is noted for the specimens aged at 1300°C except that the transformation appears to be more sluggish. The first traces of the cubic phase (at ≈14%) are present only after 10 hours and the concentration of cubic phase is only 24% after 100 hours. Interestingly after remaining constant at ≈9.% $YO_{1.5}$ in T' for which the c/a is ≈1.010 after the 100 hour anneal low and high yttria T' phases appear with c/a values of 1.0166 (4.35% $YO_{1.5}$) and 1.0094 (9.85% $YO_{1.5}$) respectively. This is a similar feature to that observed for the 1400°C anneal after 10 hours. At 1200°C there was no detectable destabilisation of the as sprayed T' phase within the 100 hour heat treatment programme. Only after a 500 hour anneal was any detectable change in structure observed.

The results are in good agreement with the equilibrium diagram presented in figure 1 and 2 and the data at 1200°C support the the observation of Andersson and Gupta that diffusional transformations are sluggish below this temperature. However at temperatures above 1200°C the original T' material transforms to a low yttria T'_1 and a high yttria phase which is probably cubic at elevated temperatures. This transformation appears to proceed in accordance with the respective tie lines on the equilibrium diagram and is more rapid the higher the annealing temperature. At 1600°C there is a very rapid separation after 1 hour into two compositions which quench to form two distinct T' phases T'_1 and T'_2. The concentration of the yttria in the low yttria phase decreases progressively with annealing time whereas the high yttria phase reaches its compositional limit more rapidly. This is also substantiated by the 1500°C data. After 100 hours at 1600°C no low yttria T'_1 was found but monoclinic was detected. The concentration of yttria is probably less than 4% after this time and would appear that the T'_1 phase transforms on cooling to monoclinic via the $M_s^{T'-M}$ transformation line shown in figure 1. For the data at 1500°C the low yttria and cubic phases both quench to form two distinct T' phases both of which appear to be within the notional composition limits for the non-transformable tetragonal phase shown in figure 2 of 4% and 13.5% $YO_{1.5}$. A more complex pattern emerges after the 1400°C and 1300°C anneals. After 3 hours at 1400°C 14% cubic phase appears. This increases to 45% after 100 hours. However this simple picture is complicated by the appearance after 10 hours of a high yttria T'_2 phase and its disappearence after 30 hours into presumably, more cubic and low yttria T'_1. The latter observation is substantiated by the increase in the low yttria phase phase after 100 hours. A similar pattern of behaviour is noted for the 1300°C annealed samples. However no cubic is detected until the 10 hours anneal and no high yttria T'_2 is detected until the 100 hour anneal. At both these temperatures the equilibrium concentrations of yttria in the cubic phase should be between 14% and 15%. These exceed the compositional limits of the non transformable tetragonal phase and cubic should be retained on cooling. However the appearance and subsequent disappearance of a high yttria T'_2 phase suggest that the real situation is more complex than this. X-rays merely give a global average and compositional fluctuations on a microscale can and do occur. For example Noma et al[13] have noted finely twinned precipitates of 3 tetragonal variants in as quenched specimens of 6-10 mole% $YO_{1.5}$. We assume that separation into high and low yttria phases is largely complete after 100 hour anneals (whether the retained phase be tetragonal, cubic or monoclinic) and have summarised the relative fractions of the two variants in figure 8. This, it can be seen, is in good agreement with the the tie lines on the equilibrium diagram (figure 2).

Conclusions

In summary, the X-ray stability study has shown that the 8.7 mole% yttria coatings consisting wholly of T' are relatively stable up to 1200°C. At 1300°C and 1400°C the coatings decompose to form a mixture of low yttria T'_1 and high yttria cubic via a diffusional process involving an intermediate high yttria T'_2 phase. These phases are retained on quenching. Ageing at 1500°C causes the formation of a low yttria T'_1 and a high yttria cubic phase which has insufficient yttria content to be retained on cooling but transforms to a high yttria T'_2 material instead. It appears that ageing at 1600°C causes coatings to decompose to either equilibrium tetragonal or low yttria T' which quenches to form monoclinic and an equilibrium cubic phase which quenches to form a high yttria T' phase in the same way as that described for coatings aged at 1500°C. What is not certain in these measurements is whether the equilibrium tetragonal phase forms or the T' is stable with respect to this transformation at this temperature. Future work using high temperature X-ray diffractometry and TEM analysis could resolve some of the issues raised in this paper.

References

1. H.G.Scott, J. Mater. Sci., **10**, 1527-35 (1975).
2. V.S.Stubican, R.C.Hink and S.P.Ray, J. Amer. Ceram. Soc., **61**, 17-21 (1978).
3. C.Pascual and P.Duran, J. Amer. Ceram. Soc., 66, 23-27 (1983).
4. R.A.Miller, J.L.Smailek and G.G.Garlick, Science and Technology of Zirconia, Advances in Ceramics, Vol.3, Ed. A.H.Heuer and L.W.Hobbs. 78-85 (1981).
5. C.A.Andersson and T.K.Gupta, Science and Technology of Zirconia, Advances of Ceramics, Vol.3, Ed A.H.Heuer and L.W.Hobbs. 184-201 (1981).
6. C.A.Anderson, J.Greggi, and T.K.Guppta., Science and Technology of Zirconia II, Advances in Ceramics, Vol.12, Ed. N.Claussen, M.Rühle and A.H.Heuer (1983).
7. P.Morrell and R.Taylor,
8. V.Lanteri, A.H.Heuer and T.E.Mitchell, Science and Technology of Zirconia II, Advances in Ceramics, Vol.12, Ed. N.Claussen, M.Rühle and A.H.Heuer. 118-130 (1983).
9. D.L.Porter and A.H.Heuer, J. Amer. Ceram. Soc., 62, 298-305 (1979).
10. P.Boch, P.Fauchais, D.Lombard, B.Rogeaux and M..Vardelle, Science and Technology of Zirconia II, Advances in Ceramics, Vol.12, Ed. N.Claussen, M.Rühle and A.H.Heuer, 488-502 (1983).
11. G.Teufer, Acta Crystallogr., 15, 1187 (1962).
12. T.Noma et al,.Extended Abstracts, 3rd International Conference on the Science and Technology of Zirconia, (1986)
13. D.S.Suhr, T.E.Mitchell and R.J.Keller,.Science and Technology of Zirconia II, Advances in Ceramics, Vol.12, Ed. N.Claussen, M.Rühle and A.H.Heuer. 503-517 (1983).

Table 1. Summary of Phase Analysis After Thermal Exposure

		Annealing Time (hrs)									
		1		3		10		30		100	
Temp.		%	(c/a)	%	(c/a)	%	(c/a)	%	(c/a)	%	(c/a)
1600	T'_1	50	1.014	40	1.015	-	-	35	1.0165	-	-
	T'_2	50	1.006	60	1.007	-	-	65	1.0056	73	1.0065
	M	-	-	-	-	-	-	-	-	27	-
1500	T'_1	75	1.0107	58	1.0126	50	1.0152	47	1.0155	45	1.0168
	T'_2	25	1.0043	42	1.0040	50	1.0047	53	1.0043	55	1.0043
1400	T'	100	1.0096	86	1.0101	-	-	-	-	-	-
	T'_1	-	-	-	-	40	1.0150	45	1.0157	55	1.0159
	T'_2	-	-	-	-	36	1.0089	22	1.0082	-	-
	C_F	-	-	14	-	24	-	32	-	45	-
1300	T'	100	1.0105	100	1.0098	96	1.0100	88	1.099	35	1.0094
	T'_1	-	-	-	-	-	-	-	-	41	1.0166
	C_F	-	-	-	-	4	-	12	-	24	-
1200	T'	100	1.011	100	1.0101	100	1.010	100	1.0112	100	1.0112

Table 2a. Data for specimens annealed at 1500°C

	Annealing time (hours) 1	3	10	30	100
Mole Fraction in %					
T'_1	75	58	50	47	45
T'_2	25	42	50	53	55
400 T'_1	5.108	5.101	5.096	5.096	5.093
004 T'_1	5.163	5.165	5.174	5.175	5.179
400 T'_2	5.124	5.123	5.122	5.124	5.124
004 T'_2	5.146	5.144	5.146	5.146	5.146
c/a T'_1	1.0107	1.0126	1.0152	1.0155	1.0168
c/a T'_2	1.0043	1.0040	1.0047	1.0043	1.0043
Mole% $YO_{1.5}$					
T'_1	8.86	7.41	5.42	5.19	4.20
T'_2	13.75	13.98	13.44	13.75	13.75

Table 2b. Data for specimens annealed at 1400°C

	Annealing time (hours) 1	3	10	30	100
Mole Fraction %					
T'	100	86	-	-	-
T'_1	-	-	40	45	55
T'_2	-	-	36	22	-
C_F	-	14	24	32	45
Lattice parameter Å					
400 T'	5.1102	5.1090	-	-	-
004 T'	5.1592	5.1604	-	-	-
400 T'_1	-	-	5.096	5.095	5.0938
004 T'_1	-	-	5.1726	5.175	5.175
400 T'_2	-	-	5.1122	5.1173	-
004 T'_2	-	-	5.158	5.1592	-
400 C_F	-	5.134	5.134	5.134	5.134
c/a T'	1.096	1.010	-	-	-
c/a T'_1	-	-	1.015	1.0157	1.0159
c/a T'_2	-	-	1.0089	1.0082	-
Mole % $YO_{1.5}$					
T'	9.70	9.32	-	-	-
T'_1	-	-	5.58	5.04	4.90
T'_2	-	-	10.24	10.78	-

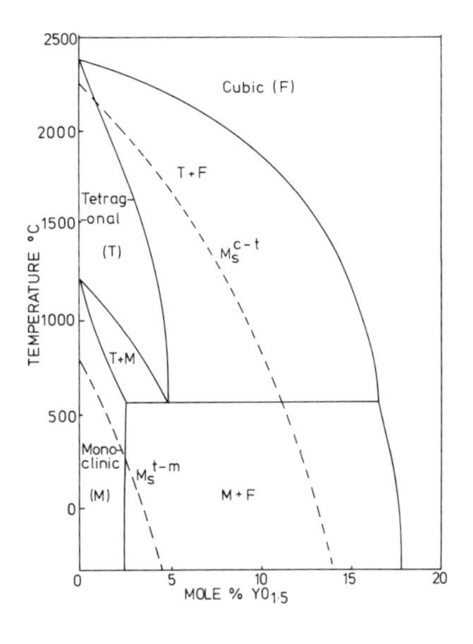

Figure 1. Equilibrium phase diagram for the ZrO_2-$YO_{1.5}$ system showing estimated martensitic start loci (Ref.6).

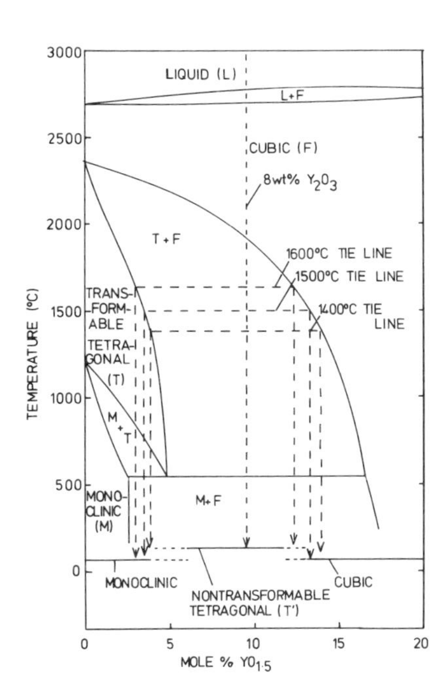

Figure 2. Equilibrium phase diagram for the ZrO_2-$YO_{1.5}$ system (Ref.1) showing tie lines and composition lines for quenched phases.

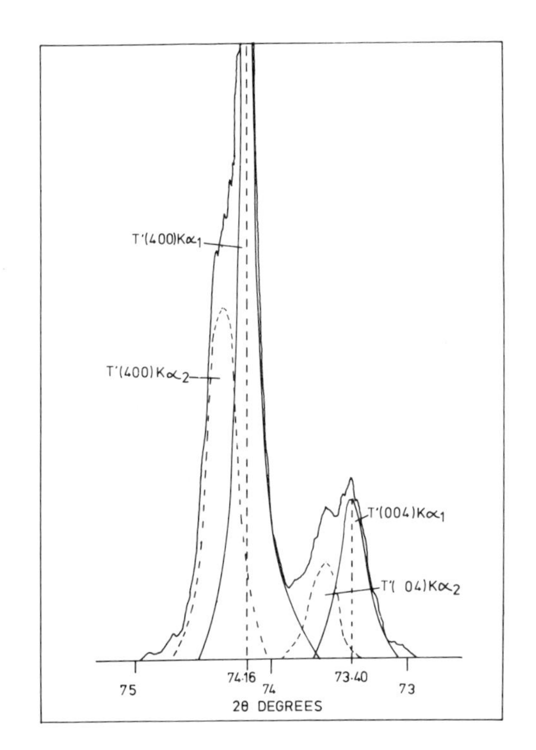

Figure 3. Typical trace and its deconvolution for an as sprayed 8.7% $YO_{1.5}$ PSZ coating.

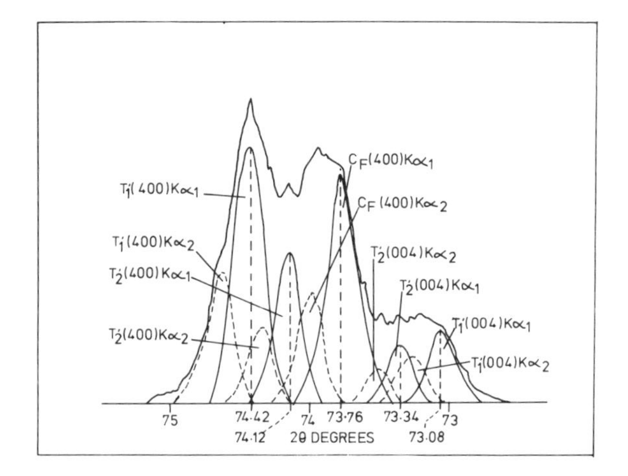

Figure 4. The component peaks forming the composite peak in the (400),(004) type region for the coating aged at 1400°C for 30 hours.

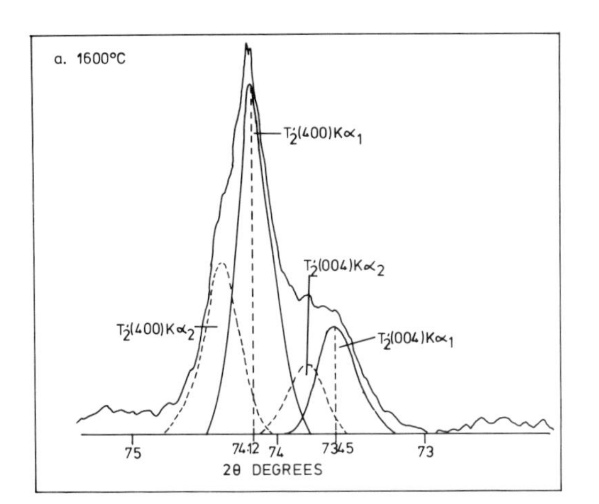

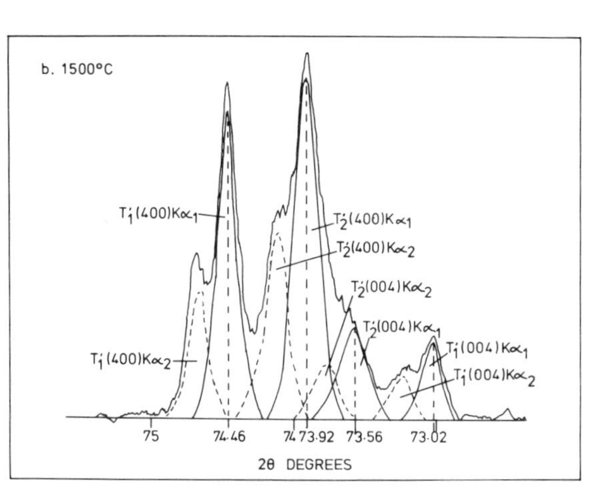

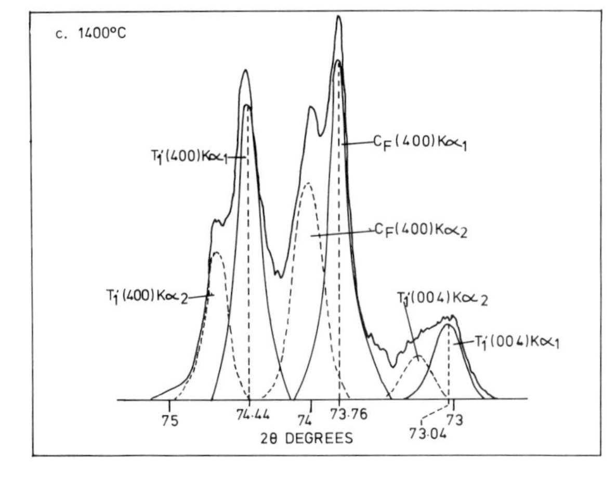

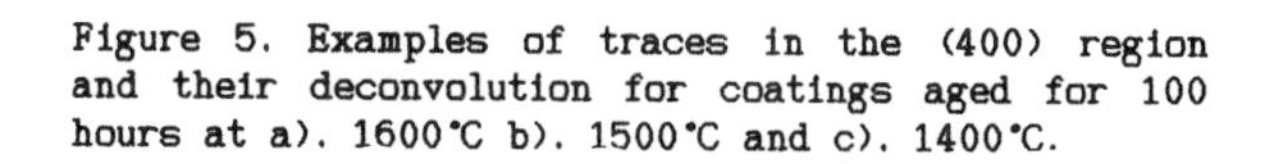

Figure 5. Examples of traces in the (400) region and their deconvolution for coatings aged for 100 hours at a). 1600°C b). 1500°C and c). 1400°C.

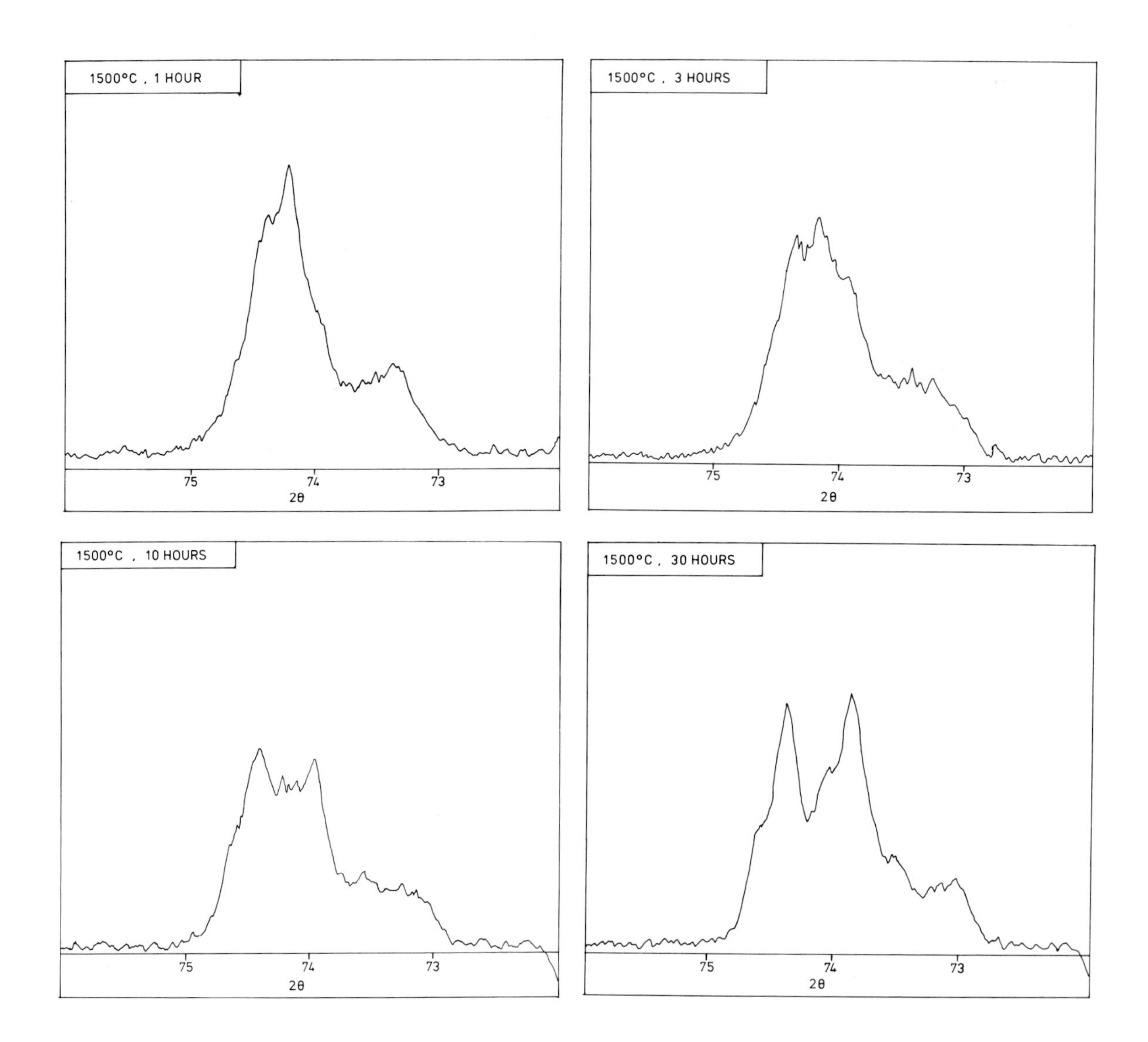

Figure 6. The sequence of (400),(004) peaks obtained for coatings after ageing at 1500°C for varying lengths of time.

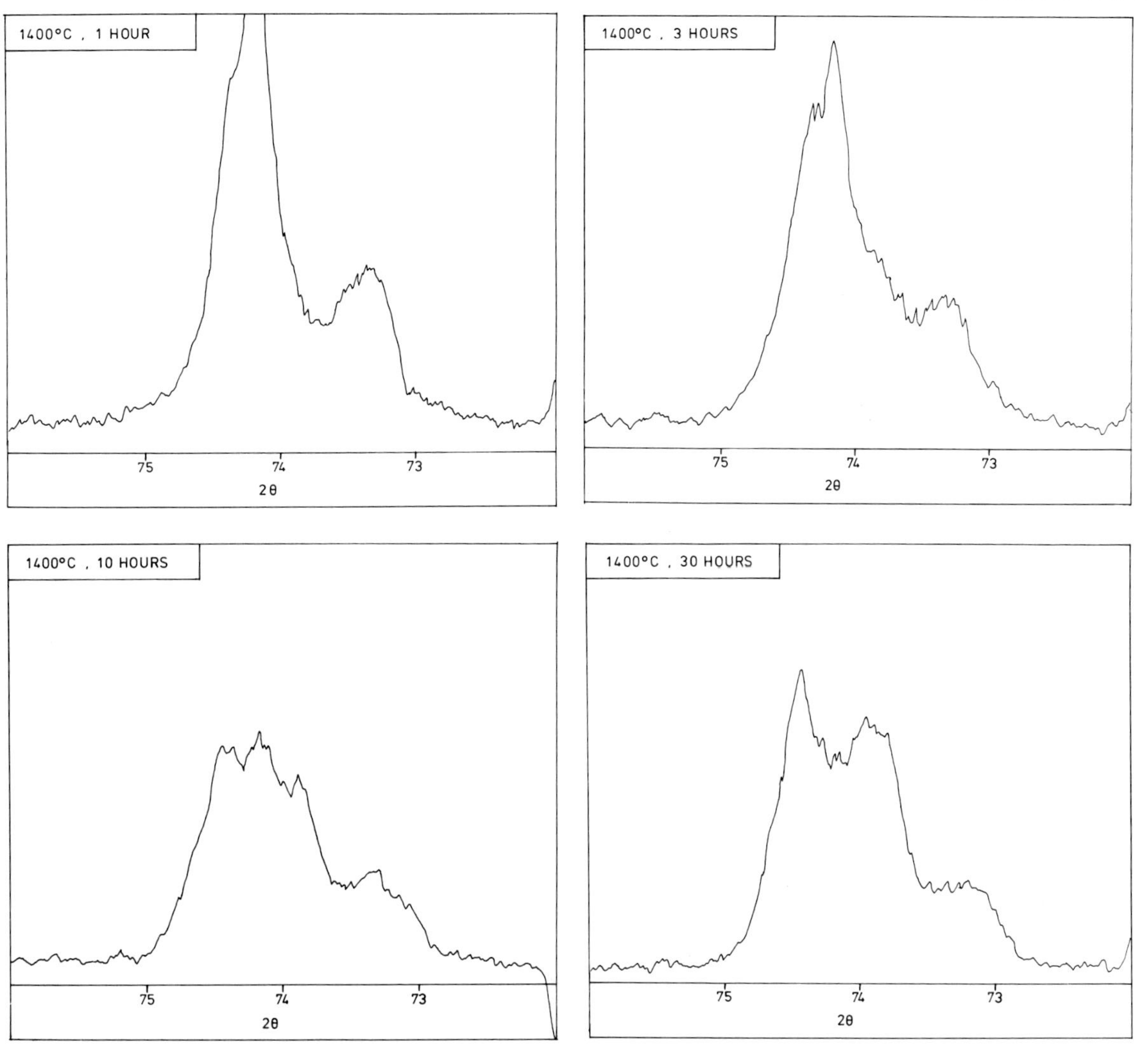

Figure 7. The sequence of (400),(004) peaks obtained for coatings after ageing at 1400°C for varying lengths of time.

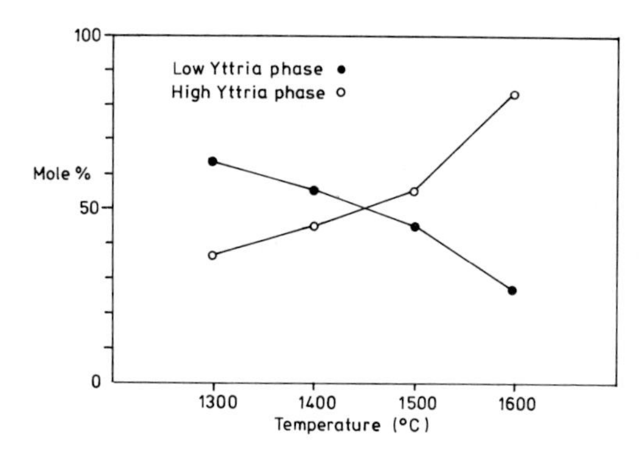

Figure 8. The relative proportions of high yttria (C_F,T'_2) and low yttria (T'_1) phases present in coatings aged for 100 hours with varying annealing temperature.

Microstructure of aluminosilicate fibres and possibilities for control

M SOPICKA-LIZER and S PAWŁOWSKI

The authors are with the Institute of Materials Science and Engineering, Katowice, Poland.

SYNOPSIS

Mullite crystallization in aluminosilicate fibres has been tested. It has been established that aluminosilicate glass is metastable and mullite crystallizes spontaneously inside the ellipsoidal precipitations at temperatures above 920°C. The viscosity of the remaining aluminosilicate glass decreases at higher temperatures leading to sintering of the fibres and cristoballite formation. Both of these processes set limits on the work temperature of the fibre applications. Three methods of improving the fibres' thermal stability are discussed. The most appropriate fibres with glass-ceramic structures are described.

INTRODUCTION

The present interest in ceramic fibres arose from two major technological needs - thermal insulation at high temperature and fibres of high strength, stiffness, and chemical stability for the reinforcement of metals. A development in fibres for insulation was the production of glassy aluminosilicate fibres by spinning a mixture of quartz sand and alumina. These fibres contain 45-50% by weight of alumina and can be used at up to 1200°C. Their thermal limit is imposed by devitrification of the glass, causing shrinkage and a fall in efficiency of thermal insulation.

Mullite appears as the first phase during heat treatment at about 940°C and cristoballite as the second one at about 1150°C.[1] It is known that most glassy fibres (glass, mineral, and slag) are used below transformation or crystallization temperatures whereas aluminosilicate fibres are used above the temperature of mullite crystallization.

The viscosity of glass falls rapidly above the transformation temperature and the fibres soften. On the other hand, the crystallization of glass is controlled by the nucleation process and appears as the growth of a few large crystals which can destroy the fine fibres. None of these processes are observed during the crystallization of aluminosilicate fibres below 1100°C. This means that mullite crystallization differs from normal crystallization of glass and the fibre microstructure is one of the main reasons for this process. This suggests that the relationship between the microstructure of fibres and devitrification process of aluminosilicate fibres should be examined.

CRYSTALLIZATION OF MULLITE

The crystallization of mullite occurs at temperatures between 940 and 1020°C and can easily be observed on a DTA curve during the heating of powder prepared from aluminosilicate fibres. The transformation temperature (Tg) of these fibres has been estimated from the DTA curve at 900°C. The attempts at the application of DTA to the kinetic study of mullite crystallization have failed because Avramy's equation

$$\ln (1 - \alpha) = k t$$

describing reactions in which nucleation is the stage controlling complex process rate could not be applied in this case.[2] It has been established that the function $g(\alpha)$ varies during crystallization and different processes control the crystallization rate at several stages. Thus mullite crystallization in aluminosilicate fibres depends on some other factors, different from those accepted by classical theories of glass crystallization.

A further study was carried out using isothermal heating of fibres. The standard aluminosilicate fibres (45 wt-% of alumina), produced in Skawińskie Zakłady Materiałów Ogniotrwałych, Poland, were tested. The specimen of fibres was placed inside the electric furnace with the temperature controlled at ±5°C. The heating time varied from 0.5 to 5 h, and the temperature from 880 to 980°C. The extent of crystallization was determined as the ratio of the amount of crystallized mullite in the heated fibres to the maximal content of mullite which can occur in this kind of aluminosilicate glass. The amount of mullite has been calculated from the X-ray data and the results are shown in Fig. 1.

The mechanism of mullite crystallization depends on temperature. There were no traces of mullite even after prolonged heating of the fibres below the transformation temperature (880°C). The fibres heated at 900 and 920°C crystallized slowly and this process could be controlled by the nucleation rate. Mullite crystallization at higher temperatures was different and a greater amount of mullite appeared during the first 30 min of heating and prolonged heat treatment did not cause any significant increase in the mullite content.

The microstructure of these fibres was examined by electron microscope at different stages of crystallization. Thick fibres with a diameter of up to 70 μm were cut with a sharp knife in preparation for an examination of their cross-sections in the scanning microscope. Thin fibres of under 2 μm dia. were tested in the transmission microscope because they were transparent to an electron beam. All the fibres were etched in fluorohydrogen acid vapour.

The unheated fibres exposed to the electron beam changed their microstructure. Rearrangement of structural units was observed over 1-2 sec and the main diffraction lines of mullite were detected (3.86, 2.55, 2.21)(Fig. 2). This phenomenon was caused by a local increase of temperature during electron beam exposure and can be defined as instant crystallization of mullite.

The microscopic studies of heated fibres revealed the details of their microstructure after crystallization. Droplets of ellipsoidal shape were observed, visible both on the fibre cross-section in the scanning microscope (Fig. 3) and inside the thin fibres examined in the transmission microscope (Fig. 4). The size of these droplets depends on the fibre diameter and was found to be about 10 μm in 60 μm fibres and about 0.2 μm in 1.5 μm fibres. The microstructure of the ellipsoidal precipitations differs from the residual glass and could be considered similar to the interconnective structure occurring in immiscible glasses. Moreover, the ellipsoidal droplets are more resistant to HF etching and thus seem to be an alumina-rich phase.

Crystallization of mullite did not cause transmutation on the fibre surface. The changes could be visible on fibres heated at higher temperatures (above 1200°C) after cristoballite formation.

The obtained results show that rapid mullite crystallization is due to the metastable miscibility of aluminosilicate glass. Binary aluminosilicate materials can form glass after rapid quenching from high temperatures and the amount of silica determines the glass-forming ability. Pure silica glass is built in three-dimensional arrays of corner-shared $[SiO_4]^{4-}$ tetrahedra. If aluminium ions were to replace silica ions in a tetrahedral network every aluminium-for-silicon replacement would require one oxygen ion bonded to three cations to maintain the charge balance.[3] The metastability of this structure, proposed by Lacy,[4] is the basic reason for the small amount of alumina allowed in silica glass solution. It is schematically drawn in Fig. 5. Beal assumed that the alumina preferred to form its own randomized octahedral structure represented by a glass approaching mullite in composition.[3] Evidence of a miscibility gap in this binary system was found and the results are discussed later. Takamori and Roy[5] assume that the microstructure of these glasses depends on cooling rate and they estimate that the minimal cooling rate to avoid phase separation in 50 mol% alumina glass is about 10^{-7} °C/sec.

Rapid quenching from high temperature during fiberization leads to local phase separation. The macroliquation precipitations crystallize spontaneously during heat treatment at temperatures above 920°C. The first stage of crystallization can be controlled by the nucleation rate and the second can be a kind of 'martensite transformation'. The internal crystallization of mullite does not destroy the fibres and, moreover, the residual silica-rich glass does not crystallize easily because of high viscosity. This answers the question why the glass devitrification does not destroy aluminosilicate fibres and can be the base for developing new kinds of these fibres. Nevertheless, the viscosity of residual glass will decrease if the heating temperature rises. This means softening and sintering of fibres parallel to cristoballite formation. Due to these processes high shrinkage and degradation of fibrous materials are observed.

FIBRE MODIFICATION

Aluminosilicate fibres heated at 900-1000°C consist of mullite grains gathered inside the ellipsoidal precipitations and residual silica-rich glass. The mullite grains are stable at higher temperatures but the properties of silica-rich glass depend on temperature and they will determine the behaviour of the fibres. The viscosity of the residual glass is the most variable property and depends on the amount of alumina in silica-rich glass. An increase in heating temperature will cause softening and sintering of the fibres due to a decrease in the viscosity of the glass. Therefore the structure of the fibre should consist of as much mullite phase as possible, preferably in the form of the crystalline network, and

almost pure silica glass with high viscosity. This kind of structure modification can be achieved in the following ways:
1. increase alumina content in order to obtain a more unstable glass with mullite as the main phase after the devitrification process. High alumina fibres will internally crystallize into a greater amount of mullite plus some residual glass that appears resistant to cristoballite formation.[6] Thermal resistance of these fibres depends on alumina content, but the melting point rises with the increase in alumina content and this causes severe equipment problems in the fibre spinning process. For this reason the high-mullite fibres are produced rather by the solution method than by melt spinning.[7]
2. a network modifier cation addition to improve the thermal stability of aluminosilicate glass. These cations can maintain charge balance in Al^{3+} for Si^{4+} replacement and the glass becomes more stable. Chromium oxide is preferred due to its high charge and thermal behaviour in aluminosilicate glass. The addition of 3-5 wt-% of chromium oxide eliminates the macroliquation precipitations and leads to mullite crystallization in the whole volume of the fibre.
3. a heterogeneous catalyst addition in order to control the mullite crystallization and to obtain a typical glass-ceramic structure with the fine-grained mullite crystals uniformly distributed in the matrix of silica glass. The mullite crystalline frame should ensure stiffness of the fibres in high temperatures when there is a decrease in the viscosity of the silica glass. Copper oxide was chosen to modify the fibre structure. It can behave like a network modifier, as chromium oxide does. However, part of the copper cations can be reduced to copper atoms during melting and then, during heat treatment, they aggregate into clusters, being the nuclei for the later crystallization of the mullite. The nuclei are uniformly distributed over the whole volume of the fibre and further mullite crystallization leads to crystal frame formation.

The aluminosilicate fibres with 1.5 wt-% of cuprous oxide were heat treated at the temperature of nuclei formation and at that of mullite crystallization. The amount of mullite approached 0.9% of the theoretical value and the desired mullite grains frame, with a grain size of 0.4 μm, was formed inside the fibres (Fig. 6).

PROPERTIES OF THE MODIFIED FIBRES

The main aim of modifying the fibres was to increase their thermal and chemical resistance. Thermal resistance was measured as the linear shrinkage of the fibrous blanket against the temperature as presented in Fig. 7. The specimen of fibrous blanket was isothermally treated for two hours. Assuming that the shrinkage value of the fibrous products does not exceed 4% at the work temperature, the permissible temperature for the use of those materials can be determined from the data in Fig. 7. The highest thermal stability was achieved for glass-ceramic fibres with the highest degree of crystallization and the most advantageous microstructure. This is 300°C higher than for the standard aluminosilicate fibres.

The chemical resistance was measured as the solubility of the fibres in the acid (0.1 n H_3PO_4) and alkaline (0.1 n NaOH) environments. The kinetic dependence of the concentration of Al^{3+} ions washing out into the solution can be seen in Fig. 8. Experiments were carried out at two temperatures. It can be seen that the glass ceramic structure guarantees a higher chemical resistance independently of the character of the environment. This result is due to the main components of the fibres - the mullite grains and the silica glass. Both are more resistant than the metastable aluminosilicate glass.

CONCLUSION

Aluminosilicate fibres crystallize spontaneously within the ellipsoidal droplets during heat treatment. The residual aluminosilicate glass, rich in silica, determines the properties of fibres at temperatures above that of mullite crystallization. It is possible to change the fibre microstructure in order to obtain higher thermal and chemical resistance. This improvement can be achieved by a change in the stability of the glass. The temperature of glass-ceramic fibre application is about 300°C higher due to the higher thermal resistance. Moreover, these fibres are about ten times more resistant in alkaline environments than the standard aluminosilicate fibres.

REFERENCES

1. B.A.Scowcroft and G.C.Padget: The structure and thermal behaviour of ceramic blankets, Trans.J.Brit. Ceram.Soc. 1, 72, 1973
2. M.Sopicka-Lizer and S.Pawłowski: The applicability of DTA to the study of the crystallization process of ceramic fibres, Therm.Acta, 38, 1980
3. G.H.Beal: Mullite glass-ceramics, High-temperature oxides, Part IV, 1971, New York, Academic Press
4. E.D.Lacy: Aluminium in glasses and in melts, Phys.Chem.Glasses, 6, 4, 1963
5. T.Takamori and R.Roy: Rapid crystallization of SiO_2-Al_2O_3 glasses, J.Am.Ceram.Soc., 12, 56, 1973
6. P.Dietricks and W.Kronert: Eigenschaften, Hochtemperaturverhalten und Einsatzbedingungen, Gas Warme International, 30, 7-8, 1981
7. J.D.Birchall: The preparation and properties of polycrystalline aluminium oxide fibres, Trans.J.Brit. Ceram.Soc., 4, 82, 1983

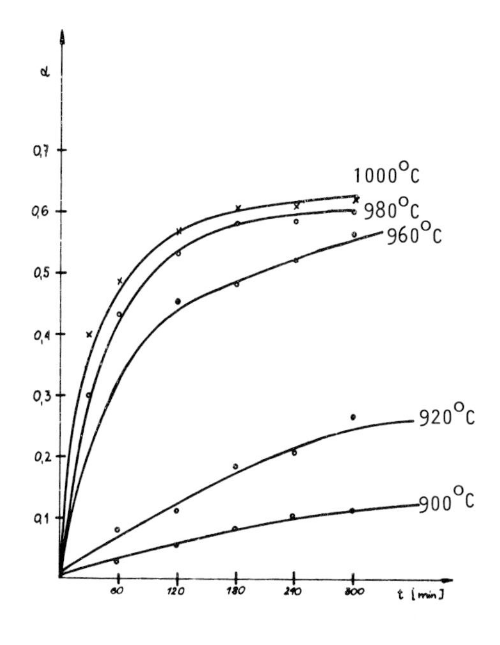

Fig. 1 Kinetic dependence of mullite crystallization in aluminosilicate fibres

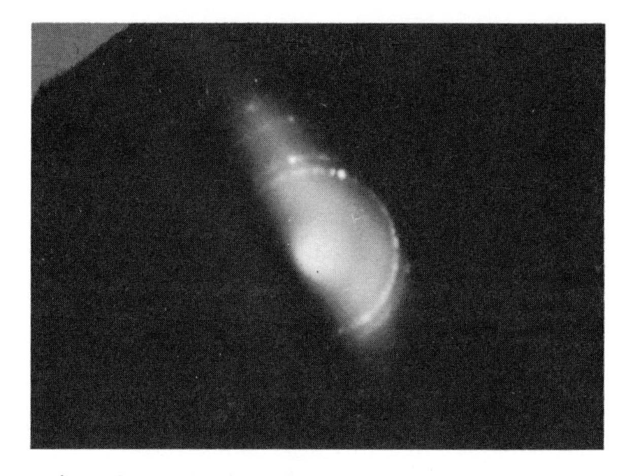

Fig. 2 X-ray examination of unheated fibre on a transmission microscope

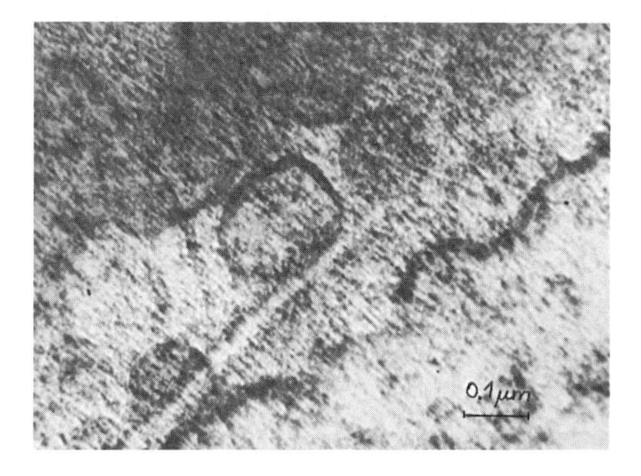

Fig. 4 Microstructure of a fibre with a diameter of 2 μm heated at 960°C: transmission microscope

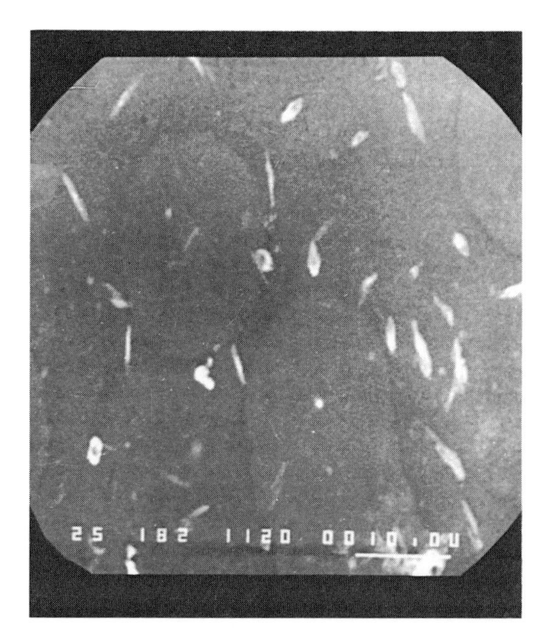

Fig. 3 Microstructure of a fibre section heated at 980°C: scanning microscope

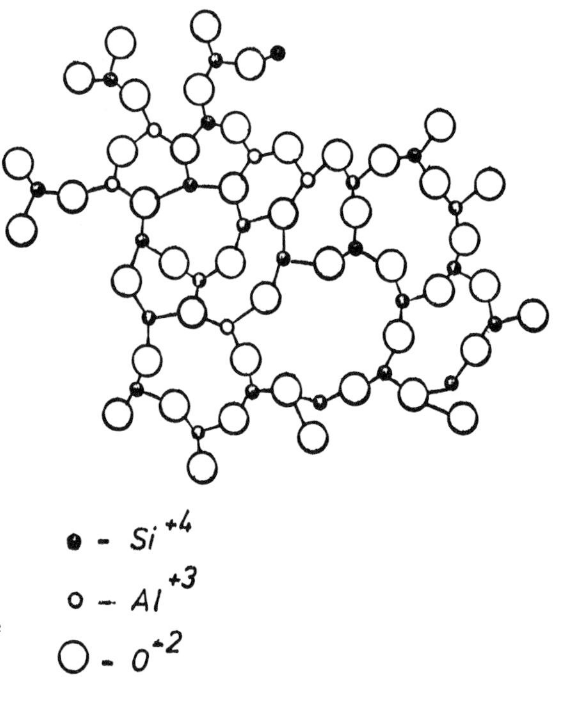

Fig. 5 Structure of aluminosilicate glass4

Fig. 6 Microstructure of section of fibre containing about 1.5 wt-% of cuprous oxide which has been treated at 890°C and then at 1040°C: scanning microscope

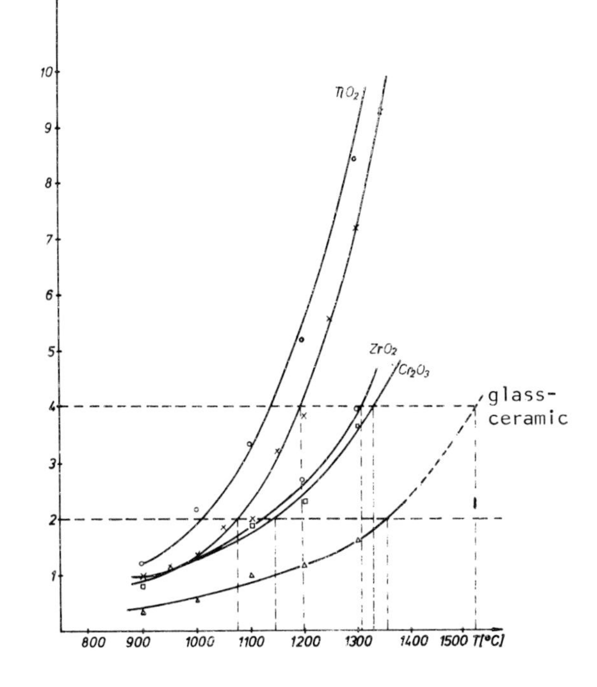

Fig. 7 Linear shrinkage of the materials of the modified fibres

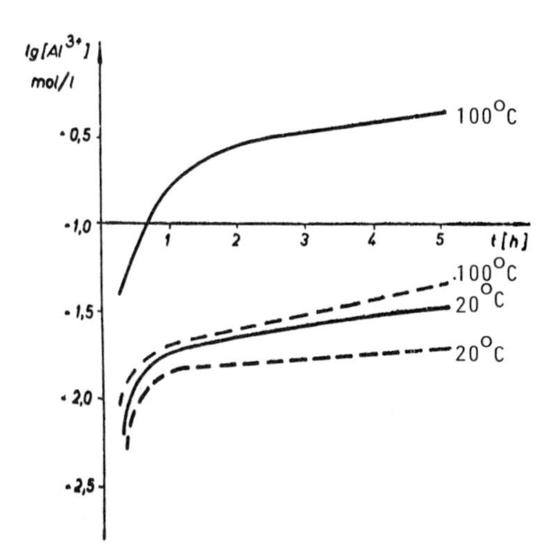

Fig.8 Kinetic dependence of Al^{3+} solubility in 0.1 n H_3PO_4

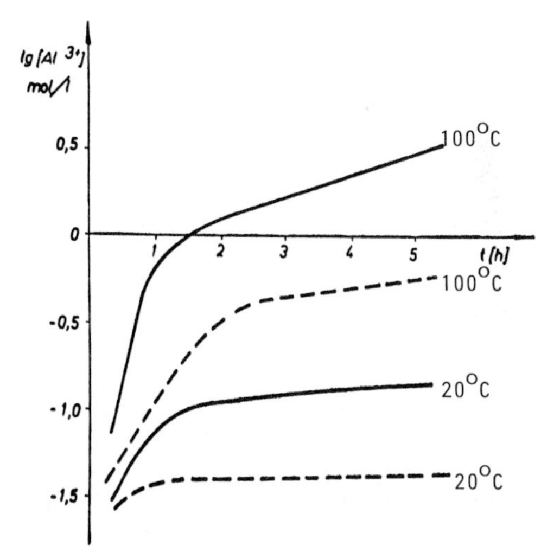

Fig. 9 Kinetic dependence of Al^{3+} solubility in 0.1 n NaOH

Development of high-manganese steels for heavy duty cast-to-shape applications in rail transportation and materials handling applications

R W SMITH, M GHORESHY, A DeMONTE, W B F MACKAY, T J N SMITH and M R VINER

Drs. Ghoreshy, DeMonte, Mackay, T. Smith and Viner are the Research Associates of Professor R. Smith of the Department of Metallurgical Engineering, Queen's University, Kingston, Canada.

SYNOPSIS

In 1882 Robert Hadfield patented the steel which bears his name. It has the nominal composition Fe-1.2%C-13%Mn. This alloy has an enormous capacity for work-hardening upon impact and is still commonly used for railway components such as frogs and crossings and also for rock-handling equipment. As such it can hardly be considered to be an "advanced material". However, work at Queen's over the last six years has shown that part of this capacity for work-hardening may be "traded" to provide alloys which are appreciably stiffer, have significantly improved wear resistance and may be cast to near net shape to produce components which still possess remarkable toughness. These materials have been produced by the addition of minor amounts of other carbide-forming and solid-solution strengthening elements and by heat-treating the ascast components. This has necessitated looking at the oxidation/decarburisation occurring during heat-treatment and the effect this has had on the results obtained earlier by other workers. In addition, because the as-cast components are battered out of shape when used in the solutionised and quenched condition, they are often explosively-hardened prior to installation. However, even then, they still deform in order to provide the necessary toughness. As a result they are often rebuilt by overlay welding. Thus part of the study has been concerned with ensuring that any new formulations/heat-treatment procedures are compatible with repair practices. A rail/locomotive wheel impact simulator was constructed to help in the selection of alloys more suitable for in-track testing. These developments are reviewed and more recent work described.

INTRODUCTION

Hadfield's steel (Fe-1.2%C-13%Mn) is a remarkable engineering alloy in that in the fully austenitic solutionised form it is soft and ductile. However, when deformed, it rapidly work-hardens and so, even though it may suffer considerable wear in non-impact abrasive conditions, impact or gouging deformation quickly causes it to work-harden. It is the standard material for railway frogs (switches) where heavy axle loads predominate e.g. N. America, Australia, Africa and the U.S.S.R., Figure 1.

There has been considerable work conducted to extend the property ranges of Hadfield's steel. One of the most detailed was by Krainer (1), who investigated variations of C, Mn, Si, Ni, Cr, Cu, W, Mo, Ta, Ti and Zr by means of hardness static and dynamic tensile tests, toughness and work-hardening characteristics. In all, he examined 75 experimental heats. These were cast as small ingots and forged into test-pieces, solutionised at 993°C and water-quenched. The dynamic tests were performed at a reported elongation rate of 16 ft/sec., a value close to the striking velocity of the usual impact machine. He found that the forged tensile specimen did not neck locally but deformed with virtually a uniform reduction in area along the gauge length. Thus, because the standard alloy was found to be relatively ductile in the as-quenched state (reduction in area of 35% to fracture) he proposed that the true tensile stress is a more accurate measure of properties. He was able to show that, for some additions, whilst the nominal (engineering) ultimate tensile stress (UTS) changed little with composition, the true UTS changed by almost 60%.

Krainer showed that the true UTS peaked at about 1.1% C (static) and 1.3% (dynamic) and fell as the Mn content exceeded about 12%. Thus, by good fortune or careful testing, Sir Robert managed to bracket the more desirable C and Mn ranges in his patent.

It should be noted that Krainer (1) tested forged samples for which the solutionising temperature was lower than that recommended for cast specimens. In addition, it is not clear whether he used any protective atmosphere when solutionising his samples His use of dynamic testing to get more informative data is open to question since the results are subject to error and the strain rate is much smaller than, for example, the impact velocity of a wheel on a railway frog.

In attempting to find additions which will increase the wear resistance and stiffness of a

Hadfield's steel, it is important to note that of interest are the surface hardness and strength after work-hardening. Since contact fatigue resistance is of primary interest in the degradation of a frog and is determined by strain capacity i.e. the difference between 0.2% yield stress (since the true point of yielding is difficult to determine directly) and the UTS of the work-hardened state, then it is most desirable that the addition(s) increase both the YS and UTS of the work-hardened state, without seriously impairing the toughness. Of the elements tested, Krainer found virtually no increase in true UTS with any element but a significant increase in nominal UTS for vanadium additions. He reports that all elements increase the work-hardening rate to about the 0.3% addition concentration. However, it should be noted that the low solutionising temperature (<1000°C) would not be adequate to solutionise some of the carbides and so lead to embrittlement and a reduced UTS with cast to shape, rather than forged components.

In particular, we are investigating applications for railway frogs and other heavy impact-wear applications. Early in the life of a frog, it experiences adhesive wear and deformation. Thus C.P. Rail uses explosive depth hardening to "preharden" the frog. Usually three "shots" from plastic explosives are used and give a surface hardness of 350-400 HB. The value falls to 220 HB about 1" from the upset surface (2). Taylor (3) also reports that shot-peen hardening was examined by C.P. Rail but was found to produce too superficial a layer to be useful in-track. Explosive depth hardening does not distort the frog and can increase life by more than 50%. In addition, it also provides an excellent quality control check on the integrity of the castings!

The standard Hadfield's steel (Fe-1.2%C-13% Mn) in the as-cast condition contains acicular carbides of $(Fe,Mn)_3C$. Present industrial practice is to heat-treat the material at 1010-1090°C for up to one hour followed by a water quench. This procedure can normally solutionise all the $(Fe,Mn)_3$ carbides. In this condition, the steel is very soft (190 HB). As noted earlier, impact deformation can increase the surface hardness to about 500 HB. The mechanism of such work-hardening depends on the local composition but it normally arises from deformation twinning/dynamic strain ageing in the standard 1.2%C alloy. However, some decarburisation usually takes place during annealing and then a low carbon matrix is present which may well show deformation-induced martensite formation (4). Thus any heat-treatment needs to be done as quickly as possible and at as low as a temperature as possible, or in a protective atmosphere.

As noted earlier, attempts have been made through the years to reduce the ease of deformation and abrasive wear of the as-quenched standard alloy by metallic addition (5-10). Often these modified alloys have been given one of a series of elaborate heat-treating cycles (5-10) to optimise precipitate form and distribution but the accompanying decarburisation/oxidation has not been reported. In our own work, in trying to find an effective heat-treatment to dissolve the carbides in modified Hadfield's steels, it was found that exposure at 1250°C to 1300°C for 5 to 10 hours in even a weakly oxidising atmosphere caused the samples to lose a considerable amount of carbon. Similarly, small specimens held at 1150°C in air readily decarburised. Thus, the microstructure no longer contained carbides and, in addition, was no longer austenitic either, but ferritic. As a result of these findings, a study of the extent of decarburisation in these alloys under the more drastic heat-treating conditions was undertaken (11,12). From this it is clear that the results obtained by earlier workers who used small specimens and no atmosphere control during heat-treatment are highly suspect (12,13).

EXPERIMENTAL DETAILS, RESULTS AND DISCUSSIONS

a) Sample Preparation

The samples were obtained by melting predetermined amounts of the components given in Table 1 in a 'Tocco' induction furnace. For these alloys, the pouring temperature was determined by the use of a Pt-Pt13Rh thermocouple. The molten metal was poured into heated investment moulds (700°C) so that only a little grinding had to be performed to obtain the required dimensions of the specimens for mechanical testing. Before grinding, the samples were heat-treated in a 'Lindberg Heavy Duty Furnace' to which a chamber had been adapted to provide an enclosure in which a flowing argon atmosphere could be introduced. The alloys examined in this phase of the project are shown in Table 2.

b) Impact Strength

Since a large impact strength was considered to be an important alloy property, it was decided to examine a number of factors
which may adversely influence impact strength, namely grain size, phosphorus and copper content. The latter is a steady increasing component in steel scrap.

It was found that (i) adequate grain size control would be obtained by using suitably heated molds (700°C) (ii) phosphorus levels below 0.05% P were permissible (Table 3) and

(iii) copper in amounts up to 1% were not beneficial (Table 4).

c) Mechanical Properties

The effects of various heat-treatments were followed primarily by metallographic and mechanical testing studies. After heat-treatment, the alloys were ground to the desired size on an automatic grinder equipped with coolant. Charpy V-notches were cut by spark-machining, following which the notched impact specimens were boiled in water for two hours to remove any residual hydrogen (14).

Deformation studies on the new alloys were carried-out in two ways: a) regular mechanical testing (Table 2); and b) using a rail/locomotive wheel impact simulator (15).

An inspection of the available phase equilibrium diagrams showed that the solubility of vanadium in the austenite matrix increased with reduced carbon content. As a result, alloys with two carbon levels were examined. Table 2 shows that the yield strength falls markedly as the carbon level is reduced in the standard alloy. Tungsten promotes a somewhat increased yield strength, toughness and hardness. A similar, but less dramatic effect has been seen with single additions of molybdenum (16). As well, additions of vanadium markedly reduce impact strength but promote a significant increase in work hardening rate (13).

The rail/locomotive wheel impact simulator is shown in Figure 2. Here the specimen, usually an undamaged section of a Charpy specimen bar, is placed in the periphery of a large flywheel. This is then rotated and brought into contact with a pressure-loaded roller. The hardness is then measured as a function of number of impacts by stopping the machine periodically and "shimming" the specimen to maintain approximately a 1mm project above the rim. Figure 3 shows how the various experimental alloys work-harden as a function of number of impacts. Figure 4 shows similar tests done on a vertical section specimen through a Hadfield's steel component which had been overlay welded. The Cr- and N-Cr-bearing welding rods were commercially available and found by C.P. Rail to be of similar utility for rebuilding frogs. The Mo-bearing rod was devised by us and is seen to offer commercial prospects.

Comparative wear tests were made by bringing a standard-size specimen, under a variety of given loads, into contact with an abrading wheel for fixed times. Reproducible results could be obtained if the wheel was re-dressed between samples. It was found that a 1% addition of vanadium to the standard alloy would reduce the abrasion rate by two thirds, presumably due to the presence of wear-resisting carbides.

d) Heat-Treatment

Significant decarburisation can occur if the standard Hadfield's solutionising treatment is followed (12) (Figure 5). As noted earlier, this is particularly serious if small castings of experimental alloys are heat-treated. In practice, with a full-sized frog, some of the decarburised layer is ground-off before fitting.

The regular heat-treatment followed for the standard Hadfield's steel i.e. one hour at 1010-1090°C is not adequate generally whenever carbide-forming additions are made. The carbides are not fully dissolved and consequently the material loses some impact strength. Various 'beneficial' treatments have been reported by others, of which a procedure mentioned by Middleham (5) and Norman (10) was claimed to be the most successful. In this treatment, the alloys were heated to 1050°C for two hours, water-quenched and then heated to 650°C for two hours, heated to 900°C for six hours and finally water quenched. This procedure, it was reported, did not dissolve carbides but dispersed them. In fact, our work indicates that the carbides are not dispersed by this heat-treatment, rather they tended to appreciably coalesce into near-spheres. The carbides were found to be still located along the grain boundaries with very few dispersed within the grains. Since the cyclic-type of heat-treatment did not prove effective, a compromise was attempted. The modified materials were heated to (1150°C) for two hours in an inert atmosphere and then water-quenched. This process did not dissolve the carbides as the present work and that of Brekel (11) indicated, but, at least, the heavily cored cast structure was partially eliminated and some grain growth occurred to leave earlier grain-boundary precipitates within the grains.

GENERAL DISCUSSION

As noted, the addition of vanadium improved the yield strength, hardness and wear-resistance qualities but the resultant loss in toughness had to be overcome. One approach that suggested itself was the possibility of reducing the carbon content thus depressing the formation of a large volume fraction of carbides. This was reasonably effective to maintain good impact strength and produce a wear resistance not much different from the 2% V addition to the standard alloy (16). Some multiple additions also produced an increase in yield strength.

In addition to improved mechanical properties it is advantageous to go to a reduced carbon content in the alloy for improved weldability since, with a larger carbon content, there is a greater chance for the precipitation of carbides in the heat-affected zone, thus lowering the toughness of the material (17). However, if alternative additions are made, then a larger carbon content can be tolerated because of the changed carbide distribution and morphology. Small amounts of molybdenum seem to help in this respect while copper additions do not. It was found that an alloy with 1.2%C-13%Mn-2%Cr-1%Mo--2%W, when heat-treated at 1150°C and water-quenched, showed no dendritic structure but large grains. This material showed significantly greater work-hardening behaviour than the standard Hadfield's steel. In addition an alloy with 1.2%C-13%Mn-1%Mo-0.3%Nb-0.3%Ti-0.3%V, when heat-treated and water-quenched produced a relatively finedispersion of carbides in a fine grained matrix (16).

One question still to be answered is what impact strength is adequate for a heavy haul railway frog. Hopefully, this will be answered shortly when we receive the results from C.P. Rail of the in-track testing of a number of modified Hadfield's steel frogs.

Of particular interest to the foundryman is the extent to which the purchaser of (say) frogs is likely to pay an appropriate premium for superior frogs. Most certainly C.P. Rail would like to procure a frog with a 50% increase in the service life from the present figure of, for example, 130 million gross tonnes. The most expensive alloying additions used in our programme to date would increase the charge costs by 200-300%. However, many of the elements are now relatively abundant, e.g. tungsten, and so the price would fall dramatically with increased consumption. A detailed cost benefit analysis of using a 200 million gross tonnes frog: as compared with the present one is being prepared. It is clear that the initial casting price is a small part of the total cost of a frog i.e. purchased casting, straightening, grinding off the decarburised layer, fitting, installation, rebuilding and, finally, replacement. We await the results with much interest.

SUMMARY

When the above results are analysed and integrated with detailed compression testing data carried-out by Sant (13), it is clear that the mechanical response of Hadfield's steels may be divided into two considerations: 1. The rate of work-hardening of the matrix is dependent primarily on the amount of carbon in solution in the austenite matrix and the presence of a fine dispersion of carbides; 2. The presence, amount and dispersion of carbides significantly influence the wear resistance of the alloy and, if the carbides form interconnected grain boundary films, seriously impairs its impact strength.

In addition, the decarburisation/oxidation analysis shows quite clearly that austenitic manganese steels should not be heat-treated in open-air or a weakly oxidising atmosphere unless the decarburised layer is to be removed before testing or use. Heat-treatment should be at temperatures as low as possible and for minimum times to solutionise the carbides and avoid excessive surface decarburisation. This in turn will affect the work-hardening characteristics of the materials.

ACKNOWLEDGEMENT

This work forms part of a continuing programme of foundry science and alloy development at Queen's University. It has been supported from a variety of sources including Queen's Advisory Research Council, Natural Sciences and Engineering Research Council of Canada, BILD (Province of Ontario), the Canadian Transportation Development Centre, C.P. Rail and ESCO. This financial and other support is gratefully acknowledged.

REFERENCES

1. H. Krainer, Archiv Eisehuttenwesen,(1937) 11 p. 279-282.
2. E.H. Taylor, Internal Report "Explosive Depth Hardening", C.P. Rail, Jan. 27, 1987.
3. E.H. Taylor, Memorandum TRCHWORK (9-2-26(35-201-CIGGT)OM00496), 9, 1986.
4. S.B. Sant and R.W. Smith, Strength of Metals and Alloys, Vol. 1, [Proc. Conf.] Montreal, Canada, Pergamon, (1985), p. 219-224.
5. T.H. Middleham, Alloys Metals Review, (1964), 12 2.
6. V.I. Grigorkin and G.V. Korotushenko, Metallov. I Term. Obrab. Met., (1968), 2, 48.
7. V.I. Grigorkin, I.V. Frantsenyuk, I.P. Galkin, A.A. Osetrov, A.T. Chemeris and M.F. Chernenilov, (1974) 4, 68.
8. G.I. Sil'man, N.I. Pristuplyuk and M.S. Frol'tsov (1980), 22, 38.
9. D.C. Richardson, W.B.F. Mackay and R.W. Smith, Solidification Technology Warwick, England, The Metals Society, 1981, p. 409-413.
10. T.E. Norman, D.V. Doane and A. Soloman, AFS Trans., (1960), 68, 287.
11. N.C. Brekel, B.Sc. Thesis (1981) Queen's University, Kingston, Canada.
12. A.I. DeMonte, Ph.D. Thesis, "Effects of Vanadium on Hadfield's Steel", (1984) Queen's University, Kingston Canada.
13. S.B. Sant, M.Sc. Thesis, "Mechanism of Work Hardening in Hadfield's Steel and the Effect of Minor Amounts of Carbide Formers" (1984), Queen's University, Kingston, Canada.
14. M. Ghoreshy, W.B.F. Mackay and R.W. Smith, Journal of Testing and Evaluation, (1985), Vol. 13, No. 2, p. 176-177.
15. J.I. Mendez, M.Sc. Thesis, "Weldability of Hadfield Steel", (1984) Queen's University, Kingston, Canada.
16. R.W. Smith, (Unpublished work).
17. H.S. Avery and H.J. Chapin, Welding Journal, (1954), 33, 459-479.

Table 1. The chemical composition (%) of each charge material.

Charge Material	Chemical constituents (balance Fe) %C	%Mn	%V	%Mo	% Si	%W
Rail stock	0.78	0.84	--	--	0.2	--
Armco Iron	0.04	0.06	--	--	--	--
Low-C Fe-Mn	0.08	86.4	--	--	--	--
High-C Fe-Mn	6.4	75.	--	--	--	--
Carbon	11.	---	83	--	--	--
Fe-Mo	---	---	--	71.2	--	--
Fe-W	0.01	0.04	--	--	0.03	91.39

Table 2. Charpy impact toughness (N-m), Brinell hardness and compressive yield (MPa) for new alloy series.

Alloy	%C	%V	%Mo	%W	Toughness	Hardness	Yield
5R3	1.2	-	-	-	261	138	327
5R16	1.2	-	-	2.0	290	152	331
5R29	1.2	0.7	0.7	2.0	91	162	448
5R30	1.2	1.0	1.0	2.0	62	207	452
5R31	1.2	1.0	1.0	1.0	64	192	478
6R6	0.8	-	-	-	258	130	266
6R35	0.8	-	-	2.0	260	137	290
6R33	0.8	0.7	0.7	2.0	108	167	386
6R32	0.8	1.0	1.0	2.0	79	186	417
6R34	0.8	1.0	1.0	1.0	73	186	399

Table 3. Charpy impact toughness (N-m) and Brinell hardness of R3 Hadfield steel (1.2% C, 13% Mn), as a function of phosphorus content.

Alloy	% P	Toughness	Hardness
4R3a	0.014	>357	170
4R3b	0.020	>357	143
3R3	0.040	351	143
3R26	0.055	342	148
3R27	0.066	266	150
3R28	0.090	296	156

Table 4. Charpy impact toughness (N-m) and Brinell hardness of Hadfield Steel (1.2% C, 13% Mn), with and without additions of vanadium and copper.

Alloy	%V	%Cu	Toughness	Hardness
2R3	---	---	190	149
2R25	---	1.0	212	149
2R20	1.0	---	76	187
2R21	1.0	0.5	88	183
2R22	1.0	1.0	85	179
2R4	2.0	---	27	217
2R23	2.0	0.5	28	207
2R24	2.0	1.0	33	207

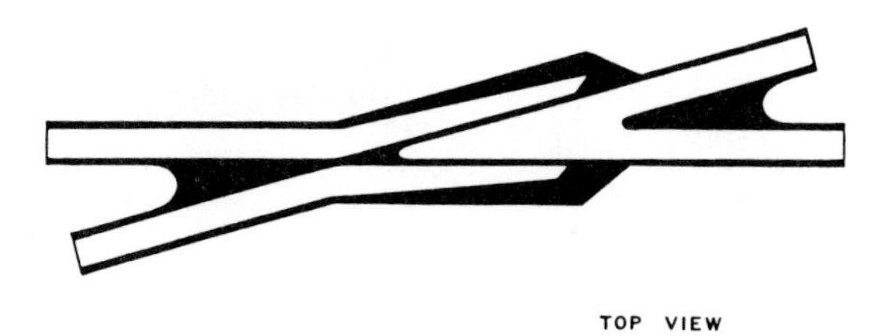

Fig. 1: Illustration of a railway crossing

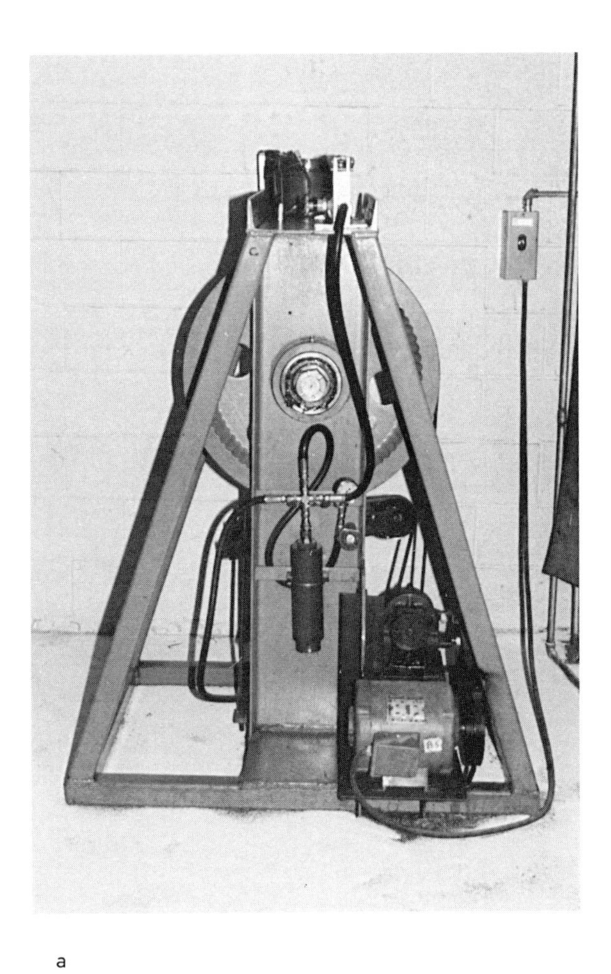

a

b

Fig. 2: Rail/locomotive wheel impact simulator
(a) General view
(b) Specimen location

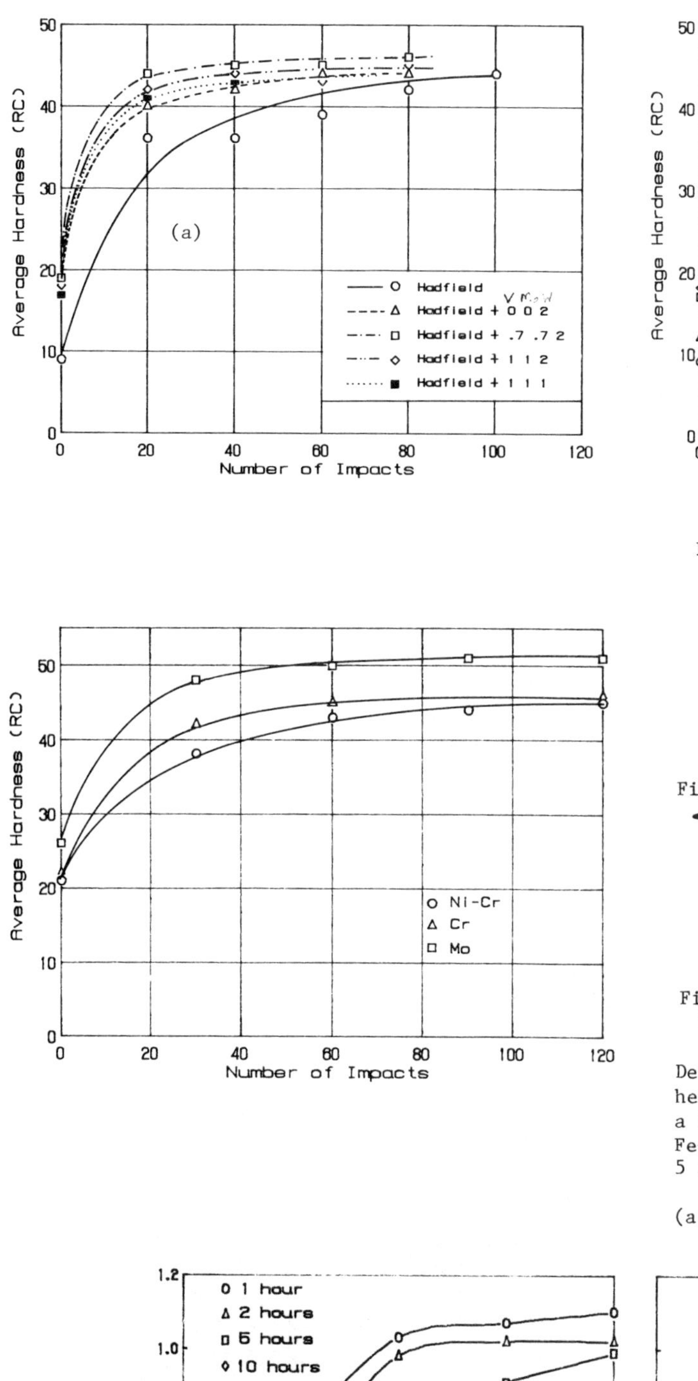

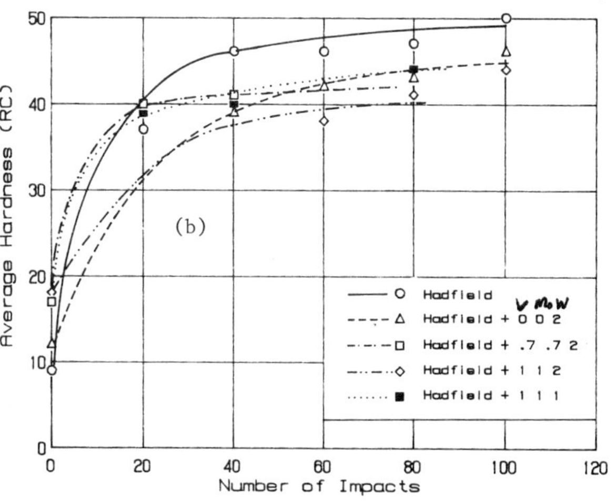

Fig. 3: Work-hardening characteristics of Hadfield steel and Hadfield steel modified with Mo, V and W
(a) 1.2%C
(b) 0.8%C

Fig. 4: Work-hardening characteristics of weld overlays on Hadfield's steel base

Fig. 5:

Decarburisation of Hadfield's steel during heat-treatment. Carbon concentration as a function of depth from surface in Fe-1.2%C-13%Mn, heat-treated at 1150°C for 1, 2, 5 and 10 hours

(a) in air and (b) in argon

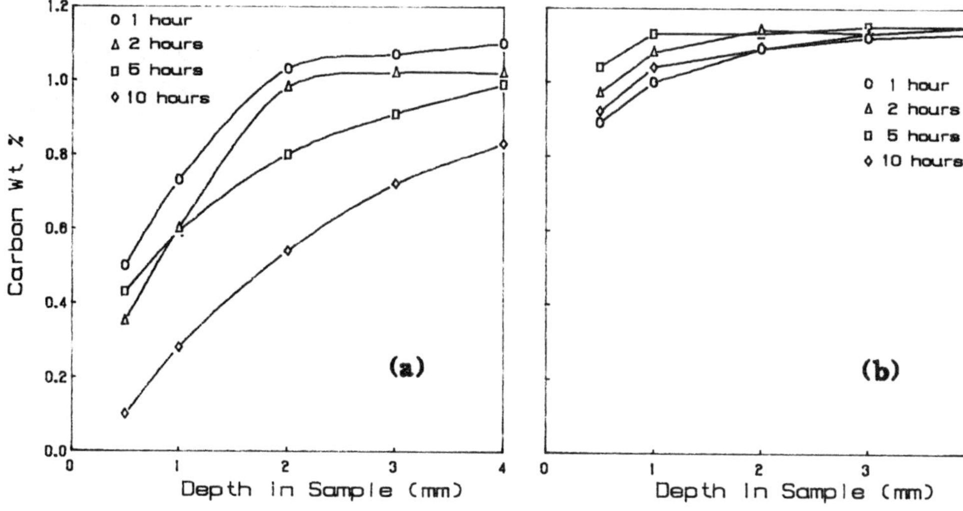

Fatigue and strain-hardening of high carbon martensite-austenite composite microstructures

M A ZACCONE, J B KELLEY and G KRAUSS

MAZ is a Research Assistant at the Colorado School of Mines; JBK is with Allison Gas Turbine Division, General Motors Corporation, Indianapolis, Indiana; GK is AMAX Foundation Professor and Director of the Advanced Steel Processing and Products Research Center at the Colorado School of Mines, Golden, Colorado, USA.

SYNOPSIS

Low cycle fatigue testing of a series of 0.8 wt. pct. C steels shows that fatigue resistance increases with retained austenite content of tempered martensite-austenite composite microstructures. At stresses below which the fatigue test were performed elastic limits of the microstructures decrease with increasing austenite content and the deformation behavior is dependent primarily on retained austenite. At high strains, strain-induced austenite transformation to martensite occurs. This transformation increases strain hardening rates which in the specimens with the highest retained austenite contents effectively retard fatigue crack initiation at prior austenite grain boundaries.

INTRODUCTION

The microstructures produced in steels by the quenching of high-carbon austenite are complex systems which consist largely of martensite and austenite in varying amounts and morphologies (1). Carbides may be retained if austenitizing is performed below A_{cm}, and fine allotriomorphic carbides may form on austenitic grain boundaries during quenching (2). When tempered below 200°C, these microstructures have high strength and hardness and good contact and bending fatigue resistance for bearing and gear applications.

A number of investigations have shown that retained austenite contents on the order of 30 pct are beneficial to fatigue resistance (3-5). Most recently, these beneficial effects of retained austenite have been attributed to the high toughness of martensite formed during plastic deformation of austenite and the possible associated increase in compressive residual stresses (5). Investigations of roller contact fatigue show increases in compressive residual stress with the transformation of austenite (6). Actual measurements of residual stress in carburized steel, however, show that compressive stresses actually decrease during cyclic stressing (7).

Fatigue progresses by cyclic hardening or softening, crack initiation, and crack propagation. In both the high and low strain regimes the material stabilizes before crack initiation mechanisms begin. In many cases the behavior which leads to crack initiation can be determined by examining monotonic deformation behavior (8). The objectives of the investigation described in this paper were to investigate the plastic deformation and strain hardening behavior, from the microstrain regime through macroscopic straining, of tempered martensite-austenite microstructures, and to relate the deformation behavior to fatigue performance.

Steels and heat treatments were selected to simulate the high carbon (0.8 pct) case structure of carburized steels. Retained austenite was systematically varied by selection of alloy steels with chromium contents ranging from zero to 1.3 pct. Increasing chromium lowered M_s and increased the amount of retained austenite.

EXPERIMENTAL PROCEDURE

Table 1 lists the chemical compositions of the five steels prepared for this study. All of the steels also contained 0.010 to 0.012 pct P, 0.006 to 0.009 pct S, and 0.04 to 0.05 pct Al. The steels were forged and hot rolled before homogenizing in vacuum for 20 hours at 1200°C, air cooling and tempering at 700°C for 10 hours prior to machining.

Two heat treatments were applied: direct quenching and reheating. The direct quench treatment consisted of holding at 930°C for 2 hours, step cooling to 850°C for 1 hour, oil quenching and tempering at 200°C. The reheat treatment consisted of heating to 930°C, air cooling, reheating to 850°C for 1 hour, oil quenching, and tempering at 200°C. The hardness of the specimens varied from HRC 55 to HRC 59 (Table II).

Fatigue testing was performed in four point bending in stress control on an MTS system at stresses of 862 MPa (125 ksi) and 1655 MPa (240 ksi) (9). Compression testing was used to determine the mechanical behavior of the heat treated specimens. Strain gauges were mounted on opposite sides of cylindrical specimens 6mm in diameter and 9mm in height to determine elastic limits and microstrain behavior. The elastic limit is reported as the stress required for a permanent strain of 2×10^{-6}. Microstrain compression testing was performed in strain control at rates of 0.00003/0.00016 s^{-1} up to 0.01 pct strain.

High strain composition testing to 15 pct strain was performed in strain control at rates of $0.0003/0.0004 s^{-1}$. The samples were placed between TiC platens within cold rolled shim stock coated with Mo_2S lubricant. The load and deflection data were acquired and stored with a DEC PDP-11 computer and were used to obtain engineering stress-strain curves, true stress-strain curves and strain hardening rates.

Microstructures were prepared with standard metallographic techniques and retained austenite was measured by x-ray diffraction. Measurements were made with molybdenum radiation at 40 kV and 20 ma. Three austenite peaks and two martensite peaks were measured while rotating the stage to reduce texture effects.

RESULTS

Microstructure

All the microstructures consisted of tempered martensite and retained austenite. The retained austenite content varied, generally increasing with increasing Cr content, Table III.

The reheat treatments had a smaller austenitic grain size (Table II) because their microstructures contained fine dispersions of carbides retained during austenitizing. Generally reheated specimens exhibit a finer grain size due to shorter times at lower austenitizing temperatures. Coupling this with austenitizing temperature below the Acm, as in the case of the reheated 0.9 and 1.3 Cr alloys, the grain size was significantly reduced relative to the other specimens.

Fatigue Effects

Figure 1 shows the results of the low cycle fatigue tests performed on the two sets of heat treated specimens. In all cases, the reheat and quench specimens had greater low cycle fatigue resistance than the direct quench specimens due to their refined microstructures, especially those in higher Cr samples. However, for the direct quench samples the increasing fatigue resistance with Cr content strongly correlated with increasing retained austenite. This effect must be present in the reheat heat treatments but is apparently overwhelmed by the change in grain size.

Selected broken specimens which were sectioned normal to the fracture surface at the point of crack initiation showed two microscopic zones associated with the initiation of low cycle fatigue failure in the martensite-austenite microstructures: (1) an initiation zone associated with severe deformation of the martensite and almost complete transformation of the retained austenite; crack initiation was always associated with intergranular fracture in all but the 0.9 and 1.3 Cr reheated specimens in which crack initiation was associated with inclusions, (2) a broad plastic zone in which much of the retained austenite has transformed to martensite. A narrower plastic zone, as indicated, would be expected to be associated with stable crack growth prior to unstable fracture. These zones are shown schematically in Figure 2, and representative micrographs are shown in Figure 3.

Metallographic examination revealed that significant plastic deformation and transformation of retained austenite accompany fatigue crack initiation in tempered martensite-austenite microstructures. With the stabilization of the cyclic substructure small cracks developed along P-cementite enriched prior austenite grain boundaries (2). The grain boundary structures were assumed to be similar for all of the alloys and therefore the fatigue crack initiation must be related to differences to strain hardening, a process where retained austenite plays a significant role. Since little evidence for stable crack growth was observed, the fatigue life was directly related to the strain hardening processes which led to crack initiation. The following sections characterize the deformation behavior of the various alloys to explain the role which retained austenite plays in the deformation and fracture processes which accompany fatigue.

Microstrain Deformation

Figure 4 shows the microstrain stress-strain behavior of the five steels in the direct quenched condition. The steels with the highest chromium contents, and therefore the highest retained austenite contents, have the lowest elastic limits and the lowest strain hardening rates in the very early stages of plastic deformation. In all of the microstructures, the initiation of plastic flow occurs at stresses well below those applied in the low cycle fatigue tests, Figure 1.

Figure 5 shows that the elastic limit of two reheated alloy specimens were higher than the same alloys in the direct quenched condition. The 0.3 Cr alloy had relatively low retained austenite in both heat treated conditions, but reheating produced somewhat finer austenitic grain size and therefore a finer martensite-austenite microstructure than did direct quenching. The reheating of the 1.3 Cr steel, because of fine retained carbides produced a significantly refined austenite grain size. As a result, the high retained austenite content of the 1.3 Cr alloy in the reheated condition was partitioned into very fine volume elements by subsequent martensite transformation. Figure 5 shows that low elastic limits in the reheated specimens correlate with high retained austenite content, and that high strain hardening rates in the microstrain regime correlate with finer dispersions of retained austenite.

With the exception of the 1.3 Cr reheat treatment, no retained austenite transformation occurred in the microstrain regime. In the 1.3 Cr specimen, the transformation of austenite to martensite in the microstrain region increased the work hardening rate leading to the crossing of the 0.3 Cr and 1.3 Cr stress-strain curves with increasing plastic strain. Despite this, differences in the elastic limits and strain hardening behavior correlate closely with the retained austenite refinement which is a function of grain size and volume fraction austenite.

Macrostrain Deformation

Figure 6 shows the compressive true stress-true strain curves of the five steels in the direct quenched condition. The flow stresses of the high Cr, high austenite microstructures are below those of the lower Cr alloys over much of the plastic strain range. However, because of higher strain hardening rates at high strains, the curves of the high Cr steels begin to cross those of the low Cr steels. Figure 7 shows quantitatively differences in strain hardening as a function of true plastic strain for the various alloys. There is little difference in strain hardening between the alloys at low strains 0.001 to 0.02. However, above a strain of about 0.02, the rate of decrease in the strain hardening rates changes sharply. The microstructures containing the highest amount of retained austenite sustain the highest strain hardening rates at high plastic strains.

The sharp change in strain hardening is accompanied by substantial austenite transformation, Fig. 8. The amount of strain-induced austenite transformation to martensite is of course greatest in the microstructures with the highest initial retained austenite contents. However, despite the substantial fraction of austenite transformation in the high Cr specimens, the amount of untransformed austenite even at the highest strains tested is still considerable. In contrast, a larger fraction of the austenite present in the low Cr steel is transformed during plastic straining. Thus at high strains the partitioning of the microstructure by strain-induced martensite formation in the high Cr steels continues to produce obstacles to deformation which increase flow stresses at rates above those in the low Cr steels. In the latter steels, the dynamic generation of barriers to deformation decreases, and strain hardening rates fall precipitously as shown by curve A in Fig. 7.

The overall rate of transformation as a function of strain appears to be equivalent in all the steels. Therefore as the austenite transforms to martensite with strain, these samples with the highest retained austenite contents maintain a higher strain hardening rate until the retained austenite is exhausted. In the case of the high Cr specimens at 0.15 plastic strain austenite continues to be available for transformation, whereas in the low Cr specimens, the available austenite has transformed and work hardening rates decreased rapidly.

DISCUSSION

Figure 9 summarizes the differences in the compressive strain hardening behavior over the entire plastic strain range for direct quenched alloys with extremes in microstructure. For purposes of discussion, the curves have been divided into three stages. Very high strain hardening rates characterize the first plastic deformation, stage 1, in the composite tempered martensite-austenite microstructures. Low elastic limits have been attributed to residual stresses caused by quenching (10,11), but in this investigation we have shown that they correlate directly to high retained austenite content. Little or no transformation of austenite to martensite occurs in stage 1, and therefore plastic deformation of the austenite must be largely responsible for differences in strain hardening behavior.

The deformation behavior in the microstrain region, correlated very well to retained austenite level and austenite refinement. The stress at which the first plastic deformation occurred was much lower than that expected for high carbon tempered martensite. Reducing the retained austenite to 5 volume percent by quenching the 0.0 Cr steel in liquid nitrogen followed by tempering at 200°C increased the elastic limit in tension and compression by a factor of 3. This increase may have been due to further refinement of the remaining austenite. However, the work hardening rate in the liquid nitrogen-quenched specimen decreased rapidly with strain, a characteristic of the deformation of a highly dislocated structure such as martensite.

In order to gain a better understanding of the deformation of martensite-austenite structures, the rule of mixtures was applied to the flow stress, described by:

$$\sigma = X_m \sigma_m + X_\gamma \sigma_\gamma \tag{1}$$

where X_m and X_γ are the volume fractions and σ_m and σ_γ are the flow stresses for austenite and martensite, respectively. Assuming that the flow curve of the liquid nitrogen quenched samples was representative of the tempered martensite, the flow stress of austenite σ_γ as a function of strain can be calculated from the composite curves in Figure 4. This calculation yields flow curves characteristic of the austenite present in the structure, with specimens which contain higher retained austenite having a lower flow stress at a given strain.

The difference in the flow curves could be due to different levels of strain hardening or differences in the mean free path between barriers to dislocation motion. The mean free path hypothesis can be modeled after theories related to grain boundary strengthening. In this case, a microstructural constituent finer than the grain size may be considered with the barriers to dislocation movement being the martensite plates. Yielding should occur when flow initiates in the largest austenite regions.

In order to evaluate the effect of refinement of the matrix austenite by martensitic partitioning, the largest dimensions of the austenite regions were measured metallographically where possible in the direct quenched and reheated Cr steels as well as in a fully austenitized and hardened 52100 steel. The elastic limit was then plotted in a Hall-Petch format with austenite region size or free mean path the structural parameter. Figure 10 shows that increasing elastic limit is directly related to the decreasing mean austenite region size in the microstructure. Generally, the higher the retained austenite content, the lower the elastic limit, but if a given amount of austenite is partitioned by martensite into finer volume elements, due to a smaller austenite grain size for example, the elastic limit increases.

After yielding occurs larger percentages of the austenite participates in the yielding process as the larger regions work harden to a level equal to that of the smallest austenite region. During this stage the martensite is assumed to undergo elastic deformation until the austenite strain hardens to a strength equivalent to the martensite at 0.001 plastic strain. In the microstrain region the deformation behavior may be described by:

$$\sigma = X_m \epsilon E + \sum_{i=1}^{\infty} X_i \sigma_i + \epsilon (X_\gamma - X_i) E \qquad (2)$$

The first term represents the elastic contribution of the martensite phase. The second term represents the contribution of successively smaller austenite islands as they begin to participate in the deformation process with i=1 the largest size. The third term represents the smaller austenite regions, strain hardened to flow stresses greater than the flow stress of the larger regions. This term decreases as deformation occurs due to more of the austenite contributing to the second term. This equation can be simplified by:

$$\sigma = \epsilon E (X_m + X_\gamma - X_i) + \sum_{i=1}^{\infty} X_i \sigma_i \qquad (3)$$

since $X_m + X_\gamma = 1$ then

$$\sigma = \epsilon E (1 - X_i) + \sum_{i=1}^{\infty} X_i \sigma_i \qquad (4)$$

In stage 2, strain hardening is independent of microstructure, perhaps because straining is accommodated uniformly by the martensite and strain-hardened austenite. The strain hardening behavior of all the steels is equivalent after the flow strength of the austenite equals that of the martensite. Here the deformation behavior is described by a simple rule of mixtures (Equation 1) where σ_γ and σ_m are the flow stresses of austenite and martensite at a given level of strain, where the accumulated strain in the austenite is greater than that in the martensite.

With further straining, stage 3 is initiated by substantial strain-induced transformation of austenite to martensite, and the microstructures with the most austenite maintain higher strain hardening rates than do the microstructures with lower amounts of retained austenite. The transformation occurs at a rate depicted in Figure 8. In this process, regions equivalent to the austenite grain size transform with strain. These regions are the grains with the most favorable orientations relative to the compression axis. With increasing strain, more grains reach the critical plastic strain necessary for deformation induced transformation. The retained austenite does not fully transform to martensite, perhaps due to austenite regions smaller than the critical nucleus size required for transformation as the austenite becomes increasingly subdivided and only grains with extremely unfavorable orientations remain.

The above observations of the differences in strain hardening behavior of the various austenite-martensite composite microstructures help to explain the differences in the low cycle fatigue performance shown in Fig. 1. Table IV presents the measured sizes of the plastic zones associated with fatigue crack initiation in several of the direct quenched specimens. The largest plastic zones, as measured by the extent of strain-induced martensite formation, were observed in specimens with low retained austenite and which failed after limited cycles of stress. Apparently, the effective surface stresses were larger in the lower retained austenite samples probably due to higher surface residual tension
This higher effective stress quickly transformed austenite to martensite. The increasing amounts of martensite hardened the surface layers reducing the effective strain at the notch of the specimen. With continuing cycles, residual stress effects and the austenite available for transformation were exhausted. With the rapid exhaustion of strain-induced martensite formation just below the notch surface in the highly strained portion of the plastic zone, the rapid decrease in strain hardening rates characteristic of highly strained low Cr specimens caused cracks to develop at embrittled prior austenite grain boundaries after a relatively small number of stress cycles.

In contrast, the higher Cr specimens had lower surface residual tensile stresses and higher work hardening coefficients. The higher value of n translate to materials which cyclically harden during cyclic stressing under load control (4,12). Thus the strain at cyclic stabilization of the higher austenite specimens is much lower than in the low austenite specimens prolonging crack initiation mechanisms. Also the larger austenite volume requires more cycles before stabilization of the microstructure occurs. Therefore from a low cycle fatigue standpoint the higher austenite samples have a better fatigue resistance due to the requirement of more cycles in the crack initiation phase of fatigue.

Although the above discussion of the effects of strain hardening or low cycle fatigue behavior appear qualitatively reasonable, it must be noted that we have related strain hardening as measured by monotonic compression testing to cyclic stressing applied by tensile bending. Cyclic stress-strain and tension tests are in progress to further evaluate the effect of strain hardening on fatigue performance of composite martensite-austenite microstructures in high carbon hardened steels.

CONCLUSIONS

1. Low cycle bending fatigue resistance of tempered martensite-austenite microstructures increases with retained austenite content. Significant transformation of austenite to martensite is present at fatigue crack initiation sites.

2. The elastic limit of martensite-austenite composite microstructures decreases with

increasing amounts of retained austenite. Refining the austenite distribution in the composite increases the elastic limit. The elastic limit correlates with the size of the largest austenite islands by a Hall-Petch type of relationship.

3. In the microstrain regime, the strain is accommodated in successively smaller austenite regions until the flow strength of the austenite matches the elastic limit of the martensite.

4. At high strains, decreases in strain hardening rates are interrupted by the transformation of austenite to martensite. The higher the retained austenite in the microstructure, the higher the strain hardening rates.

5. The high strain hardening rates of microstructures with the most retained austenite retard the mechanisms of fatigue crack initiation which develop at high strains at prior austenite grain boundaries.

ACKNOWLEDGEMENTS

This research was supported by the U.S. Army Research Office. The steels were provided by the LTV Steel Corporation (formerly the Republic Steel Corporation).

REFERENCES

1. G. Krauss: Metallurgical Transactions A, 1978, 9A, 1527-1535.

2. G. Krauss: Case Hardening Steels, Microstructure and Residual Stress Effects, 33-58, 1983, Warrendale, PA, TMS-AIME.

3. C. Razin: Harterei-Technishe Mitteilungen, 1968, 23, 1-8.

4. R.H. Richman and R.W. Landgraff: Metallurgical Transactions A, 1975, 6A, 955-64.

5. H. Brandis and W. Schmidt: Case Hardened Steels, Microstructural and Residual Stress Effects, 189-209, 1983, Warrendale, PA, TMS-AIME.

6. A.P. Voskamp, R. Osterlund, P.C. Becker, and O. Vingsbo: Metals Technology, 1980, 14-21.

7. M.A. Panhans and R.A. Fournelle: Journal of Heat Treating, 1981, 2, 45-53.

8. C. Laird: Fatigue and Microstructure, 149-204, 1979, Metals Park, Ohio, ASM.

9. J.B. Kelley: M.S. Thesis, Colorado School of Mines, Golden, Colorado, 1984.

10. H. Muir, B.L. Averbach, and M. Cohen: Transactions ASM, 1955, 47, 380-407.

11. C.L. Magee and H.W. Paxton: Transactions TMS-AIME, 1968, 242, 1766-67.

12. C.A. Stickel: Metallurgical Transactions A, 1977, 8A, 63-70.

TABLE I
Chemical Composition of Steels
(in weight percent)

Steel	C	Mn	Si	Ni	Cr	Mo
0.0 Cr	0.80	0.89	0.32	1.77	0.00	0.72
0.3 Cr	0.81	0.88	0.31	1.80	0.29	0.71
0.6 Cr	0.83	0.92	0.33	1.75	0.55	0.73
0.9 Cr	0.84	0.90	0.31	1.77	0.90	0.74
1.3 Cr	0.82	0.88	0.30	1.73	1.30	0.71

TABLE II
Hardness and Grain Size of the Heat Treated Specimens

	Direct Quench		Reheat & Quench	
Steels	Hardness (HRC)	Austenite Grain Size (μm)	Hardness (HRC)	Austenite Grain Size (μm)
0.0 Cr	59	70.0	59	38.0
0.3 Cr	58	60.0	58	32.0
0.6 Cr	57	65.0	57	19.0
0.9 Cr	55	65.0	57	11.0
1.3 Cr	56	65.0	57	9.4

TABLE III
Retained Austenite Content of Steels
(volume percent)

Steels	Fatigue Samples Direct	Fatigue Samples Reheat	Compression Samples Direct	Compression Samples Reheat
0.0 Cr	23	25	18	--
0.3 Cr	20	25	17	17
0.6 Cr	23	35	22	--
0.9 Cr	39	39	30	--
1.3 Cr	32	32	29	29

TABLE IV
Microstructural Analysis of Direct Quench
Fatigue Samples

Initial Retained Austenite (%)	Maximum Stress, ksi	Fatigue Life Cycles	Size of Plastic Zone*, mm
29	125	105,206	0.19 x 0.34
29	125	74,409	0.11 x 0.33
23	125	26,480	0.13 x 0.40
23	125	14,426	0.42 x 0.69
20	140	10,721	0.09 x 0.57

*First number represents distance measured perpendicular to the crack and the second number was measured parallel to the crack. Both were measured from the initiation point at the root of the notch.

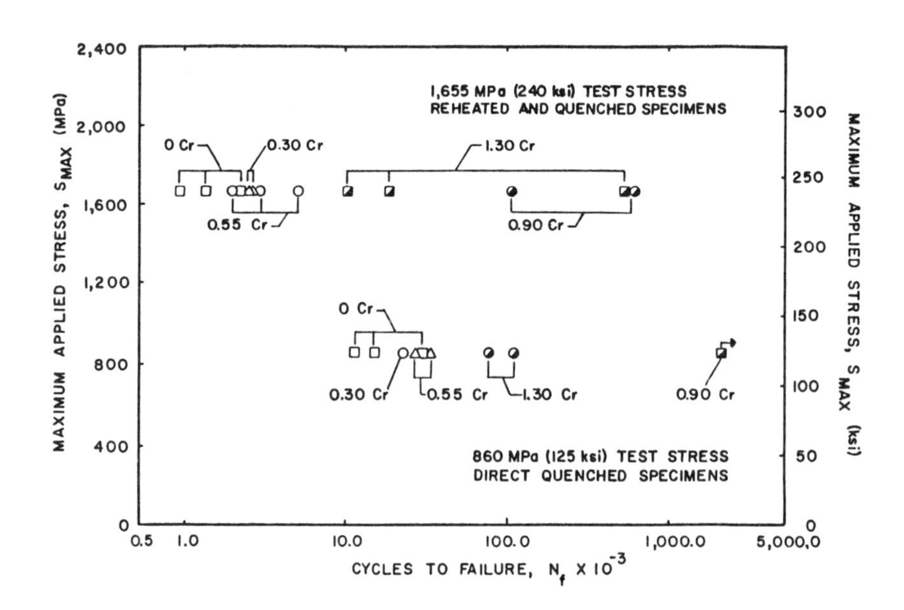

Figure 1. Low cycle bending fatigue test results of direct quench and reheat and quench specimens.

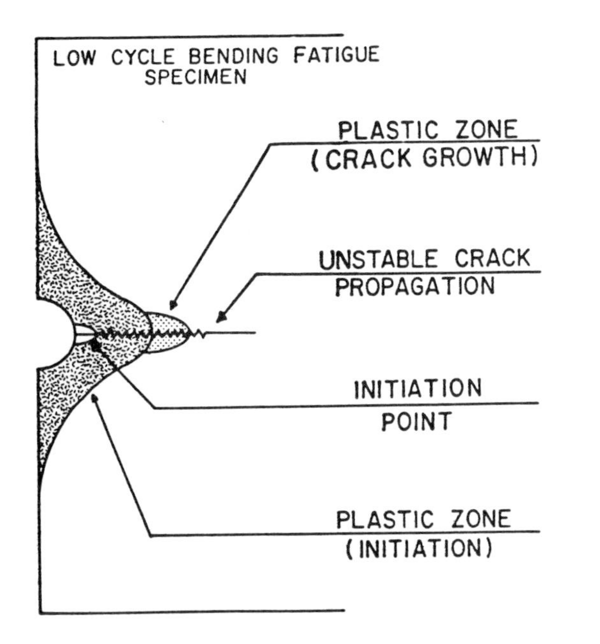

Figure 2. Schematic diagram showing regions associated with fatigue crack initiation.

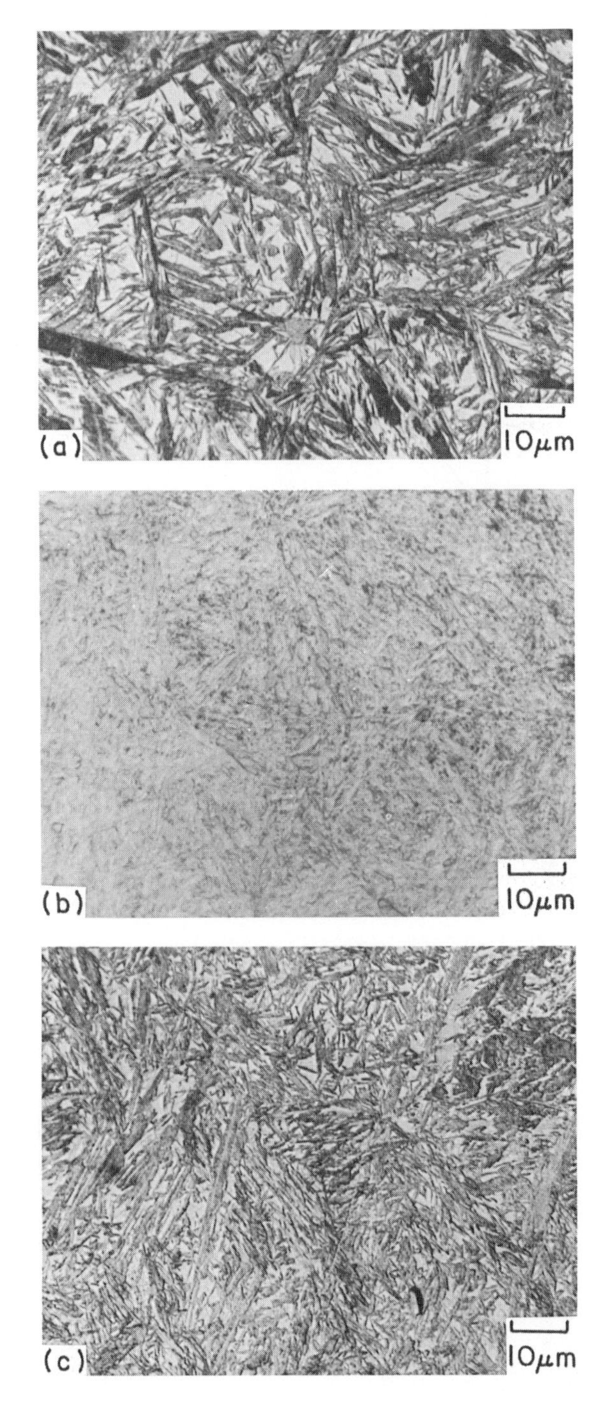

Figure 3. Microstructure of 1.3 Cr direct quench specimens associated with the crack initiation zones in the fatigue samples: a) undeformed, b) at initiation point, c) in plastic zone developed during initiation.

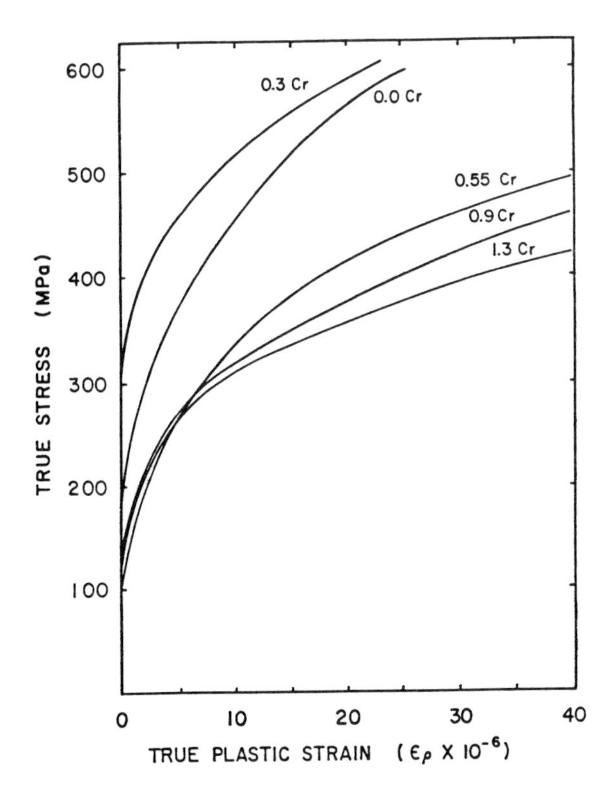

Figure 4. True stress vs true plastic strain in the microstrain region for the direct quenched specimens.

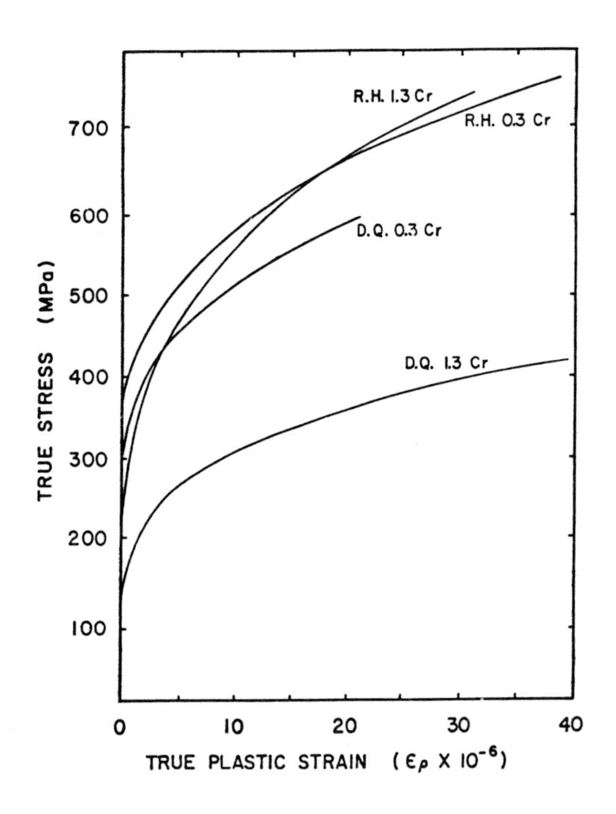

Figure 5. True stress vs true plastic strain as a function of heat treatment in the microstrain region for 0.3 and 1.3 Cr steels.

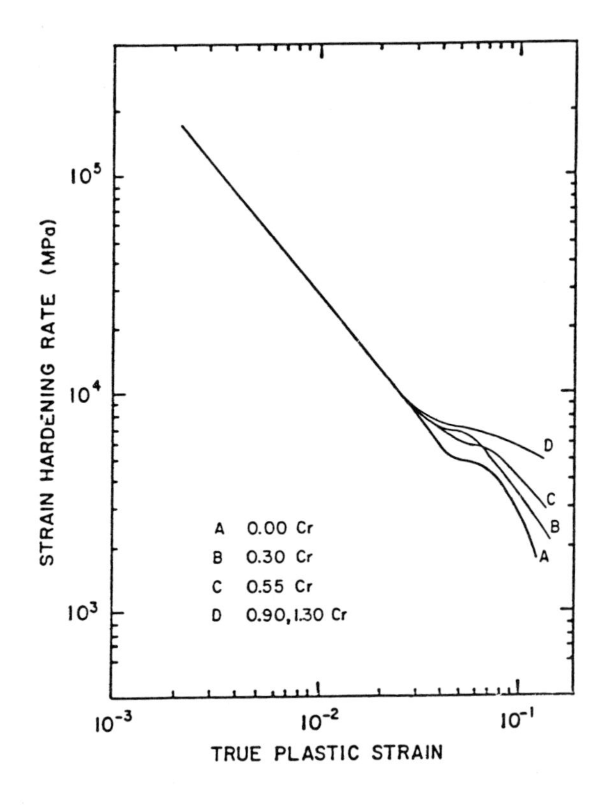

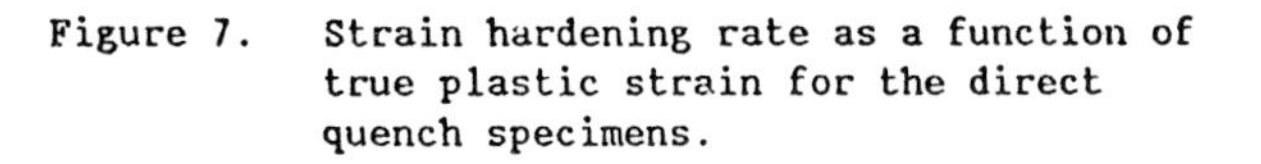
Figure 7. Strain hardening rate as a function of true plastic strain for the direct quench specimens.

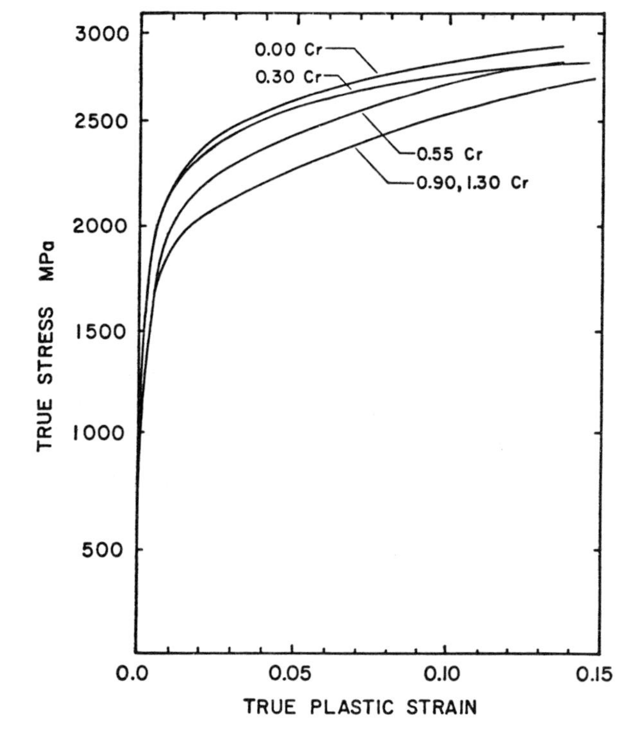

Figure 6. True stress vs true plastic strain in the macrostrain region for the direct quench specimens.

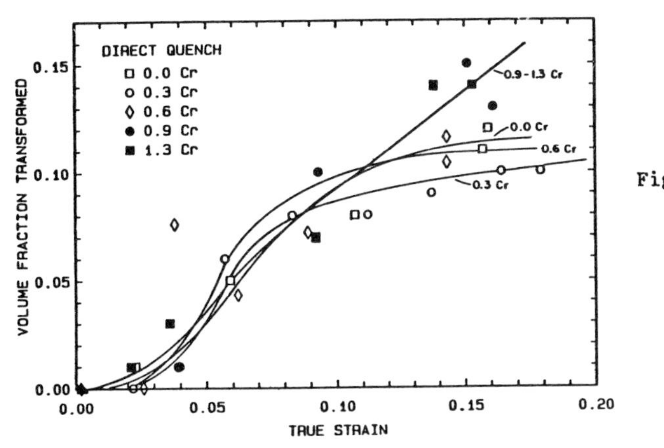

Figure 8. The volume fraction of the microstructure transformed from austenite to martensite as a function of true strain for the direct quenched specimens.

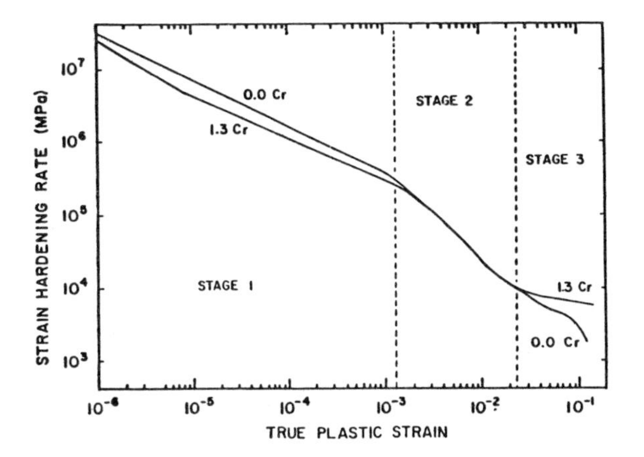

Figure 9. Summary of stages of strain hardening for 0.0 and 1.3 Cr direct quenched specimens.

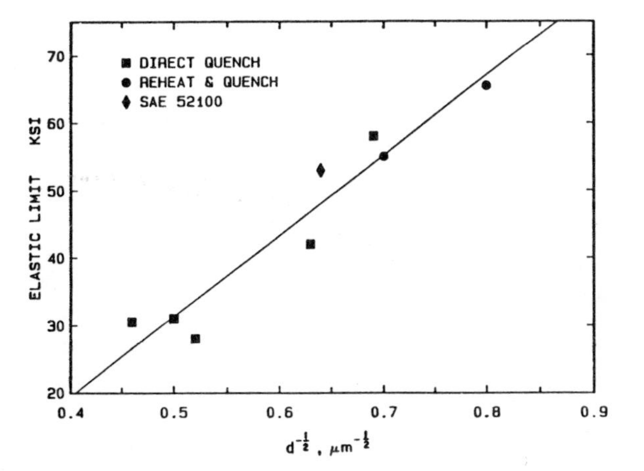

Figure 10. Elastic limit as a function of the largest austenite free mean path, d, for several specimens.

'Ultra-bar' — a new generation of quenched and tempered steels

P H WANNELL

The author is Director, Technical Services and Materials at LaSalle Steel Company, Hammond, Indiana, USA, a subsidiary of Quanex Corporation.

SYNOPSIS

Quenching & tempering heat treatment is frequently used to improve the strength, ductility, and toughness of steel. A range of mechanical properties can be produced to suit a variety of engineering applications. However, conventional heat treatment of large masses of steel requires long furnace soaking times, and process control is often difficult.

Recently, a new quenched and tempered bar has been developed which makes it possible to produce higher quality steels. New ULTRA-BAR also solves several of the problems commonly associated with conventional furnace heat treating. ULTRA-BAR employs direct electric resistance heating -- a technology pioneered at LaSalle Steel -- which is a flexible and energy-efficient method for rapidly and uniformly heating steel. The resulting product possesses high uniformity and improved properties, offering the steel consumer lower final component costs.

INTRODUCTION

LaSalle Steel began experimenting with direct electric resistance heating in the early 1970's. This technology was first applied to stress relieving (1,2) and it was soon discovered that a variety of improved steels could be produced. Continued research in this area yielded four more patents (3-6), and Quanex LaSalle is currently in the process of commercializing improved quenched and tempered products made using direct electric resistance heating. This report describes the new process, and characterizes the improved quenched and tempered products which are now available.

In order to produce quenched and tempered products, the steel must be heated to an appropriate austenitizing temperature, rapidly cooled or quenched, and then tempered to the proper strength level. For low alloy and carbon steels, the austenitizing temperature selected depends primarily upon the carbon content. Quenching is normally accomplished using an oil bath or water bath which contains special quenching additives. The quenching bath selected should cool the steel fast enough to form a predominantly martensitic microstructure. Tempering is usually conducted at temperatures below 1300 ^{o}F. (700 ^{o}C.), and the temperature selected depends upon the strength required in the finished part.

Before the advantages of new ULTRA-BAR can be fully explained, it is necessary to understand the basic differences between direct electric resistance heating and conventional furnace practice. Consequently, a portion of the discussion that follows will be devoted to explaining these differences. Following this explanation will be a comparison of the quenched and tempered steels made using the new process and using conventional furnace processing. Finally, a description of the new ULTRA-BAR products that are available will be presented.

CONVENTIONAL HEAT TREATING VS. ULTRA-BAR

a) Heating

The first step in conventional furnace heat treating involves loading the steel into a furnace which will heat it to the austenitizing temperature. Heat is transferred from the furnace to the steel by conduction, convection, and radiation. Heat transfer takes place primarily on the outside of the furnace load, and the core of the bar or bundle of bars must be heated by conduction through the steel. These heat transfer processes are relatively slow and uneven. The time for which the load must be held at temperature to ensure uniform heating affects the

* ULTRA-BAR IS A REGISTERED TRADE NAME

overall efficiency of the furnace. Other factors that affect furnace efficiency include: how well it is insulated, the efficiency of the burners, whether or not recuperators are used, and overall furnace design. The best austenitizing furnaces can achieve energy efficiencies of 30-40%, and tempering furnaces are generally poorer.

Direct electric resistance heating employs an entirely different approach which is inherently more efficient. The steel workpiece, in this case a bar, is clamped into position and a D.C. electric current is passed through the steel to cause heating. The entire cross section is uniformly heated to the desired temperature. Since only the steel is heated, efficiencies are very high. Some energy losses can be attributed to the electrical conversion equipment and radiation. The latter is minimized, however, by heating the steel rapidly to the desired temperature. Typical efficiencies for direct electric heating range from 70-90%, which is a significant improvement over the best efficiency that can be achieved by furnace heating.

b) Uniformity

One of the inherent problems with conventional furnace processing is the lack of uniformity in heating. Since the steel is heated from the outside and the heat must penetrate the furnace load, there are variations in temperature which are impossible to avoid. The steel on the outside of a large bundle will heat faster and be at temperature longer than steel in the center of the bundle. This uniformity problem can be particularly severe for the tempering operation. The tempering process is critical to product uniformity because it largely determines the final mechanical properties of the steel. A lack of temperature uniformity during tempering results in a lack of uniformity in the heat treated product.

In order to minimize the product variation which occurs during furnace heat treating, long soak times are used. However, extremely long soak times result in poor energy efficiency and poor furnace productivity. In practice, "rules of thumb" have been established to determine how long the load should be held at temperature, but these procedures are a compromise between cost considerations and product quality.

ULTRA-BAR overcomes this uniformity problem because each bar is individually heated to precisely the same temperature and for precisely the same time. There is no difference between the temperature attained during the processing of the first bar and that of the last bar processed in an order. As a result of this, a high level of uniformity is achieved in the finished product.

c) Time Effects

As was explained earlier, steel must be held at temperature in a conventional furnace for a considerable period of time to allow the core of the furnace load to come to the desired temperature. A consequence of this is that a protective atmosphere is often required during furnace austenitizing. If a protective atmosphere is not used, the steel will decarburize and scale during processing. Decarburization ruins the surface of the steel and, if too much decarburization occurs, the steel will be rejectable. Scale formation can interfere with the quenching process, and this may further degrade the final product.

ULTRA-BAR overcomes these problems by using a very short austenitizing cycle. The steel is not held in the austenitizing temperature range for more than a few minutes -- long enough to fully austenitize the steel, but short enough for neither surface decarburization nor measurable amounts of scale to form.

Often during non-productive intervals, a conventional furnace will be cycled at a lower temperature awaiting the next steel load. During this idle time, energy is wasted. In contrast, ULTRA-BAR employs instant heating. When there is no steel to be processed, the equipment is simply switched off.

d) Temperature Control

Another fundamental problem which faces the heat treater is accurately determining the temperature of the steel being heat treated. Temperature measurement is usually accomplished using a thermocouple. However, the thermocouple measures the temperature of the furnace, not that of the steel in the furnace. Consequently, there is a degree of inherent inaccuracy associated with the temperature measurement.

This problem does not exist with direct electric resistance heating because the temperature of the steel workpiece is measured directly by a radiation pyrometer which is focused on the steel. This high level of temperature-measuring accuracy contributes to the uniformity of the finished product. The use of this technique for measuring temperature also readily lends itself to automated process control; the heating process is terminated automatically when the load reaches the desired temperature. The pyrometer is focused on a region some distance from the end of the bar since, at the point where electrical contact is made, the bar is not completely heated, and this portion of the bar must be discarded. One further advantage of cutting off the ends of the bar is that the end condition, which naturally exists due to quenching rate differences, is also eliminated.

e) Distortion

When long workpieces, such as bars, are processed using conventional furnaces, there is often a straightness problem. Workpieces frequently become distorted during austenitizing because of lack of support or even heating. Special walking beam furnaces, which turn the bars during heating, have been designed to reduce distortion, however, they represent a cost premium. Further distortion occurs during quenching as a result of the stresses developed during the transformation from austenite to martensite, and a small amount of additional distortion may occur during tempering. Using a milder quenchant tends to lower the amount of distortion that occurs during quenching, but this also reduces the amount of martensite formed in the steel and thereby lowers final product quality.

Direct electric resistance quenching and tempering significantly reduces the distortion problem. During the austenitizing step, the steel is held in tension. This not only keeps the bar straight, but also straightens workpieces that were crooked prior to processing. Consequently, a consistently straight bar is delivered to the quench tank. During the quench, the bar distorts less due to the effects of rapid heating. Experiments conducted early in the development of this new process revealed that rapid austenitizing resulted in less distortion during quenching than furnace austenitizing (5). To further reduce quenching distortion, the bars are rotated during quenching. This rotation promotes a more uniform cooling which results in less distortion.

Despite efforts made to minimize distortion during austenitizing and quenching, some minor distortion is inevitable. Any residual distortion is virtually eliminated during the final tempering step in making ULTRA-BAR. During tempering, the bars are again held in tension and electrically heated. This combination of heat and tension results in straightening even though the stress applied is well below the yield stress of the steel. The data presented in Table 1 shows this temper straightening effect. It should be particularly noted that straightening occurs even if relatively low tempering temperatures are used.

f) Quenching

Quench cracking can be a major problem when heat treating steel, not only in terms of yield loss and cost, but also the adverse effect on order integrity and delivery performance.

This is another area where ULTRA-BAR possesses a major advantage. Steels which are rapidly austenitized resist quench cracking better than steels which are slowly austenitized in a furnace. The explanation for this difference in cracking tendency is based upon the strength of the austenite grain boundaries where quench cracking usually occurs. Rapid austenitizing does not provide enough time for embrittling elements to diffuse to these grain boundaries. Consequently, the grain boundaries are relatively strong and resist quench cracking. In comparison, furnace austenitizing by necessity is relatively slow, and there is ample time for embrittling elements to reach the grain boundaries, weaken them, and promote quench cracking. Tests conducted on both 4150 and 6150 alloy bars that were austenitized in a furnace and electrically resistance heated revealed that the furnace-processed steel cracked over 50% of the time, while none of the electrically austenitized bars cracked (5).

As a result of this resistance to quench cracking, a more severe quench can be used when making ULTRA-BAR than is used during conventional heat treatment. For most grades, agitated cold water becomes an acceptable quench medium, in contrast to oil or in water containing polymer-based additives. Apart from the safety and cost considerations of the latter quenchants, the ability to use a more severe quench also affords the potential to substitute leaner grades for a given end application or to provide the same grade with a higher percentage of martensite. In simple terms, this means that cost can be reduced without adversely affecting quality or quality can be improved without increasing cost.

g) Process Control

Dependable mechanical properties are increasingly demanded by design engineers and users. In a conventional furnace, it is difficult to run a pilot sample to determine the practice for a commercial order. Running individual samples through the furnace is expensive and time-consuming, and it is rarely representative of a full furnace load. In contrast, piloting of the first bar is an inherent feature of ULTRA-BAR. The process heats each bar individually so the pilot bar will be treated exactly as the bars used in the commercial order. Consequently, when a steel user needs a product in a narrow hardness range, Quanex LaSalle can obtain the steel, run a pilot test, and then set the practice for the entire order based upon the tests conducted on that pilot sample. This ability makes it possible for Quanex LaSalle to tailor the mechanical properties of the quenched and tempered steel more precisely to the customer's requirements.

h) **Costs**

Traditional heat treatment furnaces are large and expensive pieces of processing equipment. They normally are designed to perform a particular process such as quenching and tempering.

New ULTRA-BAR is made using a particularly cost-effective unit. It only takes about 25% of the floorspace of a conventional unit. The capital cost can be 33% of a conventional unit and it can be designed to perform a variety of heat treatments. The result is a very flexible unit with attractive operating costs and efficiencies.

PRODUCT ADVANTAGES

Certain advantages of ULTRA-BAR have already been mentioned, namely reduced quench cracking, no decarburization, and improved straightness. In this section, the major product advantages that can be realized by the customer will be described.

a) **Strength and Ductility**

Many quenched and tempered products are ordered to minimum tensile test specifications. For example, ASTM Specification A-193 establishes minimum levels of tensile strength, yield strength, elongation, and reduction of area. ULTRA-BAR can more easily meet these minimum specifications. The data presented in Table 2 show the results of tests conducted on 1-1/8" diameter bars of 4140 alloy that were heat treated using both conventional furnace processing and direct electric processing. The ductility of the electrically processed steel is clearly superior to that of the furnace-processed steel.

Additional tests conducted with this same heat revealed that the strength-ductility combination was consistently better for the steel processed electrically (Figure 1).

b) **Improved Product Uniformity**

It was mentioned earlier that one of the main advantages of ULTRA-BAR is improved product uniformity. Highlighted are two different comparison tests that were conducted to evaluate the relative uniformity of the new Quanex LaSalle product.

The first test compared the uniformity of a randomly selected production order from the ULTRA-BAR unit to that of quenched and tempered steel purchased from a supplier that uses conventional furnace heat treating techniques. The second test involved obtaining 60 bars from a given heat and processing 30 of the bars using the LaSalle ULTRA-BAR unit and having the remaining 30 bars heat treated to the same strength level by a commercial heat treater. Table 3 shows the results of these two uniformity tests.

c) **Surface Condition**

Another advantage of ULTRA-BAR pertains to the surface condition after heat treatment. Furnace processing can result in decarburization unless the heat treater has very good control over the furnace atmosphere. For example, the sample which was purchased and used for the uniformity test described earlier had a partially decarburized surface. Figure 2 shows the microhardness profile near the surface indicating the depth of decarburization. Many additional tests conducted on a variety of different samples indicated that ULTRA-BAR typically had almost no decarburization while furnace treated products varied widely depending upon supplier.

d) **Improved Toughness**

Quenching and tempering imparts toughness to a steel and is often an important parameter to the final end user. Several comparison tests have been conducted comparing the impact toughness of steel heat treated electrically and the same steels heat treated in a conventional furnace. The LaSalle product has been consistently tougher than the furnace treated product. Figure 3 shows graphically the difference in the toughness of steel furnace treated and processed on the ULTRA-BAR unit. This difference in toughness can be largely attributed to the finer grain size of ULTRA-BAR.

e) **Fatigue Resistance**

Fatigue comparison tests have also been conducted to demonstrate that ULTRA-BAR results in a small but consistent improvement in fatigue properties. Table 4 shows some of these results conducted on a single heat of 4140 steel.

AVAILABLE ULTRA-BAR PRODUCTS

Prior to the commercialization of ULTRA-BAR, LaSalle evaluated over thirty heats of steel over a wide range of diameters. For each heat, a variety of tempering temperatures was tested to produce tempering curves for the various steels. Figure 4 shows the results of tests conducted on ten heats of 414X steels which ranged in diameter from 0.593" to 3.500" (5). All of these steels were austenitized at 1650°F., quenched in agitated water, and tempered at the temperatures shown. It is noteworthy that the mechanical properties of these steels all fell within a relatively narrow band despite the fact that the diameters varied over a wide range.

Figures 5 and 6 show other grades that were heat treated using direct electric resistance. These grades were selected for testing because either steel users were currently using these grades or there was potential for substitution for more expensive steels.

Figure 7 shows the Charpy impact properties for some of the steels which were evaluated. It should be noted that many of the steels contained machinability additives such as sulfur, lead, selenium, or tellurium. These additives have to be avoided in conventional heat treating because they promote quench cracking. However, using the new LaSalle process, it is possible to quench these grades without this problem occurring.

The concept of substituting a leaner steel for one which is currently in use is important because it provides the potential to significantly reduce the cost of producing a given part. Consider the mechanical property data provided in Table 5 (5). These data indicate that 1045 carbon steel can be used to substitute for 4140 alloy steel in some applications. In this instance, the carbon steel processed on the LaSalle unit actually had superior mechanical properties when compared to the furnace treated alloy steel.

Clearly 1045 cannot be substituted for 4140 over the entire size range. Above 3/4" in diameter, 1045 will not completely through-harden regardless of how severely it is quenched. However, above this size, other substitutions are possible which will also provide cost reductions. For example, from 3/4" to 1-1/8", 1541 could be substituted for 4140. The grade 1547 could be used in diameters up to 1-3/8"; and above that size, 15B41 could be an alternative. When substitutions such as these are being considered, it is necessary to carefully evaluate the specific end application to ensure that the replacement steel meets all the requirements for the application.

Most of the data provided so far has pertained to round bars. However, the Quanex LaSalle process can also be used to heat treat hexagons, flats, squares, and tubes. Figure 8 shows the as-quenched and tempered hardnesses of a rectangular bar that was heat treated electrically. Note that the center of the as-quenched product is only slightly softer than the surface, even within the sharp corners where the quenching rate would have been very high. After tempering, this difference in hardness is completely eliminated.

Tests conducted on tubes and hexagons have revealed excellent uniformity in these products as well. Figure 9 shows examples of hardness tests conducted on hexagons heat treated using direct electric quenching and tempering. Additional data on tensile strength are also provided to illustrate the mechanical properties of these steels.

Case Histories

Users of high-strength steels are already taking advantage of the unique properties of ULTRA-BAR.

A major producer of large section industrial forklifts was experiencing problems of scale and crooked parts when using conventionally heat treated long-length 4140 alloy bars. A Rockwell "C" hardness of 42/47 at 3/4" radius is specified. By changing to resistance heated steel, these problems were overcome and also a less costly 1541 carbon steel was substituted.

Another company was repeatedly getting cracked ends on quenched and tempered hexagon bars used for its maintenance program. The need for costly magnetic inspection and end trimming has been eliminated by changing to ULTRA-BAR.

A manufacturer of load cells specified a 3/4" x 2-1/2" alloy bar heat treated to 32/34 Rockwell "C" hardness to become the direct load-carrying member. Uniformity of microstructure was vital in order to calibrate and stabilize the load cell. No other supplier could quote for this demanding new application.

A large machine shop produced an auxiliary power shaft for a large diesel, 1-13/16" diameter and 14 inches in length. The shaft has seven different diameters machined with several keyways and gears. The shop used both annealed and quenched and tempered aircraft quality 4140 alloy steel, the latter at lower machining speeds. A change to ULTRA-BAR quenched and tempered product gave similar machinability to the previously annealed product and several expensive finishing and inspection operations have been eliminated.

STATISTICAL PROCESS CONTROL

Statistical process control (SPC) systems provide a means to ensure the quality of the finished product by establishing techniques to monitor and control processing parameters. These techniques can be applied to conventional furnace processing. However, in furnace processing, the steel is heated in relatively large quantities, and every bar in the furnace load is not exposed to the same temperature for the same length of time due to mass effects. Similarly, quenching in fixtures does not expose every bar to the same quenching conditions. Due to these aspects of conventional processing, it is difficult to apply SPC systems which provide a means to control bar-to-bar variations.

The equipment used to manufacture ULTRA-BAR more readily lends itself to SPC techniques. Since each bar is individually heated, the processing parameters for each bar can be easily monitored, and any variations can be readily detected. Also, since the operation can be interrupted at any time, corrective action can be immediately taken. The processing parameter which is most critical to the mechanical properties of the finished product is the tempering temperature. By monitoring the tempering temperature with a radiation pyrometer and plotting the results, a continuous process control system can be established which records the tempering temperature of every bar. Other aspects of the process can also be monitored using SPC techniques. For example, the austenitizing temperature and the quench bath temperature can be checked by the operator and maintained within acceptable limits. Hence, all important aspects of the process can be incorporated in an SPC program which ensures a high level of uniformity in the finished product.

CONCLUSION

LaSalle's new ULTRA-BAR has proven to be a flexible and energy-efficient technology for producing improved quenched and tempered products. Some of the major advantages are:

* High Energy Efficiency
* Reduced Quench Cracking
* Improved Straightness
* Minimal Scaling
* No Decarburization
* Better Process Control
* Lower Costs

When the product made using the new technology is compared to steels heat treated using conventional furnaces, several product advantages are recognized:

* Improved Strength/Ductility Combinations
* Improved Product Uniformity
* Improved Surface Condition
* Improved Toughness
* Improved Fatigue Resistance

This process has been applied to a wide variety of steels, and it is now possible to consider substitutions of leaner grades for more expensive alloy steels. By being able to exploit the full hardening potential of a steel, ULTRA-BAR offers the steel user the opportunity to reduce the cost of his product while simultaneously improving quality.

REFERENCES

(1) S. H. Jones & J. M. McNeany, "Steels and Method for Production of Same," U.S. Patent #3,908,431, September 30, 1975

(2) G. W. Wilks, "Direct Electric Stress Relieving -- A New Energy Efficient Technology," Sixth Heat Treating Conference and Workshop, ASM, Cincinnati, Ohio, September 22-24, 1981

(3) E. Mudiare, "Process for Strengthening of Carbon Steels," U.S. Patent #4,040,872, August 9, 1977

(4) M. J. Rowney, "Steels Combining Toughness and Machinability," U.S. Patent #4,088,511

(5) G. W. Wilks, "Process for the Improved Heat Treatment of Steels Using Direct Electrical Resistance Heating," U.S. Patent #4,404,047, September 13, 1983

(6) G. W. Wilks, "Process for Annealing Steels," U.S. Patent #4,457,789, July 3, 1984

(7) R. J. Shipley, "Temper Embrittlement and Cold Work Effects in 5130 Steel," PhD Thesis - Illinois Institute of Technology, December 1980

TABLE 1

Temper Straightening of 4140 Alloy Bars

Bar Length - 12'4"
Bar Diameter - 2.070"
Austenitizing Temperature - 1700°F.
Quenching Media - Agitated Water at Room Temperature

Sample #	Distortion* After Quenching (in./ft.)	Distortion* After Tempering (in./ft.)	Tempering Temperature (°F.)
1	0.0355	0.0053	900
2	0.0558	0.0105	1000
3	0.0507	0.0105	1100
4	0.0202	0.0053	1200
5	0.0101	0.0053	1200
6	0.0253	0.0053	1200
7	0.0101	0.0053	1200

*Distortion was measured as the amount of bow per foot using the entire 12'4" length.

TABLE 2

Mechanical Properties of
Quenched & Tempered 4140 Alloy Steel
(Average of 6 Tests)

Property	Furnace Treated	ULTRA-BAR
Tensile Strength, ksi	168.2	168.1
Yield Strength, ksi	156.4	155.3
Elongation, %	15.8	16.5
Reduction of Area, %	53.5	57.1

TABLE 3

Uniformity Comparison Tests

Test #1	Purchased Furnace Treated Product		LaSalle ULTRA-BAR	
Samples from Different Heats	Range	Standard Deviation	Range	Standard Deviation
Tensile Strength, ksi	23.9	4.26	10.9	2.28
Yield Strength, ksi	22.7	4.25	14.4	3.70
Elongation, %	5.0	1.05	3.0	1.13
Reduction of Area, %	9.6	2.22	5.6	1.34

Test #2	Commercial Heat Treater		LaSalle ULTRA-BAR	
Samples from Same Heat	Range	Standard Deviation	Range	Standard Deviation
Tensile Strength, ksi	22.7	6.13	14.0	3.88
Yield Strength, ksi	30.0	8.17	16.1	3.73
Elongation, %	3.5	0.82	3.0	0.68
Reduction of Area, %	6.1	1.55	4.7	1.33

TABLE 4

Fatigue Test Results

	Furnace Processed	ULTRA-BAR
Tensile Strength, ksi	168.2	168.1
Fatigue Limit, ksi	88.8	91.5
Fatigue Ratio, %	0.52	0.54

TABLE 5

Alloy Substitution Example

Property	Furnace Treated 4140 Alloy Steel	ULTRA-BAR 1045 Carbon Steel
Tensile Strength, ksi	145.5	152.0
Yield Strength, ksi	129.4	129.8
Elongation, %	17.5	18.0
Reduction of Area, %	60.0	62.3

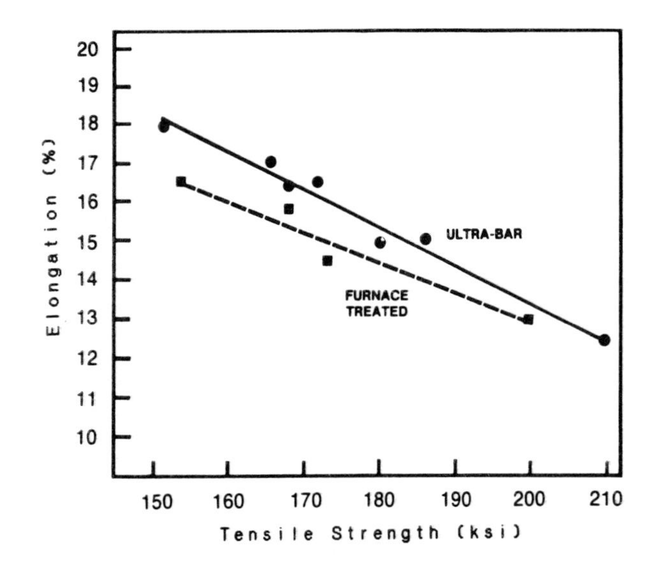

FIGURE 1
Strength-ductility combination of heat treated bars from same heat.

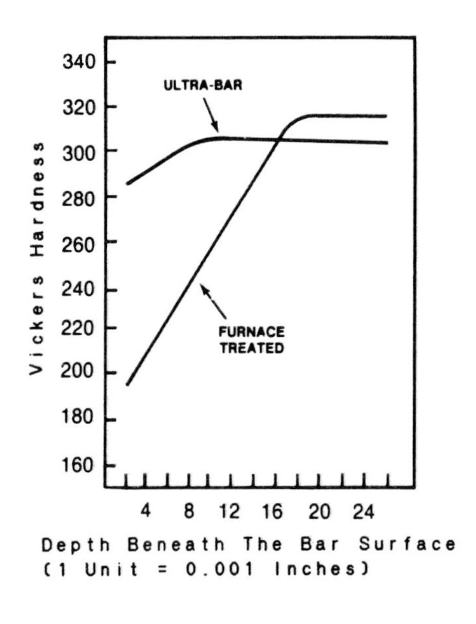

FIGURE 2
Microhardness Profile

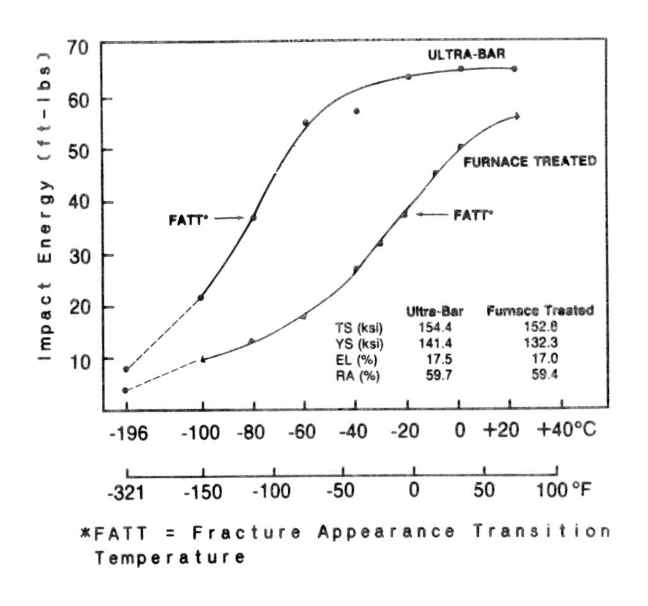

*FATT = Fracture Appearance Transition Temperature

FIGURE 3
Charpy Impact Comparison

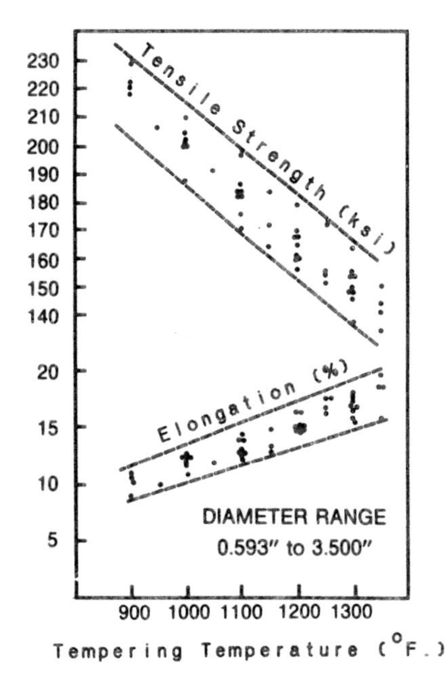

FIGURE 4
Mechanical Properties of Ten Heats of 414X Steel

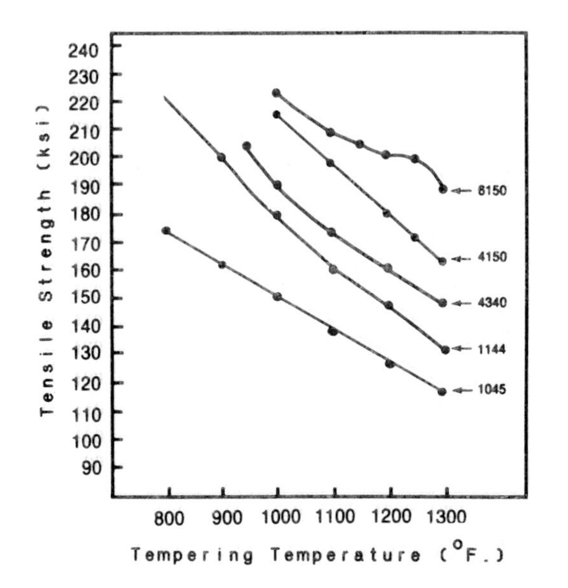

FIGURE 5
Mechanical Properties of Various Medium Carbon Steels

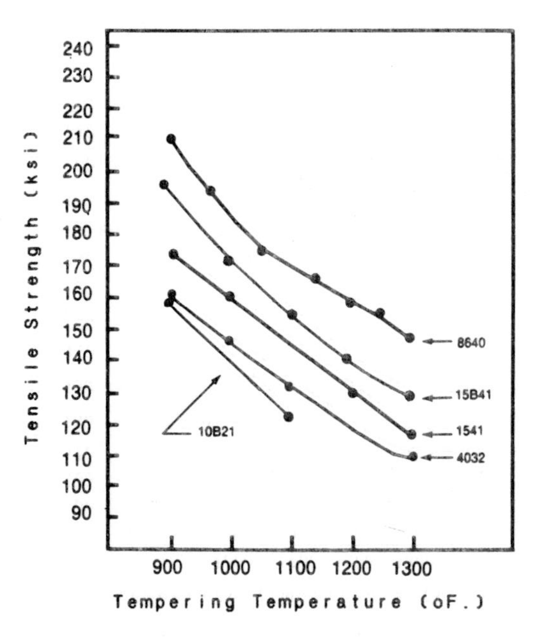

FIGURE 6
Mechanical Properties of Additional Steels

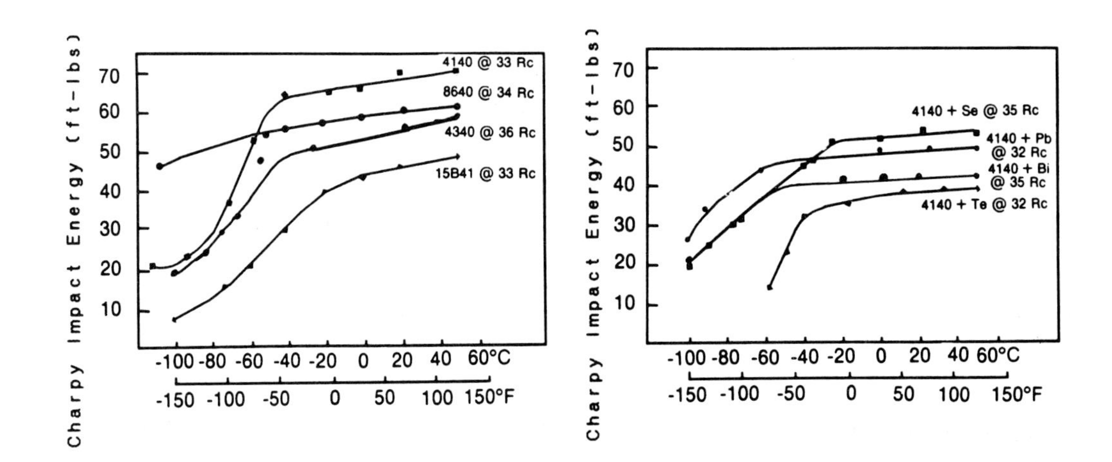

FIGURE 7
Charpy Impact Properties of Various Steels

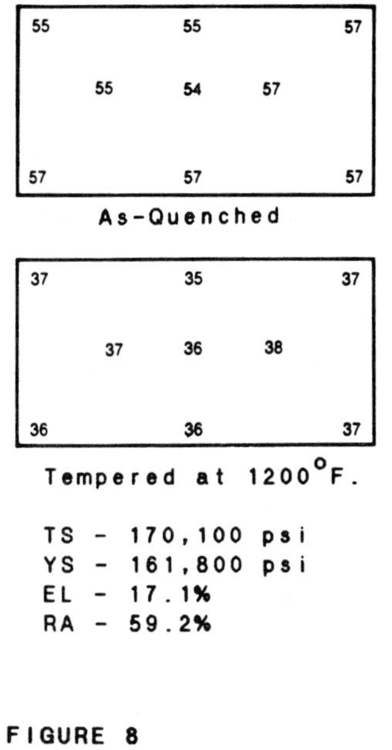

TS - 170,100 psi
YS - 161,800 psi
EL - 17.1%
RA - 59.2%

FIGURE 8
ULTRA-BAR Rectangular Bar
0.625" x 1.250"

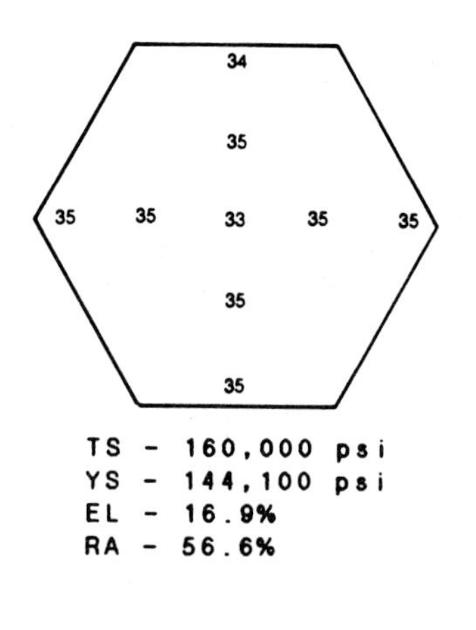

TS - 160,000 psi
YS - 144,100 psi
EL - 16.9%
RA - 56.6%

FIGURE 9
ULTRA-BAR Hexagon
2.062"

Effect of cleanness on properties of heat-treated medium carbon low alloy steels

W T COOK and D DULIEU

The authors are with BSC Swinden Laboratories, Moorgate, Rotherham, UK.

SYNOPSIS

An assessment has been made of the effects of reducing the sulphur and oxygen contents of three low alloy steels on their properties and behaviour relevant to use after heat treatment. The steels were made by modern secondary steelmaking techniques, giving sulphur levels below 0.005% with oxygen contents below 10ppm. By calcium treatment, plastically deformable inclusions were almost eliminated.

These high cleanness levels gave a significant increase in ductility in the transverse orientations of the billet materials tested, shown by both tensile and impact toughness tests on materials hardened and tempered to a range of strength levels.

Fracture toughness under ambient conditions was similarly improved, particularly where there was significant crack tip plasticity. However, tests in H_2S and 3.5% aqueous NaCl solution showed that stress corrosion crack propagation characteristics were largely unaffected by cleanness level, particularly at proof strength levels above 1000N/mm2.

Although the smooth surface rotating bending fatigue properties were improved, particularly at the higher strength levels, with increased cleanness, this effect was overridden where surface residual compressive stresses were present.

Overheating susceptibility was enhanced by reducing the sulphur content, but calcium modification of the sulphide inclusions at very low sulphur levels offset this effect.

Machinability, as assessed by turning, was generally impaired by reducing the sulphur level, this was balanced for some cutting conditions by the reduction in the level of abrasive oxide inclusions in the clean steels.

INTRODUCTION

The user of heat treatable engineering steels is now able to select the inclusion type and level to optimise properties and performance. Traditionally, enhanced sulphur levels and additives are used to improve machinability, but high cleanness steels have been made by secondary remelting. However, the rapid development of steelmaking practices using secondary vessels has allowed the bulk production of engineering steels to very high standards of cleanness. These processes were used first for special steel products such as offshore structural plates and bearing steels, but they are now generally available for the manufacture of engineering steels.

Although there have been several studies of the effects of cleanness on the properties of steels (1,2), it was thought appropriate to carry out a systematic study of the property improvements obtainable in heat treatable steels by the use of modern steelmaking methods. This paper summarises the findings of a programme undertaken with financial support from ECSC(3).

Three widely used low alloy steels were chosen to cover the range of carbon contents (0.3-0.6%) and tensile strength levels (850-1850N/mm2) for which through-hardened steels are normally used. The grades chosen corresponded nominally to the SAE 4130, 4340 and 5160 specifications. All were made under normal production conditions and were examined either in billet or in billet-derived product forms.

STEELS EXAMINED

For each steel type standard commercial material was selected, representing normal production to orders where no specific limits were set on oxide cleanness levels and where the sulphur content would fall in the normal range of 0.015-0.030wt%. The high cleanness variants were all made in VAD vessels, using a clean steel practice to prevent refractories, interactions and re-oxidation. The high cleanness CrMo SAE 4130 and one of the NiCrMo variants were made to a low sulphur content with calcium treatment for inclusion modification. The high cleanness MnCr spring steel, SAE 5160, was made to low sulphur and oxygen contents without inclusion modification.

The origins of the seven steels examined are given in Table 1, and their chemical compositions are summarised in Table 2. For reasons of supply, one of the NiCrMo steels was to the UK specification BS970:817M40, differing slightly from the intended SAE 4340 composition range. For convenience, the three steel types examined are referred to as A, B and C, as indicated in Table 1.

TEST PROGRAMME

Steel cleanness was assessed using the JK and SEP 1570 standard methods. A manual, quantitative optical technique was used to rate the projected length and the relative incidence of elongated inclusions and globular particles.

In most cases, testpieces were machined from billet samples prior to conventional heat treatment, although the spring steels were also evaluated in the form of spring flats, Table 1. After hardening to fully martensitic structures, material was tempered at selected temperatures to give a range of tensile strength levels.

The test programme is summarised in Table 3. Conventional room temperature tensile, fracture toughness and Charpy impact toughness tests were carried out, in accordance with current UK National Standards. Details of the fatigue, machinability and other nonstandard tests are given below.

EXPERIMENTAL RESULTS

1. Steel Cleanness.

An indication of the improvement in microcleanness in each grade of steel is given by the summary SEP ratings in Table 4. For the spring steel, there was a significant improvement in both the oxide and sulphide ratings between the two casts, without a major change in inclusion chemistry, the principal phases being Al_2O_3 and MnS. A similar improvement was noted between the NiCrMo steels B1 and B2. However, calcium treatment of steel B3 eliminated both elongated sulphides and oxide stringers. For this assessment, all particles in this steel were rated as globular oxides, (OG). A similar inclusion population was found for the calcium treated SAE 4130 steel, C2.

In the low oxygen content steels A2 and B2, the oxide inclusions contained some MgO, up to 28% in steel B2, with traces of calcium. In steels B3 and C2, both calcium treated, calcium was present in the oxide inclusions and up to 8% calcium was found in the sulphide phase.

2. Tensile Ductility.

It is well established that reducing the inclusion content, particularly of elongated particles, increases the transverse, ductility-related properties of wrought steels.

Tensile tests were conducted at room temperature, after tempering to various strength levels, to relate the changes in ductility to the inclusion content. The general relationship is indicated in Fig.1, for steels hardened and tempered to a common nominal tensile strength of 870N/mm2. Here the true fracture strain, calculated from the tensile reduction in area, is plotted as a function of the SEP Ko rating. A broad relationship was seen for both test orientations and, in the cleanest steels, there was an indication of a reduction in ductility as the carbon content increased between steels C, (0.3%C) and B, (0.4%C).

A more quantitative relationship may be obtained by correlating the fracture strain with the total inclusion projected length in the fracture plane, obtained from the manual optical inclusion counts. This is illustrated in Fig.2 for the B series (0.4%C) steels, at tensile strength levels between 870 and 1850N/mm2. (Test orientations were both along and across the billet product.)

Using this relationship for the 0.4%C (B Series) steels, fracture strains were calculated for the 0.3 and 0.6%C steels, as shown in Fig.3. There was a divergence in the relationship as the cleanness level increases which is consistent with a contribution to the reduction in fracture strain from the, assumed isotropic, distribution of carbide particles.

3. Impact Toughness.

The improvement in cleanness raised the transverse ductile shelf energy value, while having little effect on that for the longitudinal direction. The effect was noted in all three steels and is typified by the results for the B steels shown in Figure 4.

4. Fracture Toughness.

The results of fracture toughness tests on the three classes of steels are summarised as a function of strength level in Tables 5, 6 and 7. The nomenclature for testpiece orientation is given in Fig.5. At ambient temperature in air, under plain strain fracture conditions, there was a relatively small improvement in the ratio of the transverse to longitudinal (XY:YX) toughness values as the inclusion content was reduced for billet material. This is indicated by the results for the A and B steels at proof strength levels above 1400N/mm2, Tables 5 and 6.

The results for the A series steels, tested after hot rolling to the spring flat form show the contribution from the redistribution and deformation of the inclusions to both the level and anisotropy of toughness. The toughness values in the YX orientation were reduced on rolling from billet to spring flat for both the standard and clean materials, but particularly in the former.

At lower strength levels, there was greater crack tip plasticity in the testpiece sizes used and the toughness was assessed by COD or KQ values. Under these conditions, improving the cleanness produced a significant increase in toughness for the YX and ZX orientations, with only a small improvement for the XY orientation. However, the comparison of the results for the B and C series steels at proof strength levels below 1050N/mm2 shows how toughness in the YZ test orientation is less sensitive to inclusion content than in the YX orientation, Tables 6 and 7.

For tests on the B series steels, it was noted that the toughness in all orientations examined was improved between steels B2 and B3 at all strength levels examined. Although in part this may be related to the small differences in chemical composition between the two steel types, a contribution comes from the effect of the calcium treatment in modifying the inclusion shape.

5. Fracture in Corrosive Environments.

Tests were conducted in H_2S gas and in 3.5% aqueous NaCl solution, to determine K_1H_2S and K_1SCC respectively. The results are given for the A and B series steels in Tables 5 and 6. A similar pattern of behaviour was noted for both types of steel.

The limiting threshold stress intensity for the onset of crack growth was reduced in the corrosive environments at all the strength levels examined and for all the cleanness variants. The extent of the reduction appeared to depend mainly on the strength level and was independent of cleanness. However, for tests on the B series steels at the lowest strength level (nominally 750N/mm2 0.2%proof stress), the improvements in COD values obtained in air tests from a reduction in the inclusion content were reflected directly in enhanced values for K_1SCC.

At the lower strength levels the reduction in the limiting threshold stress intensity was small, the final values being 80-90% of the values in air.

Conversely, for materials heat treated to high strength levels, the effects of the corrosive media were severe. Both K_1SCC and K_1H_2S values were reduced to a similar very low level, around 15MPa/m, irrespective of both the inclusion content and test orientation.

6. Fatigue Performance.

Rotating bending fatigue tests were carried out using polished testpieces on all three series of steels. The limiting fatigue stress,(LFS) for longitudinal testpieces showed no dependence on cleanness for tensile strengths up to 1150N/mm2. However, a small improvement was noticed at the highest strength levels studies. In contrast, the transverse testpiece fatigue lives were improved significantly at tensile strength levels above 1150N/mm2.

The major effect was associated with the reduction in sulphide and oxide contents between the standard and cleaner steels. However, in the B series steels at the highest strength level, there was a further improvement in the transverse orientation between steels B2 and B3, table 8. The results are presented in Fig.6, in strength bands as a function of inclusion total projected length in the fatigue fracture plane. A similar general relationship was indicated for the C series steels from tests at tensile strength levels up to 1050N/mm2.

The A series spring steels were tested in three point bending after processing to simulate the conditions used in the commercial production of leaf springs, involving shot peening of the surfaces of the heat treated testpieces.

Finite life tests were conducted on testpieces which had been shot peened, starting with both as rolled and machined surfaces. No significant difference in mean lives was noted between steels A1 and A2. Similarly, S-N curves were determined, using the A1 steel as heat treated and without surface treatment as a reference material. The results are shown in Fig.7 and confirm that any difference attributable to steel cleanness was overridden by the effects of surface treatment.

7. Overheating Susceptibility of Clean Steels.

It is well established that conventional high cleanness steels are prone to embrittlement by overheating; the dissolution and re-precipitation on prior austenite grain boundaries of manganese sulphide particles under hot working conditions. The susceptibility of the B series steels to both austenite grain coarsening and overheating were assessed.

On reheating to temperatures above 850 deg.C, similar prior austenite grain sizes were found in all three steels. At higher temperatures significantly coarser grain structures were developed in steels B2 and B3. After 1h at 1350 deg.C these steels had mean linear intercept grain sizes around 1mm, compared with 0.2mm for the conventional composition steel B1.

To establish the sensitivity to overheating, the tensile and impact toughness properties and slow bend fracture characteristics were determined for material heated to progressively higher temperatures and then hardened and tempered. The tensile ductility values at a constant nominal tensile strength of 1150N/mm2 are shown in Table 9. As expected, there was a loss in ductility as the exposure temperature prior to hardening increased above 1150 deg.C, the effect being most pronounced for the low sulphur steel, B2. Although part of this loss may be attributed to variations in the prior austenite grain size, fractographic examination and the impact toughness characteristics confirmed the increased susceptibility to overheating in steel B2, Fig.8. The incidence of ductile intergranular fracture indicated clearly the benefit of the further reduction in sulphur content between steels B2 and B3, with calcium modification of the sulphide inclusions, in reducing overheating in low sulphur steels.

8. Machinability

Comparative machinability tests were carried out for both the B and C series steels, using single point turning with both carbide and high speed steel (HSS) tooling. Similar trends were noted with both steel types and the results are illustrated here with results for the B series steels, tested in both the annealed and the hardened and tempered conditions, Tables 10 and 11.

The reduction in both sulphur and oxygen contents between steels B1 and B2 was deleterious for HSS tool life, measured as the maximum cutting speed for a 20min. tool life, V20. Also, both the flank and crater depth wear rates for carbide tools were markedly increased. However, calcium treatment at the low sulphur level led to a consistent improvement in the carbide tool crater depth wear rate for all the test conditions examined. Flank wear rate was improved only for the annealed condition machined at low speed. With HSS tooling, the effect of calcium treatment was to improve the tool life over the low sulphur steel, again only in the annealed condition.

DISCUSSION

Sulphur and oxygen contents below 40 and 10ppm are readily achievable by modern bulk steelmaking methods. At these low levels, calcium treatment eliminates elongated inclusions of the MnS type. The principal effect is to improve the transverse, ductile failure-controlled properties. Thus the use of ultra-clean steels is particularly beneficial in applications where the isotropy of ductility and toughness is important.

For the environmental conditions examined, the improvements in cleanness levels studied are significant for resistance to SCC fracture propagation only at tensile strength levels below about 1000N/mm2.

It is well established that the fatigue life of ultra-high strength steels is sensitive to cleanness level under conditions where inclusions can act as failure initiation sites, as in the case of rolling contact fatigue. The results of the present work show this for rotating bending fatigue where the testpieces are of transverse orientation. The spring leaf tests indicate the importance of surface stress state in controlling fatigue initiation and, at first sight, suggest that cleanness level is of secondary importance. However, it must be remembered that significant improvements in macro-cleanness accompany the steelmaking methods used to achieve the microcleanness levels studied in the present work.

Overheating susceptibility and poor machinability have been accepted as characteristics of high cleanness steels. The present work shows that, although their grain coarsening characteristics differ, conventional and high cleanness steels can have similar overheating characteristics. A combination of a very low sulphur content and sulphide inclusion phase modification with calcium may be used to overcome the inherent overheating susceptibility of low sulphur steels. This observation is consistent with earlier work on low sulphur steels with inclusion modification (4).

Although machinability is impaired at low sulphur levels, there is a benefit from the use of calcium treatment in offsetting the effects of the reduction in sulphur content for certain machining conditions. By analogy with higher sulphur, calcium-treated steels, the good carbide tool wear characteristics noted for steel B3 may be attributed to a reduction in the number of oxide inclusions and the modification of the oxide phase to give a less abrasive product at the temperatures produced during the cutting operation.

CONCLUSIONS.

The properties and behaviour of three commonly used, heat treatable low alloy engineering steels have been examined, to determine the effects of reducing the sulphur and oxygen contents from levels of nominally 0.025% and 30ppm to 0.004% and 10ppm respectively. The lower levels represent those which can be achieved readily with modern, bulk secondary steelmaking technology.

The principal benefit from reducing the volume fraction of inclusions is the improvement in ductile failure controlled properties in test directions normal to the axis of the billet products examined. The tensile ductility can be related quantitatively to the projected length of inclusions in the fracture plane over a range of tensile strength levels for tempered martensitic structures.

For fracture under ambient, plain strain conditions, the inclusions in the process zone influence the anisotropy of toughness. However, the effect of toughness on cleanness level becomes more important as crack tip plasticity increases. In contrast, fracture resistance in aggressive environments is improved only for relatively low strength materials, where the fracture mode is by ductile tearing. Both the K_1H_2S and K_1SCC values for high strength steels, those with tensile strengths above 1150N/mm2, were insensitive to cleanness level for the tests conducted in the present work.

The limiting fatigue stress for smooth, rotating-bending testpieces in the transverse orientation was improved significantly by reducing the inclusion content for steels heat treated to tensile strengths of 1150N/mm2 or above. Under these conditions a high cleanness steel is essential to achieve the maximum fatigue performance. However, for applications where the surface stress state can be controlled, as in leaf springs, the results indicated that any effects of improvements in cleanness can be dominated by variations in the surface condition.

It has been demonstrated that very low sulphur steels with calcium modification of the sulphides have a low susceptibility to overheating.

In general, machinability, as assessed by turning tests, is impaired by a reduction in sulphur content. However, at 0.002%S with calcium modification carbide tool crater wear for single point turning of annealed material was generally better than for a conventional 0.026%S steel. This improvement is attributed to the reduced abrasion from the low volume fraction of calcium-modified oxide inclusions.

Calcium treatment of low sulphur, high cleanness materials was found to be of benefit, giving improvements in transverse ductility and offsetting the problems of reduced machinability and increased overheating susceptibility.

ACKNOWLEDGEMENTS.

This research work was carried out with the financial support of the European Coal and Steel Community (ECSC). Several of our colleagues in BSC Swinden Laboratories have contributed to the programme; in particular Drs. R. Amin and A. P. Hirst. Thanks are due to them and to Dr. R. Baker, Director of Research BSC, for permission to publish this paper.

REFERENCES.
----

1. Gladman, T, in 'The Effects of Inclusions on Mechanical Properties' Instn. of Metallurgists Monograph No. 3 No 3. London, 1979.

2. Spitzig, W. A. Met Trans. A 14A p.471 (1983)

3. Cook, W. T, Dulieu, D, Amin, R.K, & Hirst, A. P.

   ' The Evaluation of Clean Steel Practices and Resultant Property Improvements in Alloy Engineering Steels.'

   ECSC Publication No. EUR 10340 EN, (contract No.7210 MA/805.) Brussels 1987.

4. O'Brien, R. N, Jack, D. H. and Nutting, J, in; Proceedings of Heat Treatment '76' p.161. The Metals Society, London, 1977.

TABLE 1 STEEL TYPES EXAMINED

Steel Type	Identity	SAE Grade	Process Route	Cast Wt. Size t	Ingot Wt. t	Rolled Product Size mm sq.
Spring Steel	A1	5160	Standard	120	4	108*
	A2	5160	Low S VAD treated	120	4	108*
Med. C NiCrMo	B1	4340	Standard	120	4	90
	B2	4340	Low S VAD treated	120	4	90
	B3	BS 970: 817M40	Low S VAD+Ca treated	90	4	90
Med. C CrMo	C1	4130	Standard	120	4	90
	C2	4130	Low S VAD+Ca treated	90	3	90

*Also spring plate 76.5 mm x 12.7 mm section, was hot rolled from the billet

TABLE 2 CHEMICAL COMPOSITION (PRODUCT ANALYSES)

	Chemical Analysis, Wt. %													
	C	Si	Mn	P	S	Cr	Mo	Ni	Al	Ca	Cu	N	Sn	O_2
SAE5160 - Spring Steel														
A1	0.63	0.23	0.92	0.015	<u>0.023</u>	0.88	0.03	0.15	0.018	<0.0005	0.15	0.007	0.014	0.0033
A2	0.60	0.24	0.80	0.018	<u>0.005</u>	0.81	0.07	0.23	0.025	<0.0005	0.18	0.010	0.016	<0.001
Medium Carbon NiCrMo - General Eng. Steel - SAE4340, BS970:81740														
B1	0.43	0.15	0.62	0.007	<u>0.026</u>	0.79	0.26	1.64	0.016	<0.0005	0.22	0.009	0.018	0.0021
B2	0.43	0.24	0.67	0.014	<u>0.004</u>	0.82	0.23	1.91	0.035	<0.0005	0.12	0.008	0.011	<0.C01
B3	0.42	0.36	0.57	0.004	<u>0.002</u>	1.14	0.24	1.45	0.039	0.0017	0.12	0.007	-	0.0014
Medium Carbon CrMo - General Eng. Steel - SAE4130														
C1	0.32	0.25	0.56	0.017	<u>0.016</u>	1.00	0.19	0.05	0.005	<0.0005	0.043	0.009	-	0.0031
C2	0.30	0.37	0.54	0.008	<u>0.001</u>	0.95	0.24	0.31	0.036	0.0023	0.13	0.010	-	0.0011

TABLE 3 SUMMARY OF TESTING

Identity	Cleanness Assessment	Machinability Turning	Mech. Props.	Fatigue	
				Rotating Bending	Three Point Bending
A1	✓	-	✓	✓	✓
A2	✓	-	✓	✓	✓
B1	✓	✓*	✓**	✓**	-
B2	✓	✓*	✓**	✓**	-
B3	✓	✓*	✓**	✓**	-
B4	✓	-	✓**	-	-
C1	✓	✓	✓**	✓*	-
C2	✓	✓	✓**	✓*	-

* - Tested at two strength levels
** - Tested at three strength levels

TABLE 4 SEP 1570 CLEANNESS RATING

Steel	Inclusion Type	Summated K_o Rating per 1000 mm^2	Total Summated Index, per 1000 mm^2				n for K_n=0
			K_0	K_1	K_3	K_4	
A1	SS OA OG	300 400 30	740	730	650	500	8
A2	SS OA OG	20 40 30	90	70	40	15	5
B1	SS OA OG	150 30 70	250	240	100	15	5
B2	SS OA OG	20 30 50	100	80	20	7	5
B3	SS OA OG	0 0 40	40	20	0	0	3
C1	SS OA OG	80 10 15	105	90	35	15	5
C2	SS OA OG	0 0 50	50	35	15	0	4

TABLE 5 FRACTURE TOUGHNESS AND ENVIRONMENTAL FRACTURE - 'A' SERIES STEELS (0.6% C)

Steel	Product Form	Orient-ation	0.2% PS N/mm^2	K_{IC} $MPa\sqrt{m}$	K_{max} $MPa\sqrt{m}$	Ratio YX:XY K_{IC}	K_1 H_2S $MPa\sqrt{m}$	K_{ISCC} $MPa\sqrt{m}$ (100h)	Ratio K_{ISCC}: K_{max}
A1	Billet	YX		61	61		-	-	-
		XY	1400	87	90	0.70	-	-	-
A2	Billet	YX		69	72		-	-	-
		XY	1430	74	80	0.93	-	-	-
A1	Spring Flat	YX	1420	45	45	0.58	16	15	0.34
		XY		78	84		16	15	0.18
A2	Spring Flat	YX	1420	64	64	0.89	14	15	0.23
		XY		62	73		15	15	0.21

TABLE 6 FRACTURE TOUGHNESS AND ENVIRONMENTAL FRACTURE - 'B' SERIES STEELS (0.4% C)

0.2% PS & Tempering Temp.	Steel	Orien- tation	COD mm	K_Q/K_{IC} MPa$\sqrt{m}$	K_{max} MPa$\sqrt{m}$	Ratio YX:XY COD/K_{IC}	K_1 H_2S MPa$\sqrt{m}$	K_{ISCC} MPa$\sqrt{m}$ (4)	Ratio K_{ISCC} : K_{max}
750 N/mm^2 660 °C (1)	B1	YX XY	0.04 0.24	84 112	93 172	0.17	87 159	73 154	0.78 0.90
	B2	YX XY	0.21 0.25	108 106	166 187	0.84	154 174	130 157	0.78 0.84
	B3	YX XY	0.25 0.27	111 115	174 180	0.93	153 169	148 163	0.85 0.91
1050 N/mm^2 560 °C (2)	B1	YX XY	0.02 0.12	68 130	68 170	0.17	57 58	50 60	0.71 0.35
	B2	YX XY	0.11 0.16	132 144	160 190	0.69	47 64	50 60	0.30 0.32
	B3	YX XY	0.14 0.18	140 151	180 200	0.78	64 67	50 60	0.28 0.30
1550 N/mm^2 250 °C (3)	B1	YX XY	- -	50 53	51 53	0.93	16 16	15 13	0.30 0.25
	B2	YX XY	- -	52 53	53 54	0.98	16 16	15 13	0.28 0.24
	B3	YX XY	- -	73 73	73 74	1.0	16 16	15 13	0.21 0.18

All specimens heat treated in test piece size.

1 & 2 - Dimensions 25 mm x 62.5 mm x 60 mm ⎫ CT specimens
3 - " 18 mm x 32.5 mm x 31 mm ⎭

3 - All results valid K_{IC}

4 - 1000 h test duration for conditions 1 & 2, 100 h for condition 3

Heat Treatment:- OQ 850 °C, Temper as shown

TABLE 7 FRACTURE TOUGHNESS - 'C' SERIES STEELS (0.3% C)

HEAT TREATED IN TEST PIECE SIZE:- 860 °C OQ, TEMPER AS SHOWN

Nominal 0.2% PS and Tempering Temp.	Steel	Orientation	COD mm	COD Ratio YZ:XY
720 N/mm² 650 °C	C1	YZ XY	0.11 0.25	0.4
	C2	YZ XY	0.33 0.35	0.9
870 N/mm² 590 °C	C1	YZ XY	0.05 0.18	0.3
	C2	YZ XY	0.19 0.25	0.8
1050 N/mm² 530 °C	C1	YZ XY	0.04 0.10	0.4
	C2	YZ XY	0.09 0.10	0.9

Bend Test Piece Dimensions: 10 mm x 20 mm x 90 mm

TABLE 8 FATIGUE PROPERTIES (ROTATING BENDING)

Steel	Nominal Tensile Strength (and 0.2% PS) N/mm²	Orientation	LFS N/mm²	Ratio LFS:TS	Ratio of LFS T:L
B1	900 (750)	L T	475 440	0.53 0.48	0.92
B2		L T	510 440	0.55 0.47	0.86
B3		L T	475 475	0.53 0.52	1.0
B1	1150 (1050)	L T	650 540	0.56 0.47	0.83
B2		L T	600 600	0.53 0.53	1.0
B3		L T	590 570	0.50 0.49	0.97
B1	1850 (1550)	L T	700 500	0.38 0.27	0.71
B2		L T	850 625	0.46 0.34	0.73
B3		L T	850 700	0.46 0.38	0.82
C1	850 (750)	L T	435 415	0.51 0.50	0.85
C2		L T	435 435	0.50 0.50	1.0
C1	1130 (1050)	L T	590 540	0.52 0.49	0.92
C2		L T	595 595	0.52 0.52	1.0

L - Longitudinal T - Transverse
LFS - Limiting Fatigue Stress
Heat treated in near test piece size (see Tables 6 & 7 for heat treatment)

TABLE 9 EFFECT OF OVERHEATING

Steel	Property	Reheating Temperature Prior To Hardening & Tempering °C			
		As-Rolled	1150	1250	1350
B1	TS N/mm^2	1150	1180	1180	1180
	0.2% PS N/mm^2	1090	1080	1080	1070
	% El.	17	15	14	14
	% R of A	60	55	48	47
	True fracture strain	0.92	0.8	0.65	0.63
B2	TS N/mm^2	1140	1160	1155	1130
	0.2% PS N/mm^2	1060	1060	1050	1030
	% El.	16	15	13	10
	% R of A	61	53	44	22
	True fracture strain	0.94	0.76	0.58	0.25
B3	TS N/mm^2	1150	1130	1140	1140
	0.2% PS N/mm^2	1060	1020	1020	1020
	% El.	17	17	15	14
	% R of A	65	60	52	44
	True fracture strain	1.05	0.92	0.73	0.58

Heat Treatment:- 850 °C OQ, T 560 °C

TABLE 10 SINGLE POINT TURNING DATA FOR B SERIES STEELS - ANNEALED

Steel	Carbide Tool Wear Data for Cutting Speed of:-				HSS V_{20} m/min	Hardness HV30
	225 m/min		150 m/min			
	FWR µm/km	CDWR µm/km	FWR µm/km	CDWR µm/km		
B1	218	138	142	41	42	202
B2	Failure	Failure	145	50	22	206
B3	214	106	45	22	34	226

FWR - Flank wear rate

CDWR - Crater depth wear rate

TABLE 11 SINGLE POINT TURNING DATA FOR B SERIES STEELS - HARDENED AND TEMPERED

Steel	Carbide Tool Wear Data for Cutting Speed of:-				HSS V_{20} m/min	Hardness HV30
	225 m/min		150 m/min			
	FWR µm/km	CDWR µm/km	FWR µm/km	CDWR µm/km		
B1	175	173	106	62	35	294
B2	263	223	118	66	24	300
B3	259	104	160	46	25	307

FWR - Flank wear rate

CDWR - Crate depth wear rate

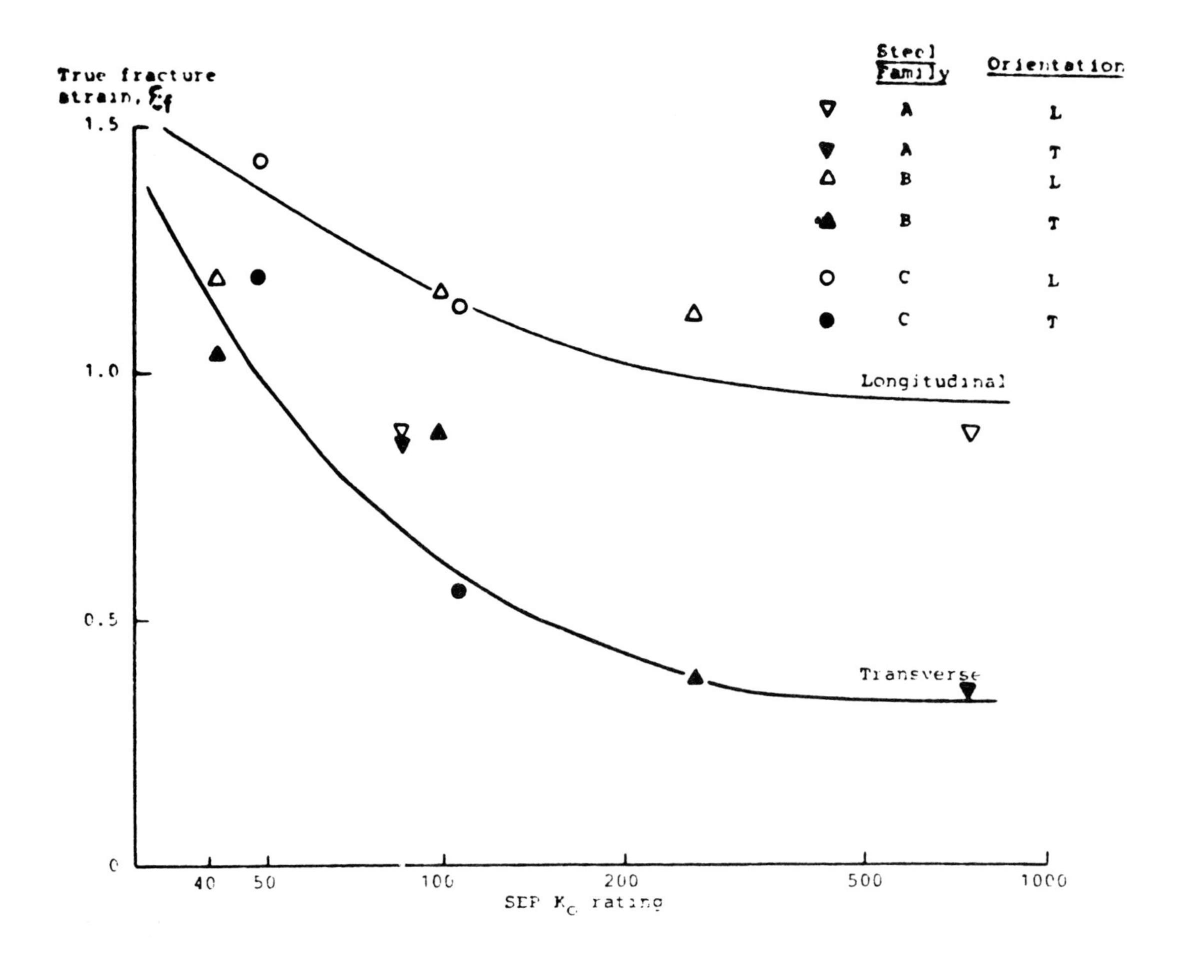

1. Tensile true fracture strain in the longitudinal and transverse orientations as a function of K_o SEP rating for all steels at a common nominal TS of 870N/mm2.

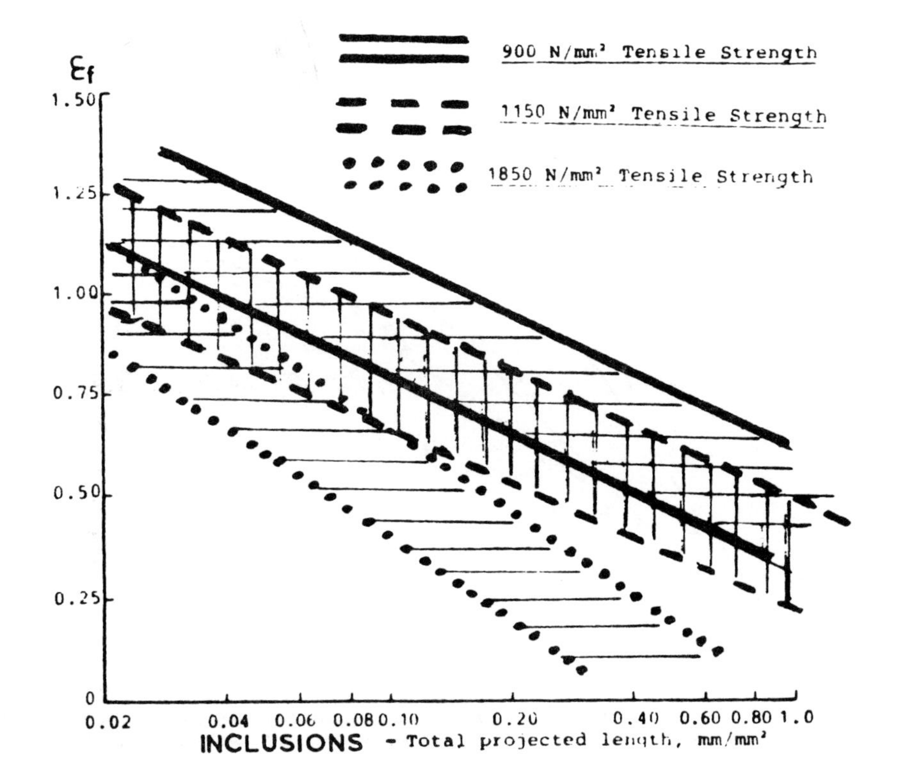

2. Scatterbands for the tensile true fracture strain of the B series (0.4%C) steels at three strength levels as a function of the total inclusion projected length in the fracture plane.

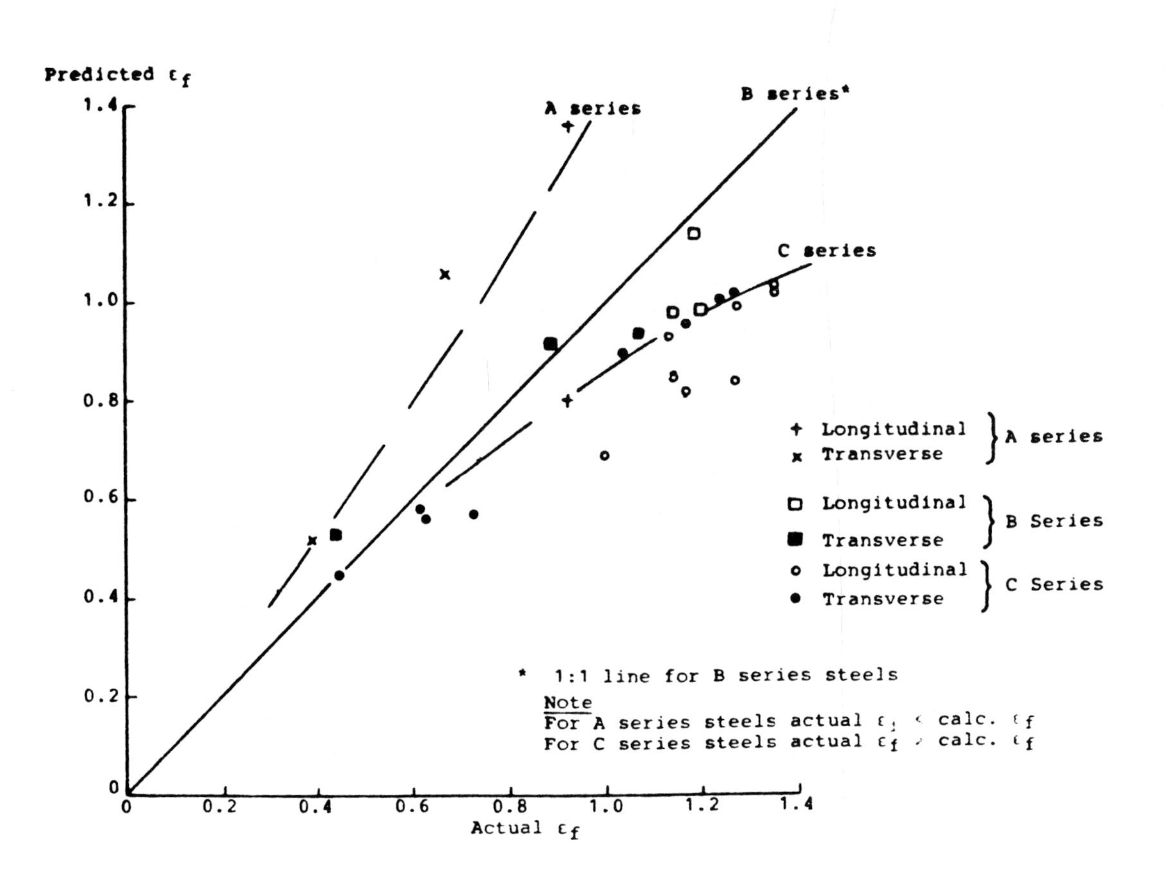

3. The relationship between the observed tensile true fracture strain for the three steel families examined and that calculated for the dependence of fracture strain on inclusion projected length and tensile strength, using the relationship derived for B series steels, Fig 2.

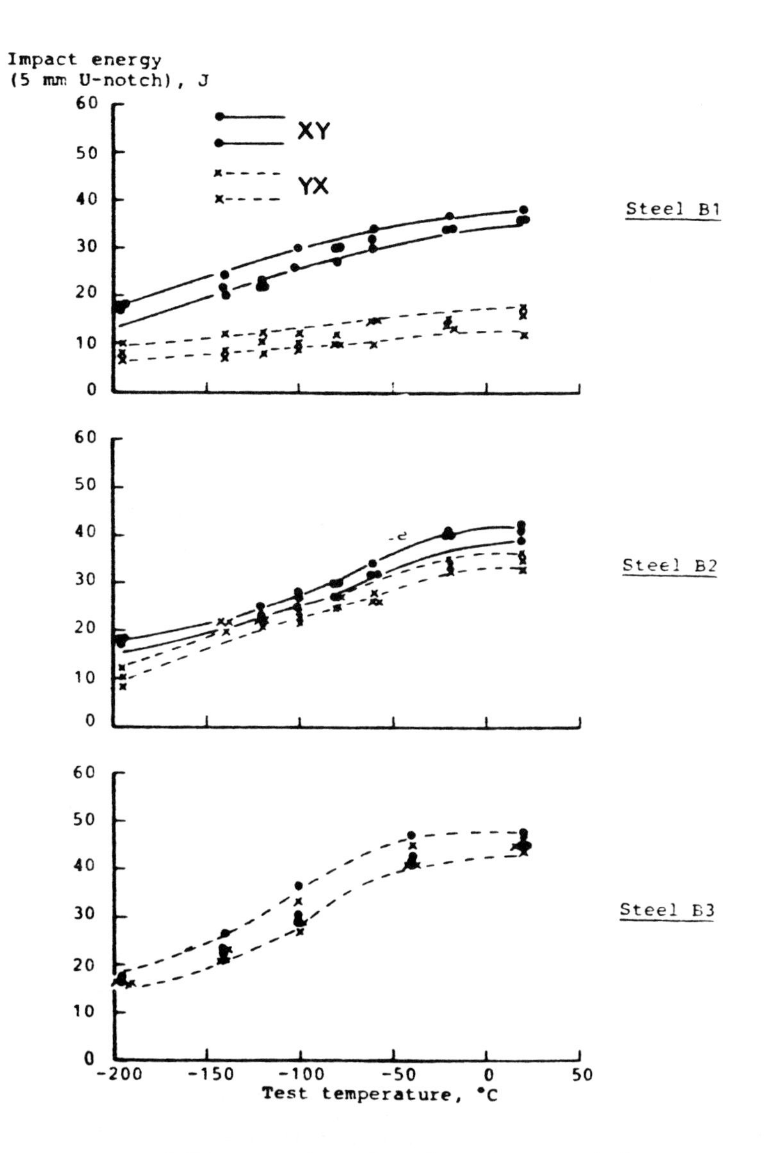

4. Charpy 5mm U notch impact transition curves for the B series steels, at a tensile strength of 1150N/mm2, showing the progressive improvement in transverse energy values in the order B1-B2-B3.

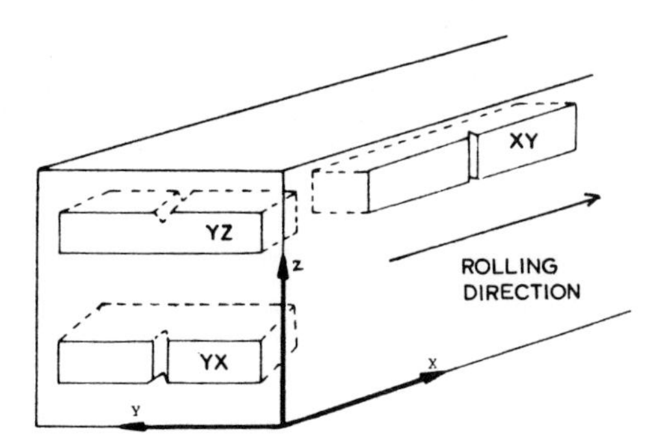

5. Nomenclature for the orientation of impact and fracture toughness testpieces.

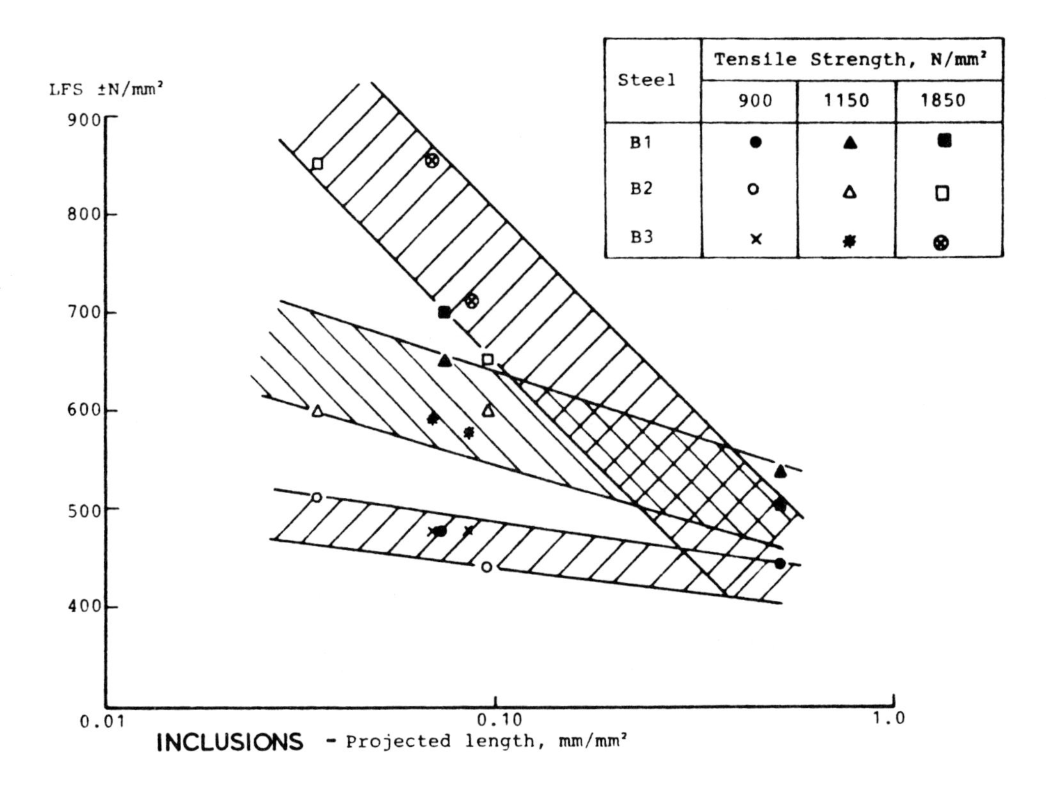

6. The relationship between the Limiting Fatigue Stress (LFS) in smooth, rotating bending tests and total inclusion projected length. Results for two test orientations and three strength levels are shown.

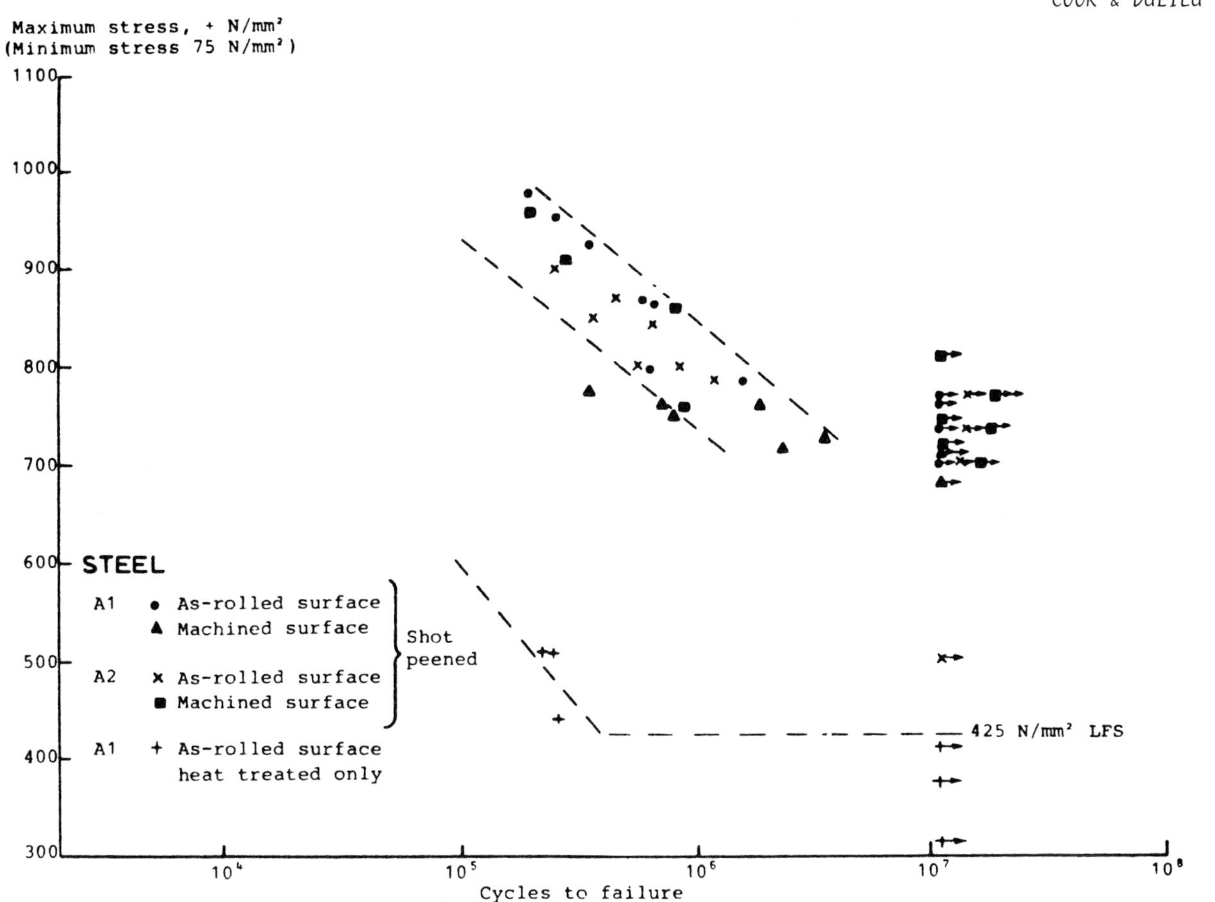

7. Stress-Life relationships for the A series (0.6%C) steels, tested in 3-point bending after shot-peening. (Tests were also conducted on steel A1 without shot-peening, for reference.)

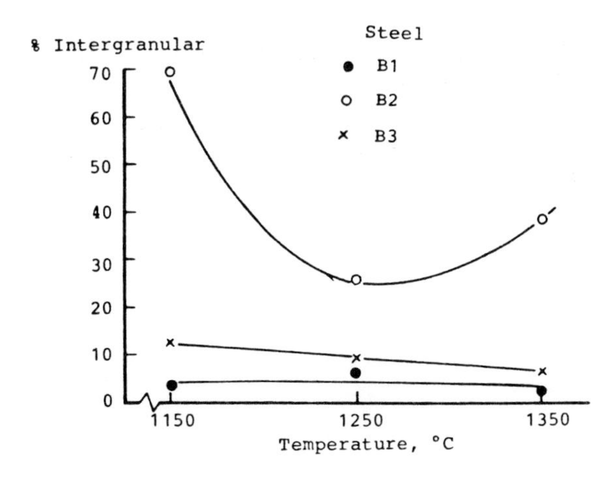

8. The relative overheating susceptibility of the B- series steels. The effect of heating temperature prior to hardening and tempering on the % intergranular fracture observed after slow bend fracture.

Generation of thermal stress and strain during quenching of low carbon steels

A J FLETCHER and B HOBSON

Dr. Fletcher is with the Department of Metals and Materials Engineering, Sheffield City Polytechnic; Miss Hobson is with BSC Swinden Laboratories.

SYNOPSIS

The generation of thermal stress and strain during the direct quenching of low-carbon steels from the forging temperature is compared with the same processes during the quenching of medium carbon steels from the austenitisation temperature, by means of a visco-elastic-plastic mathematical model. The results obtained from a set of four such alloys suggests that direct quenching from 1150°C should not create thermal stresses greater than those produced by a conventional quenching treatment of a medium carbon steel from 850°C. However, the former is likely to produce an increased amount of residual thermal strain.

Symbols

E Young's Modulus;
ℓ Number of time steps to end of quench;
α Coefficient of linear expansion
ε Strain
θ Temperature
σ Stress
ν Poisson's Ratio

Suffices

xx, yy and zz principal directions in plate. The last named is in the transverse direction.
n current element number.
tp transformation plasticity.
t time step reach during quench.

INTRODUCTION

Medium carbon low-alloy steels are quenched from a temperature where the austenite phase is stable to produce a martensitic phase of high strength, which may then be tempered to give the required combination of mechanical properties. Prior to this heat treatment the material will have been heated to a temperature considerably higher than that required for austenitisation, in order to carry out the necessary hot working operations. The austenitisation treatment is an additional expense which might be avoided by a direct quench from the working temperature. This cost reduction is accompanied by a lower toughness and a large austenite grain size[1]. This may be overcome by a reduction in the carbon content of the steel, although this involves a lower tensile strength and poorer hardenability. These latter disadvantages could be overcome by an increase in the alloy content of the material[2][3]. The mechanical properties of this type of material has been subject to considerable study[4][5], but there has been little investigation of the effect of direct quenching on the internal stress generation. That the change from a conventional quench and temper to a direct quench from the hot working temperature would affect the stress generation process is evident for the following reasons:

(i) The initial temperature is higher with a direct quench.
(ii) The low carbon steels possess a small expansion during the formation of martensite.
(iii) The yield strength of the material is reduced by a reduction in carbon content.

Considerable interest has been shown in the development of mathematical models that may be used to calculate the stresses set up during a quench[6-10]. One of these[9], a plane-stress infinite plate model, has been used to compare the generation of thermal stress and strain during the direct quenching of low carbon steels from the hot-working temperature with that produced during the conventional quenching from the normal austenitisation temperature of a medium carbon material. The model used is a visco-elastic-plastic type with transformation plasticity included. This has been the subject of a number of publications during its development and the most recent[11][12], suggest that the predicted residual stresses compare well with those obtained by experiment.

Method of Calculation

The method of calculation was the same as that used in previous publications[9][11][12]. However, the essential stages in the method are repeated here. The geometry of the specimen was considered to be an infinite plate, which allowed the stress and strain to be represented by single parameters, since the principal stresses in the plane of the plate were equal and the transverse principal stress was zero.

Viscosity, elasticity and plasticity were all considered, although the first named was not completely integrated into the calculation, as described below. The stress generation was carried out in a series of small incremental steps. There were at least 600 in number and in some cases more than twice as many were required. Half the plate only was considered on account of the symmetry of the heat flow, which was considered to occur at equal rates through each face. The region between the central plate and one face was divided into ten elements whose plane lay parallel to the face. The temperature was considered to be constant within an element, as were the physical and mechanical properties of the material.

In each step the following procedure was followed:

(a) The stress increment was obtained in each element from the temperature change introduced in the current time interval, assuming conditions of maximum restraint and elasticity. Thus:

$$\Delta\sigma_t^n = \frac{E\alpha}{1-\nu}\Delta\theta_t^n \quad \ldots\ldots \quad n = 1,10 \quad t = 1,\ell$$

(b) These stress increments were added to the existing stresses, which were adjusted uniformly to satisfy the boundary condition of zero net force on the plate section.

(c) The stresses were now used to determine the amount of stress relaxation introduced during the time interval using empirical equations based on experimentally determined data and a standard linear solid model[14]. The same data was used to determine the associated viscous strain increment by the method described previously[11]. The stress relaxation increments were then added to the stresses obtained at the end of stage (b) and these values were subsequently adjusted uniformly to satisfy the zero net force boundary condition. Since this adjustment modified the stresses relevant to the stress relaxation calculation this was repeated using the modified values and the boundary conditions re-established.

(d) The von Mises yield criterion was now applied and in the plastic regions of the plate the stress produced at the end of stage (c) was equated with the appropriate stress on the yield surface. In the plane stress condition this equated with the current flow stress. The application of the yield criterion produced only a first estimate of the position of plastic zones since it was necessary to adjust the elastic stresses to produce compliance with the zero net force condition. The iteration was continued until the stress on each element due to the unbalanced force was <0.1 MPa. The associated plastic strain was obtained from the Prandtl-Reuss equation.

(e) The changes in stress in the elastic region produced by the introduction of plasticity led to a change in strain, in accordance with Hooke's Law.

Although the sequence described above was sufficient in the case of these time intervals where no decomposition of the austenite occurred, the introduction of additional strains was necessary when any part of the specimen underwent transformation. The transformation strain was incorporated into the elastic strain increment by means of a composite coefficient of expansion obtained directly from the relationship between the length of a dilatometer specimen and temperature[10]. In addition, within the 40°C temperature range immediately below M_s there was an additional strain due to transformation plasticity. The incremental strain due to this effect introduced in any element during the current time interval was obtained from empirical data determined previously[13]. Within the temperature range specified above the amount of strain introduced by this process was proportional to $\theta_{M_s} - \theta_{n-1}^m$. Hence:

$$\varepsilon_{tp} = (\varepsilon_{tp})_{total}\frac{(\theta_{M_s} - \theta_{t-1}^n)}{80} \quad \ldots \quad n = 1, 10 \quad t = 1, \ell$$

This equation involves a factor of 0.5 on account of the presence of plane stress conditions.

Experimental Procedure

Four air-induction melted casts, with compositions given in Table 1, were forged and rolled to bar of 12.5 mm diameter. Tubular dilatometer specimens 10 mm x 5 mm were produced from this bar stock. Each specimen was heated at 1100°C for 10 minutes in a dilatometer and the length change in the specimen determined while it cooled in a stream of helium to simulate 20 mm thick plate water quenched. This data (Table 2) was used to determine the coefficients of linear expansion.

Physical and Mechanical Property Data

It was assumed that the physical and mechanical properties of the austenite phase were effectively the same as those obtained previously[10][14][15] from a medium carbon nickel steel 835M30. The exception to this were the coefficients of expansion referred to above. The yield strength of the martensite in the as-quenched condition was determined in a uniaxial tensile test, but the thermal properties of this constituent were again assumed to be the same as those of the same phase in 835M30[14]. The proportion of martensite to austenite was obtained from the dilatometer curve, and the yield strength and the contribution of each consistuent to the strength of a mixture of the two phases was assumed to be linear.

Results

All four sets of calculations showed similar relationships between stress and strain at specific positions in the plate. Therefore, these are only described once (alloy of cast number 1027), see Figure 1. However, the more significant events are indicated by letters in this figure, and the characteristics of these events in the case of the other calculations are shown in Table 3. Figure 1 shows that as the temperature in the surface of the plate began to fall relative to the centre a tensile stress was set up near the surface and a compressive stress was generated at the centre. The temperature at the start of the quench was 1100°C and the yield strength of the material was, in consequence, very low in the early part of the cooling process. Hence, plastic flow was extensive at both surface and centre and the rise in the thermal stress was restricted by the low yield stress. The surface temperature was still falling more rapidly than the centre when the M_s temperature was reached at the surface. The expansion produced by the formation of martensite caused the unloading of the stress at both surface and centre (points 'a' and 'd' respectively in Figure 1). This involved an approximately linear relationship between stress and strain, the small departure from linearity being due to viscous flow and transformation plasticity. The high M_s temperature and the low strain associated with the martensitic reaction caused transformation plasticity to be limited to positive strain.

The surface stress eventually became compressive, under the influence of the martensite transformation. However, the transformation front gradually moved towards the centre of the plate as it cooled further and caused the associated expansion to generate compressive stress at progressively greater depths below the surface. Eventually, the compressive stresses already generated at points behind the transformation front unloaded, to maintain net zero force in the section (point 'b' in Figure 1) while the initiation of the martensite transformation at the centre caused the unloading of the tensile stress, indicated by point 'e' in Figure 1. Further reversals in the direction of loading occurred at both positions (points 'c' and 'f' in Figure 1) once the transformation had been completed throughout the section.

At intermediate positions the relationships between stress and strain followed a similar sequence to that described above. However, the surface region was unusual in that no tensile stress developed after the martensite transformation began. At depths below 1 mm the tensile stresses were produced at some stage while the temperature was below M_s, even though a compressive stress was produced at the end of the quench at depths of less than 3 mm.

The variation in the most significant steps in the stress and strain generation processes produced by the changes in alloy composition are shown in Table 3. The temperature at point 'a' varied between 568°C and 500°C and the associated tensile strain (i.e. the maximum tensile strain at the surface) varied between 0.75% and 0.64%. The temperature in each case coincided with the M_s point of the alloy. There was, however, no direct correlation between the amount of strain at 'a' and the M_s temperature: the absence of such a correlation was probably due to the variations in the coefficient of expansion, which were also produced by the changes in alloy composition. The importance of this change in the coefficient of expansion is shown by point 'd' which corresponds, at the centre, to point 'a' at the surface. Thus the maximum compressive strain occurred at the centre at higher temperatures, where the relative value of the coefficients of expansion of the various alloys were rather different from those relevant to point 'a'. The maximum strains at 'b' occurred in alloy 1027, whereas the maximum strain at 'a' occurred in alloy 1028.

The next stage was the reversal of the stresses as a consequence of the transformation to martensite. It was terminated at the surface by the progress of the transformation front towards the centre of the specimen, (point 'b') This was associated with a substantial negative stress, which varied significantly from one composition to another. There was no clear correlation between this stress and the coefficient of expansion, because the transformation plasticity that occurred during the early part of this stage also had an effect on this stress.

The negative stresses introduced at the surface during transformation were sufficient to cause a second period of plastic flow at the hotter centre, so the stresses shown at point 'e' (Table 3) reflected the current flow stresses at the centre. The positive strain component introduced at the centre while the surface was transforming varied between 0.96% and 0.79% with the largest values associated with alloy No.1030.

The subsequent transformation of the centre of the plate to martensite caused the unloading of the tensile stress present there at this stage in the quench (point 'e' in Table 3). After points 'b' and 'e' in Table 3 had been passed the surface and centre respectively were subject to unloading, followed by a stress reversal at the centre (see point 'f' in Table 3). At the surface this stress reversal occurred in only one case (alloy 1030, point 'c' in Table 3). Thus it was alloy 1030 (the material that produced the highest compressive stress at point 'b'), which produced the greatest change in stress during subsequent unloading.

The variations in residual stress across the plate is shown in Figure 2. In all cases the stress at the surface was compressive and possessed values in the range -140 MPa to -280 MPa. As the distance below the surface increased the stress became less compressive, until it reversed at a point which varied with the composition of the alloy. Thus, the width of the surface compressive zone varied, in a 20 mm thick plate, from 1.5 mm to 2.5 mm; it was thickest in the case of the steel of lowest carbon content (1027) and smallest in the case of alloy 1030, which contained both titanium and vanadium. Below the point where this reversal occurred the stress rose rapidly to a maximum tensile value, before falling as the centre of the plate was approached. In the case of one steel (1030), the reduction in stress was sufficient to cause a second stress reversal, with a second compressive zone at the centre. The maximum tensile stress, which occurred at approximately a point mid-way between surface and centre, was never very high.

In all four specimens the residual strains were tensile at all points, with the maximum values always at the surface (Figure 3). All the residual strains were more tensile than that produced at the same position in an identical specimen of a medium carbon steel subject to a conventional water quench from 850°C. Amongst the specimens quenched from 1150°C, the steel with no titanium or vanadium additions and with the lowest carbon content (1027) produced the greatest residual strains, while the steel with the lowest residual strains possessed the high carbon content and the titanium addition. This situation, reflects the amount of strain introduced in the early stages of the quench where plastic flow was present (see points 'a' and 'e' in Table 3).

DISCUSSION

The presence of a zone of compressive residual stress in the vicinity of the plate faces is an advantage with respect to the prevention of cracks at the surface. The direct quenched steel appear to have an advantage over higher carbon material subject to a conventional austenitisation at 850°C, since the zone of compressive stress at the face is in the latter case very narrow, if it exists at all. The alloy that contained no additions of titanium and vanadium but possessed the lowest carbon content (1027) possessed the thickest surface compressive zone at the end of the quench. However, it is not certain that the differences between these zones in the four directly quenched alloys are significant.

The maximum strains introduced during cooling were, in all cases greater than that produced during the quench of a medium carbon steel from 850°C. This is reflected in the greater amount of residual strain produced by direct quenching, in comparison with that associated with the latter. Nevertheless, this is not a major disadvantage to the use of direct quenching. There were small variations in the amount of residual strain at a particular point in the directly quenched alloys, which was not directly correlated with a single parameter, although the M_s temperature, the transformation plasticity and the coefficient of expansion all affected these strains.

Although the early stages of the martensite transformation generated a compressive stress at the point in question, a tensile stress was subsequently produced at most points as the transformation front moved on. The level of this stress in the specimen under consideration was always relatively low, although in specimens with other geometries, stress raisers and greater temperature gradients could initiate cracks at this stage. None of the alloys considered compared unfavourably in this respect with medium carbon material subject to a quench from 850°C.

The very high M_s temperature in the low-carbon alloys led to an interesting interaction between strain generation and the transformation. Hence, approximately equal amounts of transformation plasticity occurred before and after the stress reversal at the surface. Consequently the strains were not displaced to significantly lower values, as would be the case in a higher carbon material, where most of the transformation plasticity occurred after the stress became compressive.

CONCLUSIONS

1. The analysis of the stress and strain generated during direct quenching of low-carbon steels from the forging temperature suggests that such a procedure should not lead to severe difficulties as far as fracture and distortion is concerned. There will, nevertheless, be greater distortion than would be produced by a quench of a medium carbon steel from 850°C.
2. The relationships between stress and strain during direct quenching of each of the low carbon materials showed that they were associated with differences in the M_s temperature and the coefficients of expansion of the various alloys. However, these differences are not of major significance.

ACKNOWLEDGEMENTS

The authors wish to thank the British Steel Corporation and Sheffield City Polytechnic for provision of research facilities and Dr R Baker of BSC for permission to publish this paper.

REFERENCES

1. R G Williams, W T Cook and D Dulieu, Heat Treatment '84, p27.
2. OVAKO Technical Information. Imacro high-strength weldable hardened and tempered steel.
3. OVAKO Technical Information. Imacro HY weldable hardened and tempered steel.
4. A Brownrigg and G G Brown, Materiaux et Techniques, Dec.1977, p61.
5. H Martenson. Scan.J.Met. 1972, v1, p319.
6. I Inoue, S Nagaki, T Kishino, M Monkawa, Ing-Arch, 1981, v50, p315.
7. S Sjostrom, Materials Science and Technology, 1985, v1, p823.
8. S Denis, A Simon and G Beck, Harterei - Tech.Mittelungen, 1982, v37, p18.
9. A J Fletcher and R F Price, Metals Technology, 1981, v8, p427.
10. J A Burnett, Mat.Sci.& Tech., 1985, v1, p863.
11. F Abbasi and A J Fletcher, Mat.Sci.& Tech., 1985, v1, p770.
12. A J Fletcher and A B Soomro. Materials Science and Engineering.
13. F Abbasi and A F Jletcher, Mat.Sci.& Tech., 1985, v1, p830.
14. R F Price and A J Fletcher Met.Tech., 1980, v7, p203.
15. A J Fletcher and A B Soomro. To be published, Materials Science and Engineering.

Table 1 The Compositions of the Alloys

	C	Si	Mn	Mo	Al	B	Ti	V
1027	0.08	0.22	1.90	0.06	0.032	0.0003	--	--
1028	0.12	0.26	2.01	0.05	0.039	0.0003	--	--
1029	0.08	0.27	2.01	0.05	0.036	0.0026	0.044	--
1030	0.08	0.21	1.95	0.06	0.033	0.0023	0.042	0.056

Cr: 0-11, Ni: 0.09 in each case
Bal: Iron

Table 2 The Effect of Temperature on the Dialtometric Properties of the Alloys

Cast No	Coefficient of Expansion in Austenite	M_s temperature	Coefficient of Expansion in Martensite	M_f temperature
1027	$1.8x10^{-5}$ K^{-1} >900°C $2.7x10^{-5}$ K^{-1} <900°C	531°C	$1.5x10^{-5}$ K^{-1}	179°C
1028	$1.7x10^{-5}$ K^{-1} >900°C $2.5x10^{-5}$ K^{-1} <900°C	502°C	$1.6x10^{-5}$ K^{-1}	180°C
1029	$2.3x10^{-5}$ K^{-1} >900°C $2.1x10^{-5}$ K^{-1} <900°C	509°C	$1.7x10^{-5}$ K^{-1}	200°C
1030	$2.0x10^{-5}$ K^{-1} >900°C $2.4x10^{-5}$ K^{-1} <900°C	505°C	$1.1x10^{-5}$ K^{-1}	315°C

Between M_s and M_f Coefficient of Expansion was zero in all cases

Table 3 The Relationships between Stress, Strain and Temperature at the Surface and Centre of Quenched Plates of the Various Alloys

Cast No	Point	a	b	c	d	e	f
1027	Temperature °C	568	314	137	968	570	180
	Temperature Difference between surface and centre °C	438					
	Stress MPa		-340	-25	-28	139	-138
	Strain %	0.74	0.58	0.68	-0.54	0.37	0.33
1028	Temperature °C	500	316	113	998	506	149
	Temperature Difference between surface and centre °C	499					
	Stress MPa		-492	-179	-27	161	-31
	Strain %	0.75	0.56	0.60	-0.56	0.24	0.25
1029	Temperature °C	532	296	135	973	530	196
	Temperature Difference between surface and centre °C	470					
	Stress MPa		-416	-76	-30	149	-100
	Strain %	0.64	0.42	0.53	-0.59	0.20	0.16
1030	Temperature °C	532	296	135	973	530	310
	Temperature Difference between surface and centre °C	475					
	Stress MPa		-572	101	-31	153	184
	Strain %	0.72	0.43	0.65	-0.64	0.31	0.28

Letters refer to points indicated in Figure 1

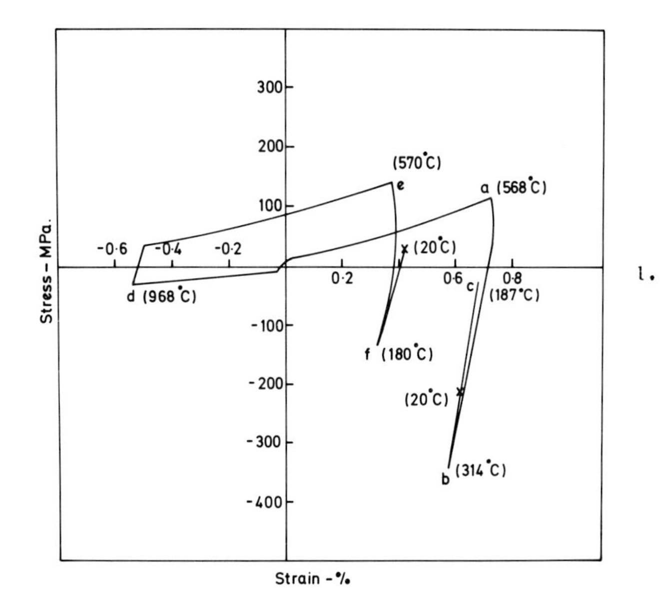

1. The Relationships between Stress and Strain at the Surface and Centre of a Plate of Alloy 1027.

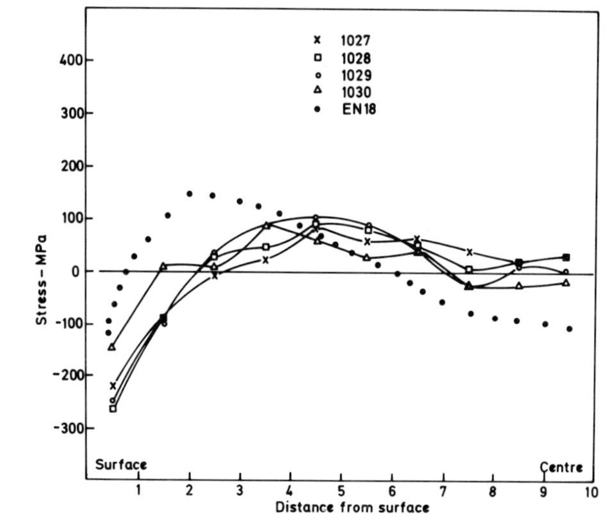

2. The Distribution of Residual Stresses at the End of the Quench.

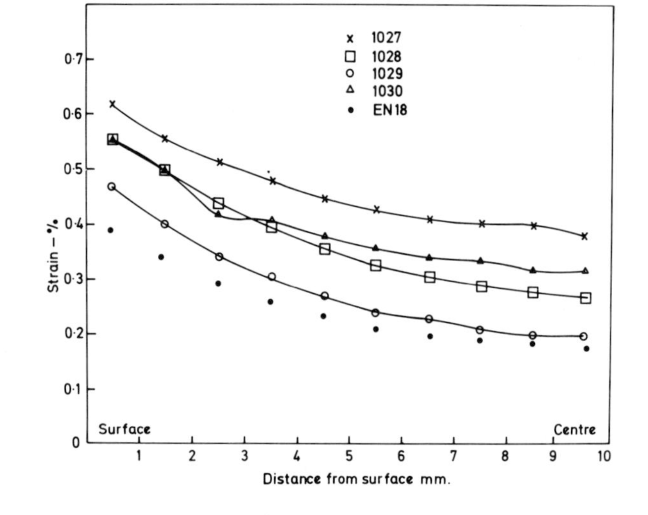

3. The Distribution of Residual Strains at the end of the Quench.

Effect of heat treatment atmospheres on structure and properties of some nickel base alloys

R P H FLEMING

The author is in the Department of Metallurgy and Materials Engineering, City of London Polytechnic.

ABSTRACT

This paper is mainly concerned with the preparation, structure and properties of some binary, ternary, and more complex nickel base alloys, and their heat treatment in controlled atmospheres as compared with their heat treatment in vacuum. The alloys were prepared using three grades of pure fine carbonyl nickel powder combined with the refractory metals Ta, W, Mo and Nb to give solid solution strengthening, and with the addition of stable carbides, such as TaC, NbC and TiC to give dispersion strengthening and sometimes precipitation hardening. The alloys were prepared using standard P/M techniques of mixing, die or isostatic compaction, followed by heat treatment/ sintering at temperatures in the range of 1250–1450°C for periods of 1–16 h or longer. Some alloys were subjected to hot impact extrusion at 1150°C or hot isostatic pressing (HIP) at 1330°C and at 7.5 tonf in^{-2} in order to increase the density (i.e. to theoretical density) and improve the metallographic structure and properties of the alloys. All the alloys were tested for hardness, density, UTS, %El, %RA, hot hardness, stress-rupture life, and oxidation resistance at 1150°C in air. It was demonstrated that, in general, vacuum heat treatment is superior to heat treatment in argon, cracked ammonia or other N_2/H_2 mixtures. This is shown by superior mechanical properties of some alloys heat treated in vacuum as compared with their heat treatment in other controlled atmospheres. Some interesting metallographic structures are illustrated in the paper.

INTRODUCTION

The history of the development of Nimonics has been adequately recorded by Betteridge[1] and, more recently, that of the P/M nickel base alloys has been dealt with by Gessinger[2] and others in this field.[3-25] The Nimonics are based on the Ni–Cr, 80-20 alloy, called Nichrome, to which were added suitable amounts of Ti and Al to give precipitation hardening.

These alloys were used at relatively low temperatures, 650–750°C. More advanced alloys were further strengthened by additions of the refractory metals Ta, W, Mo, and Nb to give solid solution strengthening of the matrix plus precipitation hardening.

These alloys are cast and forged to shape, but are limited in use to about 980°C as this is the temperature at which the γ' precipitate goes back into solution, thus making these alloys unstable at temperatures above about 980°C. One solution to this problem is to use powder metallurgy techniques to produce alloys of high purity, high density, fine grain size and superior mechanical properties.

Powder metallurgy also lends itself to the development of materials that cannot be produced by conventional metallurgy. This investigation was designed to include a number of special design parameters, such as dispersion strengthening using stable oxides, carbides, nitrides, borides, etc. in a nickel base solid solution to give improved high temperature creep resistant alloys that can perform above 1000°C and having good built-in oxidation resistance.

All these objectives have been achieved in some of the alloys that were developed, and the results show that alloys having good oxidation resistance combined with good creep resistance at temperatures up to 1150°C for over 100 h in air. Finally, the use of hot isostatic pressing for some of these alloys has shown that nickel base alloys produced by powder metallurgy techniques can have theoretical density combined with maximum mechanical properties.

SOME DESIGN CONSIDERATIONS

When new nickel base alloys are being developed it is likely that various aspects of design will be considered and included, as below:

1) Solid solution strengthening
2) Dispersion strengthening
3) Precipitation hardening
4) Solid phase sintering
5) Liquid phase sintering
6) Vacuum sintering
7) Heat treatment
8) Hot working (hot impact extrusion)
9) High temperature brazing
10) Microstructural

It should be noted that in designing new alloys some, if not all, of the above design parameters should be included in the development programme. In the present work the author has reviewed all of the design aspects mentioned above, and has also published a series of papers[3-25] which deal specifically with each parameter. In this paper the author can attempt a brief summary only of his work over the past forty years.

VACUUM HEAT TREATMENT AND INFLUENCE OF SINTERING ATMOSPHERE

The green metal compact usually contains a large number of voids, the total number and volume of which are determined by the powder characteristics (i.e. particle size, shape, and size distribution), the plastic deformation of the particles, and the compacting pressure. During sintering the total pore volume decreases, the compact shrinks, and densification occurs.

It is known that sintering affects the porosity in the following ways: (a) the total number of pores decreases progressively during sintering at all temperatures; (b) the average pore diameter existing in the compact increases with the sintering time; (c) for each condition of sintering there is a pore size which occurs in maximum number in the size scale, and this maximum shifts towards larger pore size as sintering proceeds; (d) late in the sintering process there exist pores which are larger than any present at the start of sintering.

These facts are independent of whether sintering occurs in a reducing or neutral atmosphere, or in a vacuum. The effect of vacuum sintering vs sintering in a reducing atmosphere for a single-component metal cannot be described in a simple manner as it depends on the surface condition of powder particles. All that can be stated is that there is a considerable variation in the activation energy which determines the rate of sintering, and this depends strongly on the sintering atmosphere.

However, it may also be stated that: (a) sintering in vacuum results in minimum porosity; (b) the vacuum-sintered compacts have the minimum number of fine pores, and the maximum number of coarse pores; whereas in the hydrogen-sintered compacts the maximum number of pores lies in the medium size pore range. It is also known that the pore size distribution determines the grain structure of the sintered compact, and it may therefore be concluded that the sintering atmosphere directly affects the final grain structure of the sintered compact.

A small grain size gives higher strength than a coarse grain size (i.e. Petch relationship). It is also known that hydrogen has a serious embrittling effect in some metals and alloys, for example, in iron base alloys (i.e. steels), tungsten base alloys (i.e. heavy alloys), and aluminium base alloys. On the other hand, vacuum heat treatment has a number of beneficial effects on the mechanical properties of these alloys, namely, the level of hydrogen is reduced or eliminated and this permits further densification of the alloys, particularly in the presence of a liquid phase leading to 100% theoretical density. This in turn results in a much higher level of strength and ductility for vacuum sintered materials as compared with hydrogen sintered alloys.

EXPERIMENTAL

A series of binary alloys of pure Ni, mainly type 123 (supplied by Inco Ltd), with appropriate additions of pure fine refractory metals powders (supplied by Murex Ltd) namely W, Ta, Mo and Nb, up to 50 wt.%. These alloys were made by mixing in a small stainless steel ball mill for a standard period of 1 h. The alloys were compacted in latex sleeves by cold isostatic pressing (CIP) at pressures of 10–20 tonf in^{-2}.

Binary, ternary, and more complex alloys were sintered in vacuum and in other controlled atmospheres, such as argon and N_2/H_2 mixtures. Ternary alloys were prepared using Nichrome 80-20 powder with appropriate additions of the refractory metals and also stable oxides (i.e. Al_2O_3, ZrO_2, MgO, etc.), borides, nitrides and carbides (i.e. TaC, TiC, and NbC), both singly and in combination.

Heat treatment temperatures in the range 1250–1450°C were used for both vacuum and other atmospheres. Alloys containing residual porosity were subjected to hot isostatic pressing (HIP) at 1330°C and 7.5 tonf in^{-2}. All the alloys were tested for hardness and density, and each alloy structure was examined using the optical microscope and recorded using photomicrographs. All alloys were also tested for UTS, %El, %RA, hot hardness, and creep-rupture life. Some alloys were also tested for oxidation resistance at 1150°C in air, these alloys included Ni–Cr–Ta and W–Mo–Cr–Ni–TiC alloys.

RESULTS

The results of this investigation are presented in the form of a series of graphs and photomicrographs of the observed structures. The author wishes to note that he has published a series of papers over the past twenty five years which include many aspects of the development of nickel base alloys for high temperature use,[3-25] these papers include the role of metallography, vacuum sintering, hot working, heat treatment, dispersion, and solid solution strengthening aspects.

The results of this investigation are presented as graphs in Figs. 1–8 and give some data for some binary, ternary, and more complex alloys. Fig. 1 shows the beneficial effect of vacuum heat treatment on Ni–W alloys. Fig. 3 shows the results for vacuum vs N_2/H_2 for Ni–Mo and more complex alloys. Fig. 5 shows results for similar alloys. Figs. 2, 4 and 6 show the correlation between UTS and hardness. Figs. 7 and 8 show some results for the oxidation of Ni–Cr–Ta and Ni–Cr–W–Mo–TiC alloys at 1150°C in air. Figs. 9-16 are typical microphotographs of some of the structures observed in the investigation.

DISCUSSION OF RESULTS

The results presented in this paper show quite clearly that vacuum heat treatment of binary, ternary, and more complex nickel base alloys produced by powder metallurgy techniques produces alloys with improved mechanical properties and also improved density and metallographic structures.

The advantages of heat treating binary Ni–W alloys in vacuum as compared with the use of N_2/H_2 atmospheres are clearly shown in Fig. 1. Similar results were obtained for Ni–Mo and some more complex Ni–Cr–Mo–TaC alloys shown in Figs. 3 and 5. Hardness–UTS correlations are shown in Figs. 2, 4, and 6 respectively. Oxidation resistance curves are presented in Figs. 7 and 8 and clearly demonstrate that these alloys, namely, Ni–Cr–Ta and Ni–Cr–W–Mo–TiC are extremely oxidation resistant at 1150°C in air. Typical microphotographs of the alloys are presented in Figs. 9-16.

CONCLUSIONS

It is clear from the results presented in this paper that a number of conclusions may be drawn from the work, namely:

1) The results presented in Figs. 1 and 2 clearly show that the hardness and strength of Ni–W alloys are considerably increased by sintering in vacuum as compared with the same alloys sintered in nitrogen–hydrogen mixtures. Looked at another way, alloys heated treated in vacuum are less brittle and more ductile. Thus alloys heat treated in nitrogen–hydrogen mixtures are brittle and of lower strength and ductility. Generalising this statement it is clear that nitrogen and hydrogen embrittle metals and alloys during heat treatment; especially Fe and steel, W and heavy alloys, Al and Al alloys, and nickel base alloys.

2) Hardness tests made on the Vickers machine are very reliable and consistent (i.e. using a diamond pyramid indenter). Other mechanical tests such as UTS, %El, %RA, and Charpy are extremely variable due to a wide range of defects such as segregation, porosity (i.e. micro and macro), cracks, surface irregularity, foreign particles, nitrides, oxides, slag, sand, etc. All of these defects are common in all metals and alloys and tend to reduce their strength. Charpy and fracture toughness testing (i.e. K_{1C}) are based on machining a notch (i.e. V, round, key-hole, etc.) in a cylindrical specimen. The notchings supplied by Charpy are quite inadequate for the job as the tool quickly becomes blunted and thereafter produces non-standard notches of variable depth, root radius, etc., not to mention temperature variations during testing. Thus it is clear that fracture toughness testing is based on extremely variable data, not to say unreliable results, therefore is there any point in all the mathematical calculations for estimating K_{1C}?

3) It is also clear that nickel base alloys can be substantially strengthened by additions of the refractory metals (i.e. Ta, W, Nb, and Mo) to give solid solution strengthening and stable carbides such as TaC, TiC, and NbC to give dispersion strengthening. In particular, the Ni–Cr–Mo–TaC alloys are very hard, strong, tough and oxidation resistant.

4) Finally it should be noted that the author has produced a wide range of new alloys including nickel base alloys, stainless steels, high speed steels, and sintered carbides. Many of these alloys have been designed to have oxidation resistance up to 1150°C in air, combined with high strength and good stress-rupture properties.

These observations are based on forty years experience in the field of metallurgical engineering, and scientific research.

ACKNOWLEDGEMENTS

The author wishes to thank his colleagues at the City of London Polytechnic for their continued help and encouragement in his research work.

REFERENCES

1 'The nimonic alloys', (ed. W Betteridge and J Heslop), 1974, London, Edward Arnold
2 G H Gessinger: 'Powder metallurgy of the superalloys', 1984, London, Butterworths
3 R P H Fleming: PhD thesis, 1971, University of London
4 R P H Fleming: Powder Metall., 1963, 12, 179
5 R P H Fleming: 'Modern developments in powder metallurgy', 1977, 11, 13
6 R P H Fleming: Planseeber. Pulvermetall., 1977, 25, 214
7 R P H Fleming: 'European symposium on powder metallurgy', 1978, Stockholm, p216, Jernkontoret
8 R P H Fleming: International powder metall. conf., Met. Soc., 1978, Coventry
9 R P H Fleming: 1st Int. Isostatic Pressing conf., 1978, Loughborough University
10 R P H Fleming: in 'Heat Treatment '79', 1980, London, Met. Soc., p147
11 R P H Fleming: in 'Hot Working and Forming Processes', 1980, London, Met. Soc., p254
12 R P H Fleming: Corrosion seminar, 1979, Bath University
13 R P H Fleming: in Proc. BABS/MS Conf., 1979, London
14 R P H Fleming: Modern developments in powder metallurgy, Vol 12, 1981, NJ., MPIF/APMI, p439
15 R P H Fleming: 'Microstructural Science', Vol 9, 1981, NY, Elsevier, p231
16 R P H Fleming: Planseeber. Pulvermetall., 1978, 26, 252
17 ibid., 14
18 ibid., 1980, 28, 55
19 ibid., 74
20 R P H Fleming: Proc. Conf. on insitu composites, Boston, 1981
21 R P H Fleming: in 'Trends in refractory metals, hard metals and special materials and their technology', Vol 2, 10th Plansee Seminar, 1981, Austria, Metallwerk, p465
22 R P H Fleming: in 'Mechanical behaviour and nuclear applications of stainless steel at elevated temperatures', 1982, London, Met. Soc., p15
23 R P H Fleming: Powder metallurgy in Europe, International Powder Metallurgy Conf., 1983, Milan, Assoc. Italiana di Metallurgia
24 R P H Fleming: in 'Towards improved performance of tool materials', 1982, London, Met. Soc., p107
25 R P H Fleming: 11th International Plansee Seminar '85, Vol 3, 1985, Reutte, Tyrol, Austria, 407

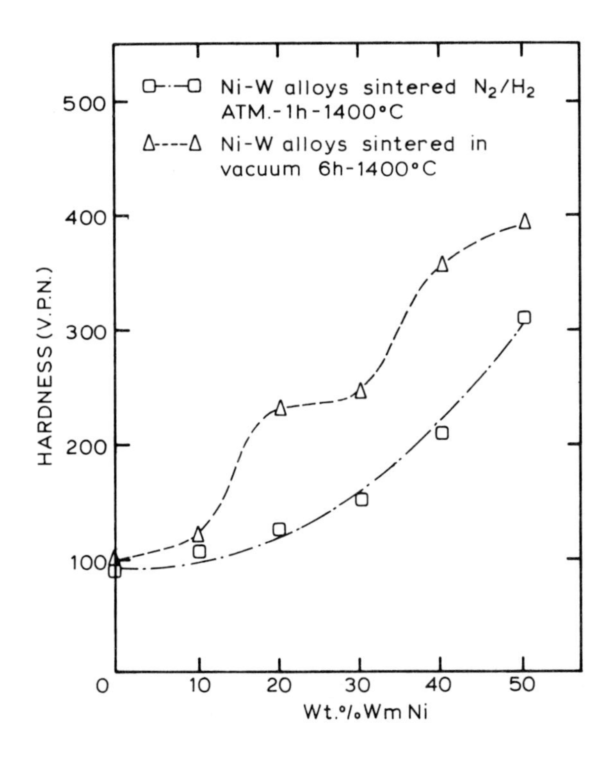

Fig. 1 Hardness vs composition Ni—W alloys

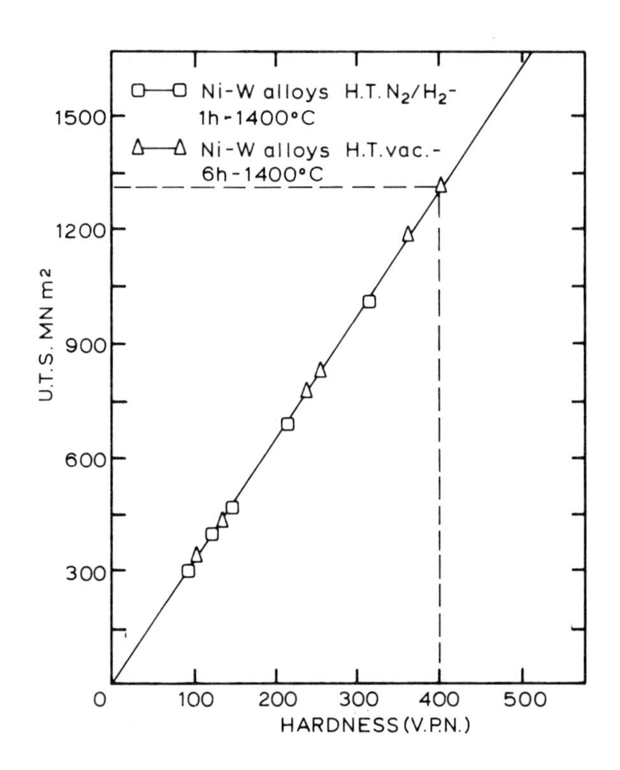

Fig. 2 UTS vs hardness for Ni—W alloys

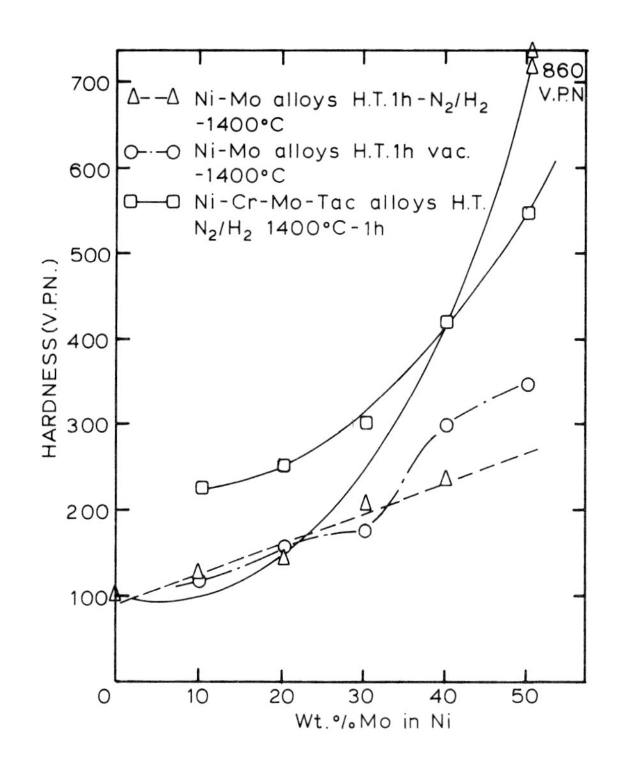

Fig. 3 Hardness vs composition Ni—Mo alloys

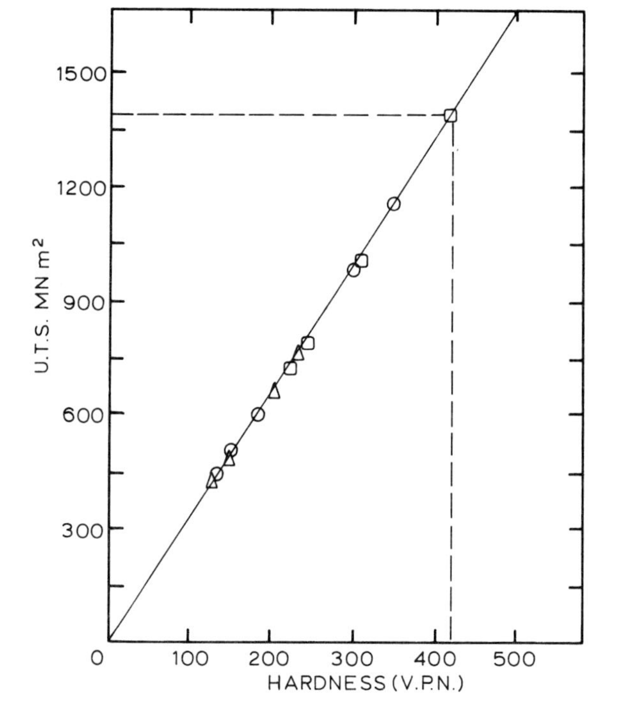

Fig. 4 UTS vs hardness of Ni—Mo alloys

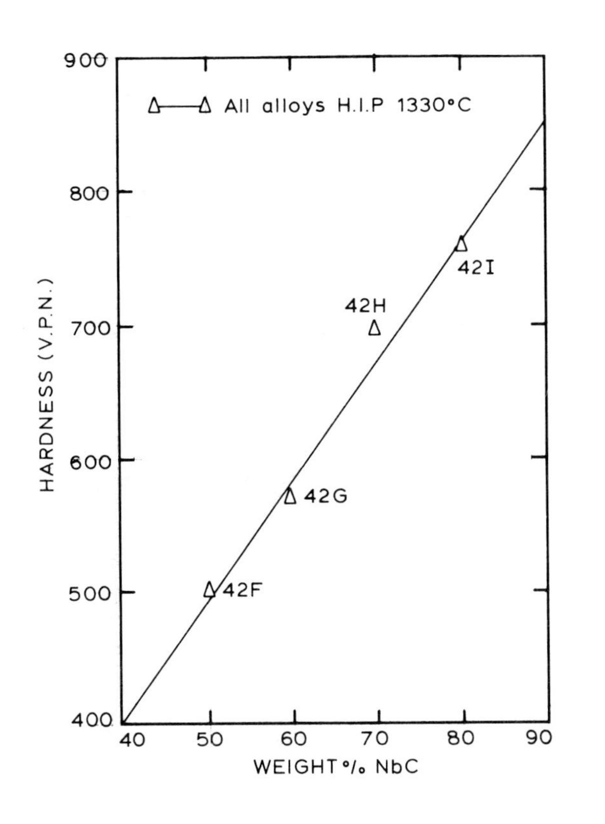

Fig. 5 Hardness vs composition NbC—Ni alloys

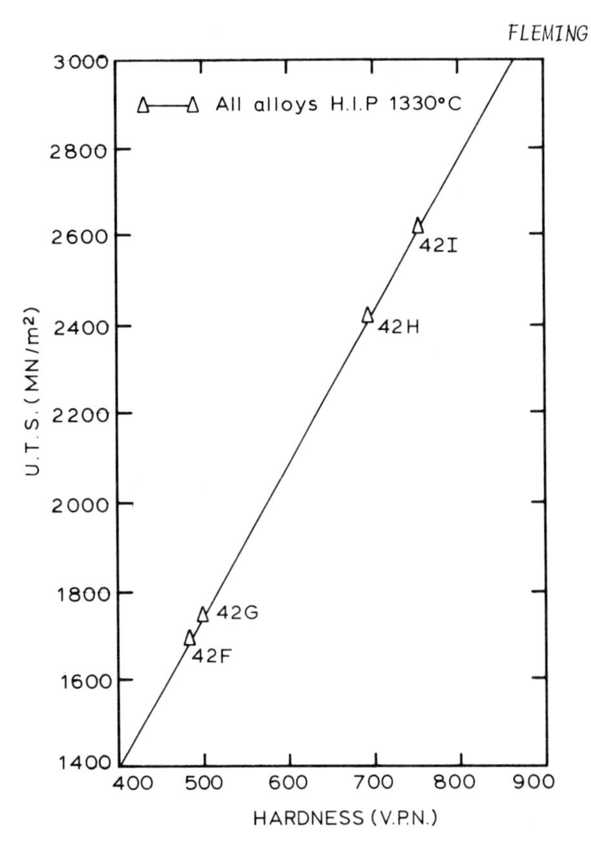

Fig. 6 UTS vs hardness for NbC—Ni alloys

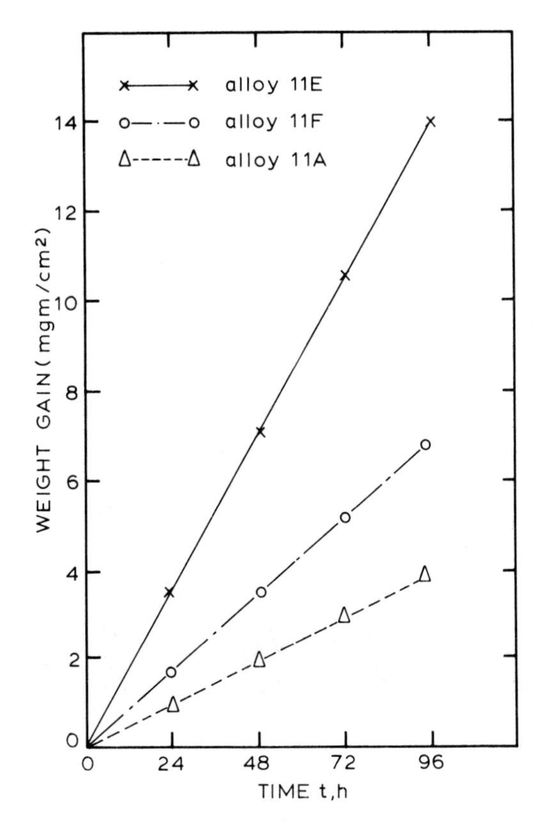

Fig. 7 Oxidation of Ni—Cr—Mo—W—TiC alloys at 1150°C in air

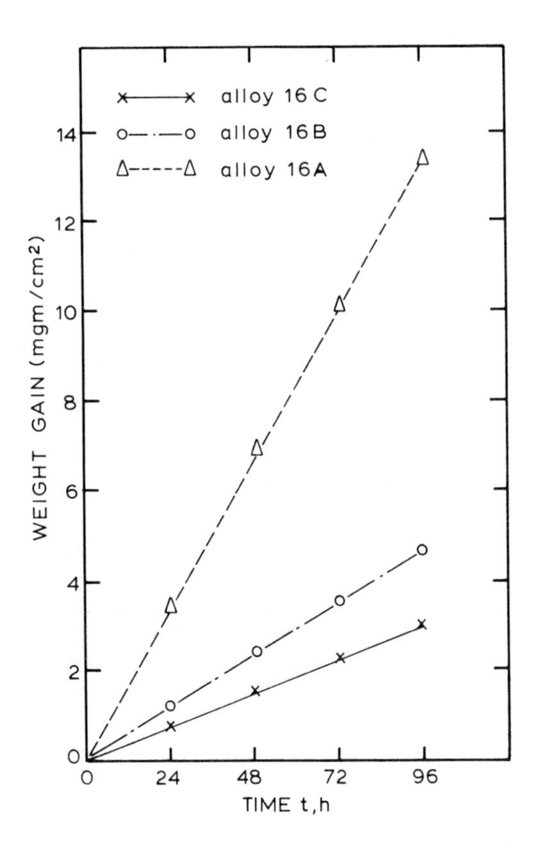

Fig. 8 Oxidation of Ni—Cr—Ta alloys at 1150°C in air

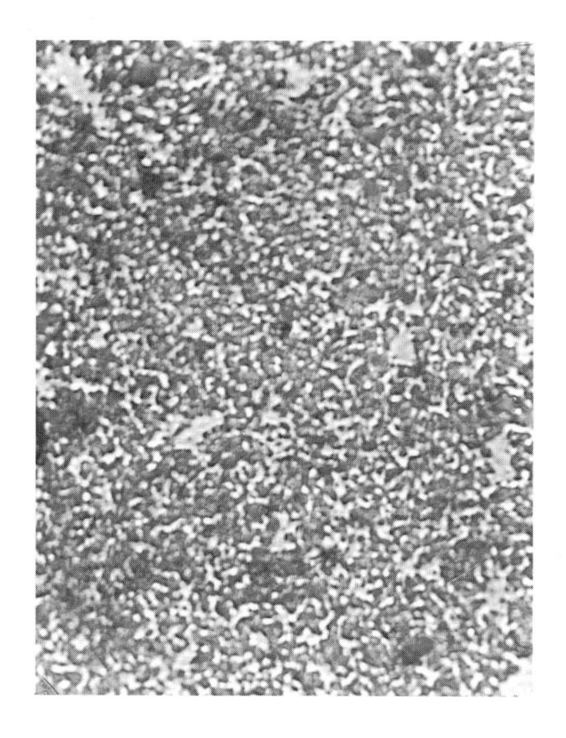

Fig. 9 NbC + Ni (60/40) sintered at 1350°C — 1h — Ar x150

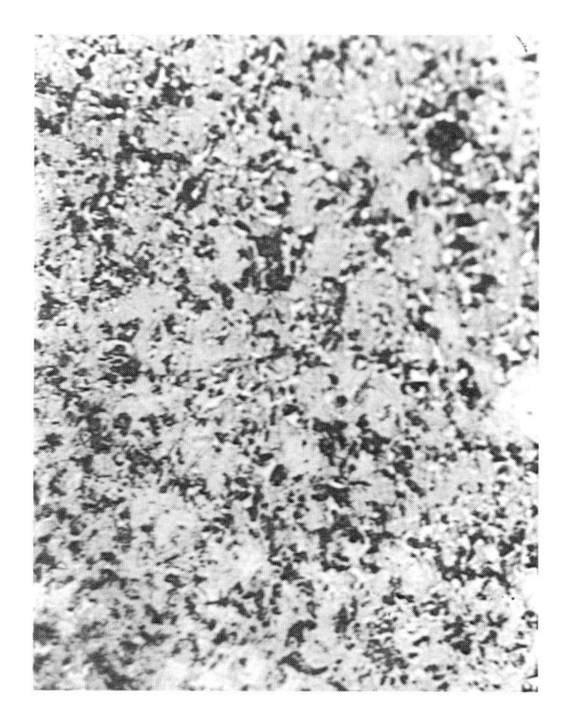

Fig. 10 NbC + Ni (60/40) sintered at 1350°C — 1h — Ar x375

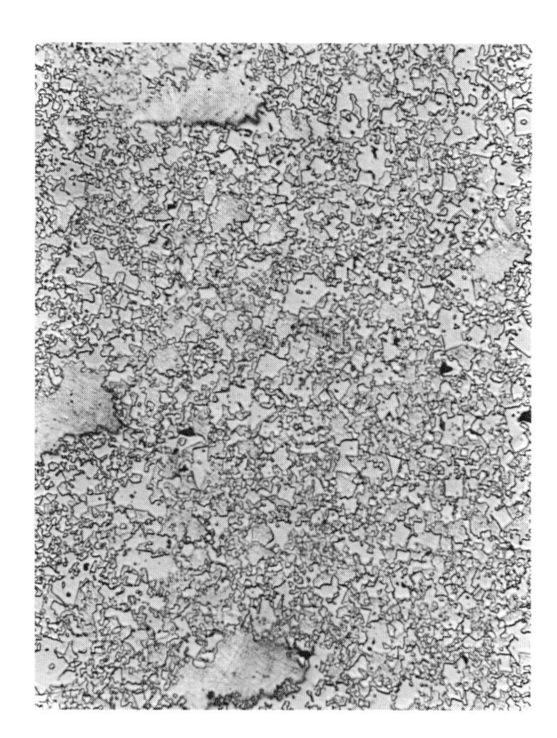

Fig. 11 NbC + Ni (70/30) sintered at 1350°C — 1h — Ar x150

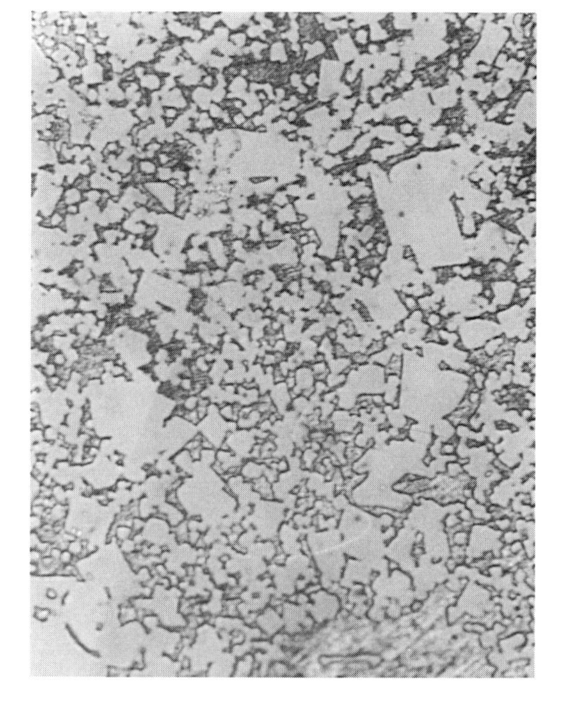

Fig. 12 NbC + Ni (70/30) sintered at 1350°C — 1h — Ar x375

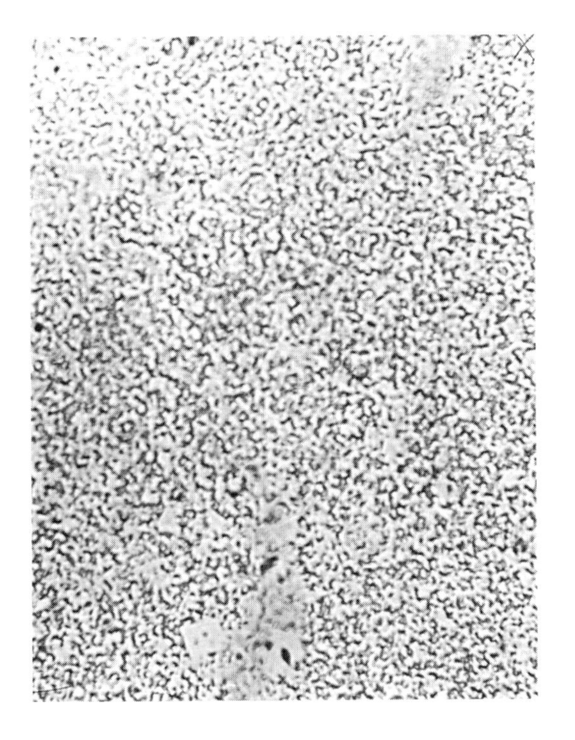

Fig. 13 NbC + Ni (80/20) sintered at 1350°C – 1h – Ar x150

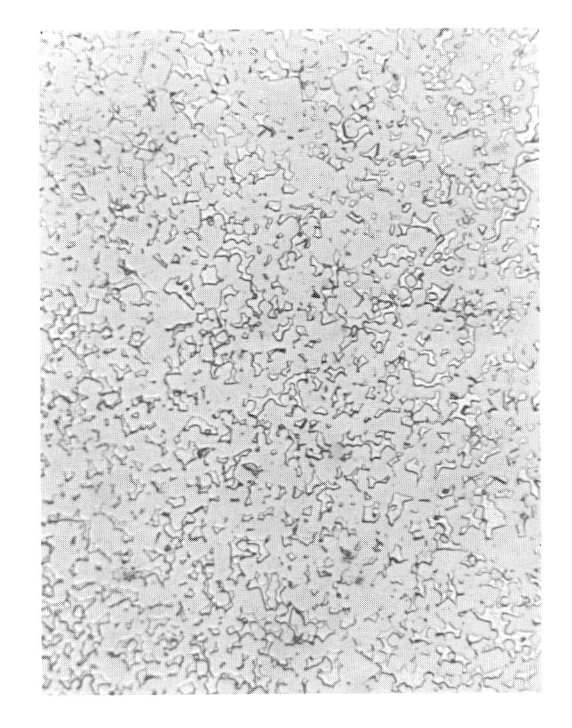

Fig. 14 NbC + Ni (80/20) sintered at 1350°C – 1h – Ar x375

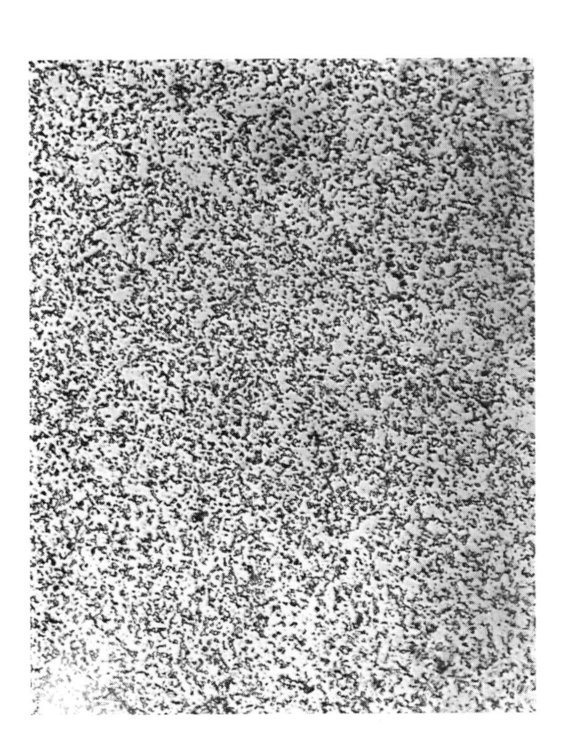

Fig. 15 NbC + Ni (90/10) sintered at 1350°C – 1h – Ar x150

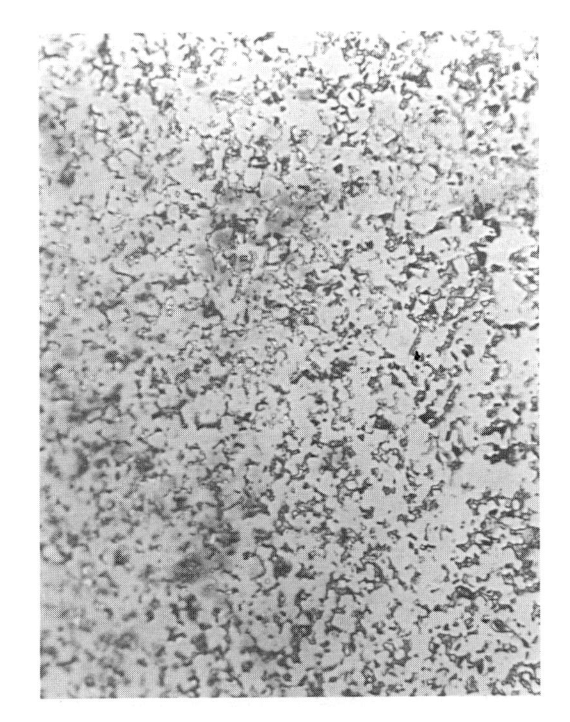

Fig. 16 NbC + Ni (90/10) sintered at 1350°C – 1h – Ar x375

□

Comparison of heat treatment response of P/M and conventionally produced T15 and M35 high speed tool steel inserts

J M GALLARDO and E J HERRERA

The authors are in the Department of Metallurgy at the University of Seville, Spain.

SYNOPSIS

A comparison of heat treatment response of P/M (directly sintered) and conventionally produced T15 and M35 high-speed steels is made. ISO RNGN-090300 cutting tool inserts were prepared either by machining from conventional HSS bar or by cold pressing of powders followed by vacuum sintering. Hardness, grain size and specific volume of both types of inserts were measured after hardening and tempering at various temperatures, using, eventually, samples with different percentage of porosity. In general, both steels show a similar behaviour so far the degree of porosity is equal or inferior to 2%. Nevertheless, P/M steel hardens on quenching at a slightly lower austenitizing temperature and its austenitic grain size is smaller than its conventionally obtained counterpart. Explanations of differences in behaviour are suggested.

INTRODUCTION

The properties required in cutting tool high speed steel are hot hardness, wear resistance and toughness. These properties are ultimately determined by heat treatment. In recent years, extensive research and development on HSS cutting tools prepared by P/M methods have been made (1). P/M processing offers several advantages, among them a uniform and fine microstructure. It cannot be assumed that P/M steels will behave in an identical manner to wrought ones. Several authors have investigated heat treatment behaviour of P/M high speed steel, mainly fabricated by the hot pressing method (2-5). In general, results indicate that carbides dissolve more readily in P/M steels and, therefore, austenitizing temperature has to be modified.

In this paper a comparative study of heat treatment response of commercially wrought high speed steels and P/M material is made. P/M steels were prepared in the laboratory by cold compaction and direct sintering.

EXPERIMENTAL PROCEDURE

Heat treating experiments were performed with cylindrical inserts, aproximately in the form of ISO RNGN 090300 (9.9 mm diameter x 3.5 mm height). They were prepared from M35 and T15 steels. Samples of wrought steels were machined from commercial annealed bar of 12.7 mm diameter. P/M samples were obtained from isothermically annealed directly-sintered blanks. Direct sintering was carried out in the laboratory (6). Annealed prealloyed water--atomized powders were cold compacted in a cylindrical die, using zinc stearate die wall lubrication and unidirectional pressures of up to 960 MPa. Sintering was performed in vacuum (<6.6 Pa) in a tubular furnace. Sintering temperatures ranged from 1200 to 1300°C and sintering times from a few seconds to three hours.

The basic chemical composition and annealed hardness of P/M and conventional (C) steels are given in Table I. T15 steels are richer in carbon and alloying elements, mainly vanadium, than M35 steels. P/M samples contain slightly more carbon (≈0.07%) than their wrought counterparts.

Heat treatment experiments were carried out in two stages. In the first stage, P/M inserts processed under various conditions were given a standard heat treatment, i.e., austenitization for 3 min in salt bath at 1205° (M35) and 1210°C (T15), respectively, interrupted quench at 550°C followed by air cooling and, finally, triple tempering (1 h) at 550°C. These P/M specimens had a sintered porosity ranging between 0.5 and 7%. After cleaning, hardness, grain size and specific volume were determined. Heat treated samples with a residual porosity equal to or lower than 2% -and not oversintered (6)- showed the best performance in continuous cutting tests.

In a second stage, P/M specimens were prepared at process conditions giving the best cutting results in the above tests. These P/M inserts, with a porosity ≤2%, were heat treated at various austenitizing and tempering temperatures. Austenitizing temperatures ranged from 1200 to 1260°C. The austenitizing cycle was carried out in vacuum (<6.6 Pa) in a calibrated tubular furnace, followed by oil quenching. Soaking time was three minutes for all tested temperatures. Inserts were triple tempered (3 x 1h) in salt bath over the temperature range

500-580°C. Samples of conventionally produced tool steels were also subjected to the above heat treatments. Hardness, grain size and specific volume were measured after each treating step.

Grain sizes were measured using the Snyder-Graff intercept technique (7). Specific volumes were determined by weighing in air and carbon tetrachloride.

RESULTS AND DISCUSSION

Hardness

Figures 1 (M35) and 2 (T15) show the hardness of samples heat treated under standard conditions as a function of residual porosity. Figures 1 and 2 also show the corresponding values of conventional wrought steels (zero porosity). Heat treated hardness greatly increases with decreasing porosity. P/M specimens with relative density ≥98% reach the highest hardness values. These values are, nevertheless, slightly lower than those of M35 (63.5 ± 1.0 HRC) and T15 (64.8 ± 1.5 HRC) commercial wrought steels. In short, hardness differences between conventional and P/M steels can be neglected in comparison with those arisen from the presence of more than 2% residual porosity.

Hardness results of second stage experiments are presented in figures 3 and 4. The effect of austenitization temperature on as-quenched hardness is shown in figure 3. Curves can be divided into two zones. At first, the increase in the austenitizing temperature produces a higher dissolution of carbides and hardness increases, but, at a certain point, dissolved elements favour greater amounts of retained austenite and, consequently, there is a reduction in hardness (8-11). Concerning the relative behaviour of P/M and conventional steels, these ones need higher temperatures (5-10°C) to reach maximum quenched hardness, which is lower (0.5 HRC) than that of P/M samples. T15 steel hardens somewhat more (≈ 1 HRC) than M35. These results may be explained in terms of the percentage of carbon in martensite, the amount of retained austenite and content, composition, size, shape and distribution of carbides. So, the higher carbon and vanadium content of T15 steels, the slightly higher carbon and alloying content of P/M samples and their more uniform carbide distribution could be responsible for differences in hardness.

To study tempering behaviour, inserts hardened at maximum as-quenched hardness were triple tempered (1 h) at various temperatures. Figure 4 shows the secondary hardening region of the tempering curves of wrought and sintered alloys austenitized at four different temperatures. Under the treating conditions indicated in figure 4, conventional steels reach a higher tempered hardness than P/M steels, since secondary hardening is weaker in P/M samples (ΔHRC≈0.5) than in wrought ones (ΔHRC≈2). T15 steels maintain their superior hardness. Peak hardness of figure 4 is attributed to the attainment of a critical dispersion of fine alloy carbides; as the carbide dispersion slowly coarsens, with increasing temperature, hardness drops (12).

Microstructure

The microstructure of P/M steels (density ≥98%) in the heat treated condition is rather similar to that of conventional high-speed steels, taking into account that P/M samples have a small amount of porosity. Figure 5 shows the typical microstructure of a P/M T15 steel specimen quenched from 1210°C.

Austenite grain size dependence on austenitizing temperature, for the range 1200-1260°C, is shown in figure 6. Grain size in P/M steels is finer than in wrought commercial steels. These results are in disagreement with the findings of Jagger et al. (13), probably due to the fact that these authors austenitized without a previous post-sintering annealing treatment. T15 samples have also a finer grain size than M35 ones. Austenite grain size increases with increasing temperature, as expected. This behaviour could be explained, because grain growth is mainly controlled by temperature and second particle barriers. The higher the content of carbon and alloying elements (P/M and T15 steels) and the finer and more uniform second particle are (P/M steels), the smaller is grain size. In addition, pores (P/M steels) can be considered "second particles", that drag grain boundary migration.

Microstructures developed in P/M samples sintered in non-optimum conditions -first stage experiments- are similar to those described above. Nevertheless, some oversintered inserts show an ultrafine grain size (Snyder-Graff >25) and coarse primary carbides, after hardening and tempering (2,6,14) These features are illustrated in figure 7, regarding P/M M35 steel. On the contrary, undersintered specimens (porosity >2%) present, after hardening and tempering, a coarse grain structure (Snyder-Graff <12), as shown in figure 8, for P/M M35 steel.

Specific volume changes

Volume changes from annealed to full treated condition were studied in P/M (density ≥98%) and wrought steels. Figure 9 presents the measured values of specific volumes of samples in the annealed, quenched (at peak hardness) and triple tempered (at peak secondary hardness) states. P/M steels, in spite of residual porosity, are slightly heavier than their conventionally produced counterparts. The higher alloying content of P/M steels could explain this difference. The specific volume of wrought steels increases from the annealed to the tempered state. P/M steels also increase volume from the annealed to the tempered condition, but, in general, to a lesser extent. This behaviour could be related to the fact that P/M samples were not fully annealed after the post-sintering isothermal annealing, as can be seen by the hardness values in table I. The damping effect of porosity cannot be disregarded as an influencing factor either.

REFERENCES

(1) Metals Handbook, 9th. Ed., Vol. 7, ASM, Metals Park, Ohio, 1984, p.370.

(2) DULIS,E.J. and NEUMEYER,T.A.: in Materials for Metal Cutting, ISI P126, Iron and Steel Institute, New York, 1971:p 112.

(3) KASAK, A., STEVEN, G. and NEUMEYER, T. A.: SAE Technical Paper 720182, SAE, New York, 1972.

(4) BORNEMAN, P. R.: Met. Eng. Q., (1) 1973:50.

(5) BEISS, P. and PODOB, M. T.: Powder Met., 25 (2) 1982: 69.

(6) GALLARDO, J. M. y HERRERA, E. J.: Téc. Metal. (Barcelona), (274) 1986: 9.

(7) SNYDER, R. W. and GRAFF, H. F.: Metal Progr., (33) 1938: 377.

(8) GULYAEV, A. P.: Kachestvennaya Stal, 5 (1) 1937: 41.

(9) GORDON, P., COHEN, M. and ROSE, R. S.: Trans. ASM, 31 1943: 161.

(10) GROBE, A. H. and ROBERTS, G. A.: Trans. ASM, 45 1953: 475.

(11) ROBERTS, G. A. and CARY, R. A.: Tool Steels, ASM, Metals Park, 1980, p. 683.

(12) HONEYCOMBE, R. W. K.: Steels (Microstructure and Properties), Edward Arnold, London, 1981, p. 153.

(13) JAGGER, F. L., PRICE, W. J. C., WALKER, P. I. and SMITH, P.: Powder Metallurgy, 20 (3) 1977: 200.

(14) GROBE, A. H., ROBERTS, G. A. and CHAMBERS, D. S.: Trans. ASM, 1954: 759.

TABLE I
Chemical composition and annealed hardness of tool steels

Steel	%C	%Cr	%Mo	%W	%Co	%V	HRC
T15 (P/M)	1.64	4.24	0.48	12.41	4.91	4.84	26*
T15 (C)	1.58	4.15	0.50	12.34	4.91	4.63	23
M35 (P/M)	0.93	3.90	4.93	6.30	4.69	1.89	27*
M35 (C)	0.85	4.25	5.00	6.25	4.75	2.00	20

(*) Hardness of P/M samples with porosity <2%.

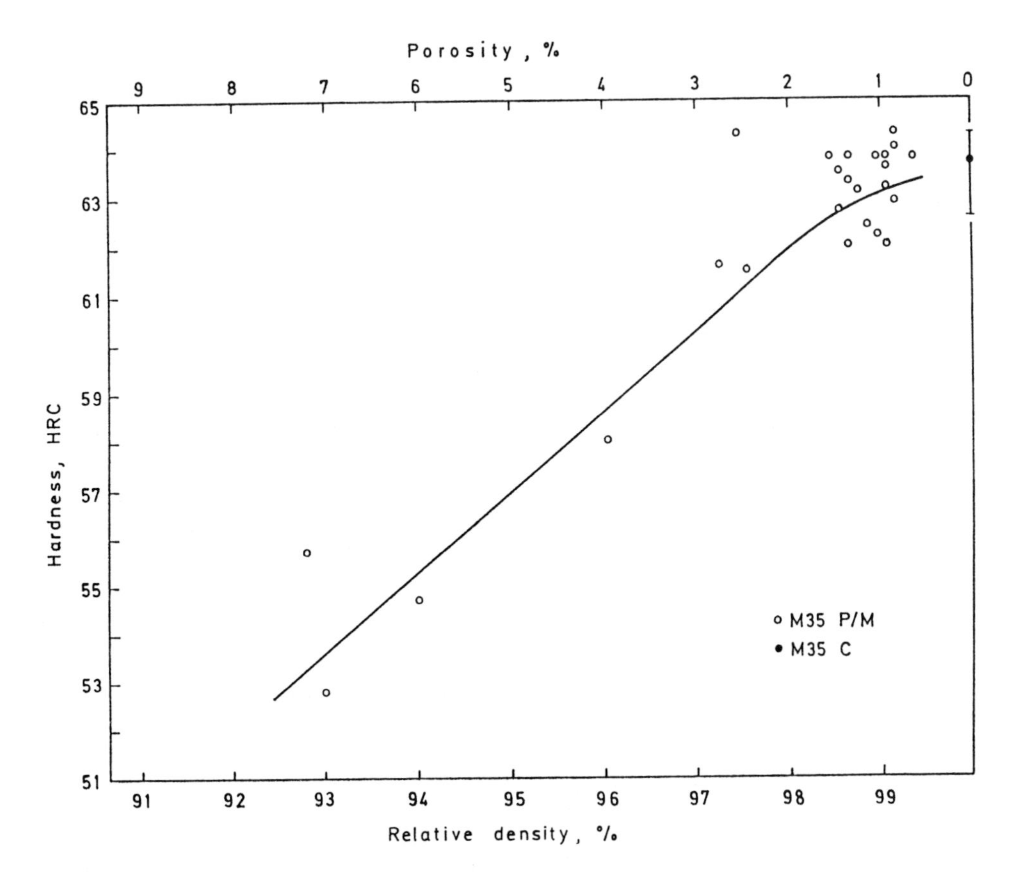

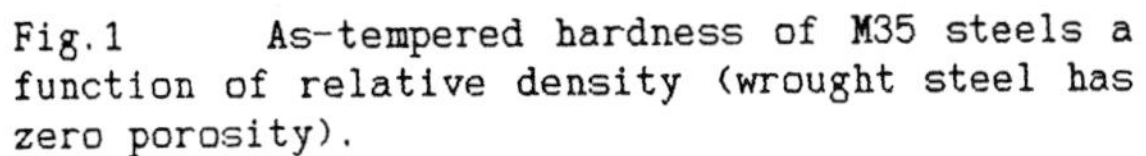
Fig. 1 As-tempered hardness of M35 steels a function of relative density (wrought steel has zero porosity).

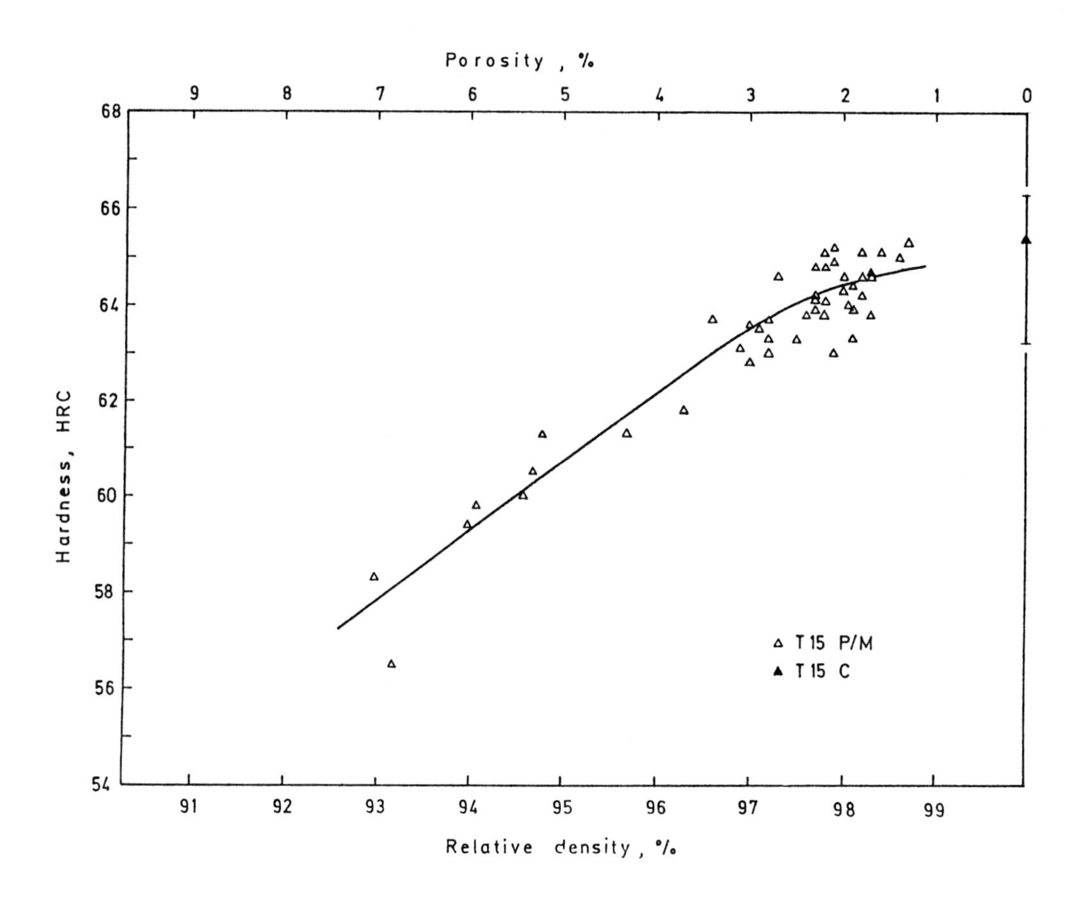

Fig.2 As-tempered hardness of T15 steels a function of relative density (wrought steel has zero porosity).

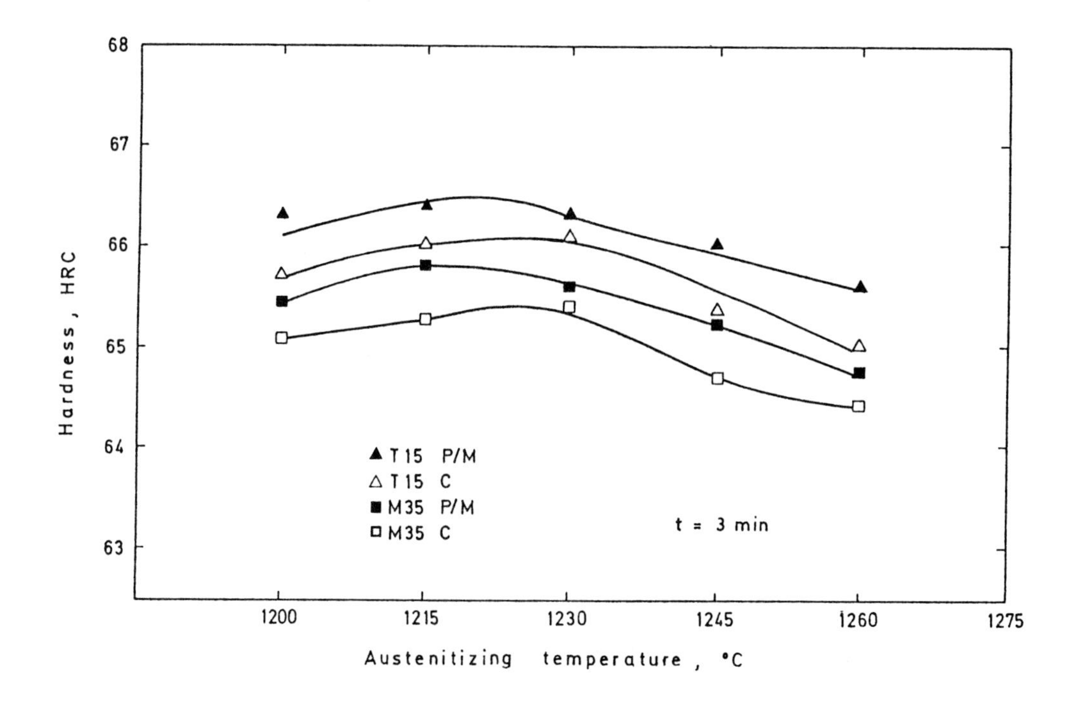

Fig.3 Effect of austenitizing temperature on as-quenched hardness.

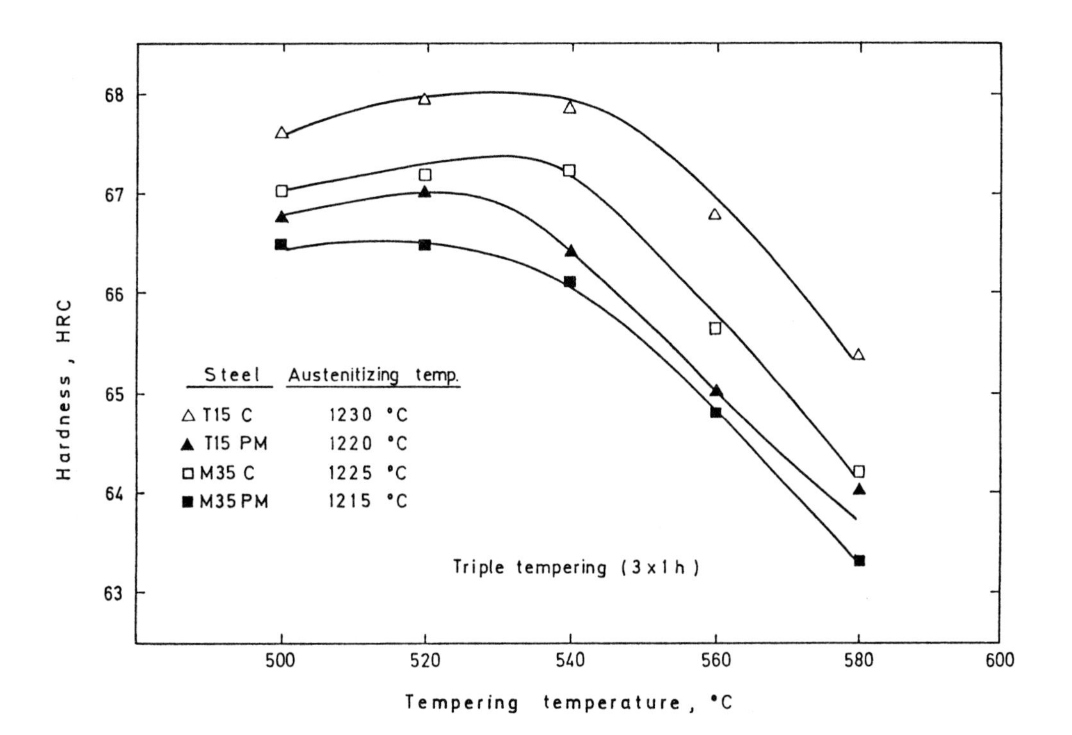

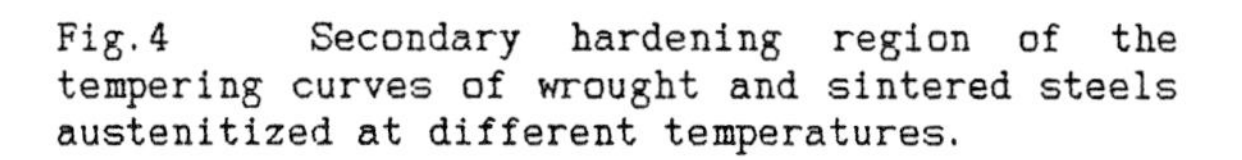
Fig.4 Secondary hardening region of the tempering curves of wrought and sintered steels austenitized at different temperatures.

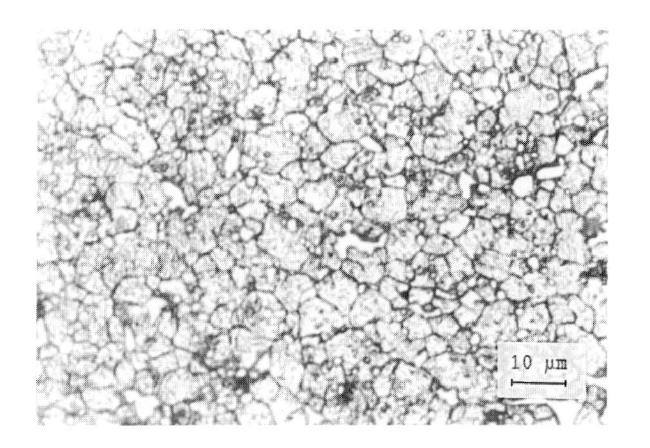

Fig.5 Microstructure of PM T15 steel austenitized at 1210°C and quenched.

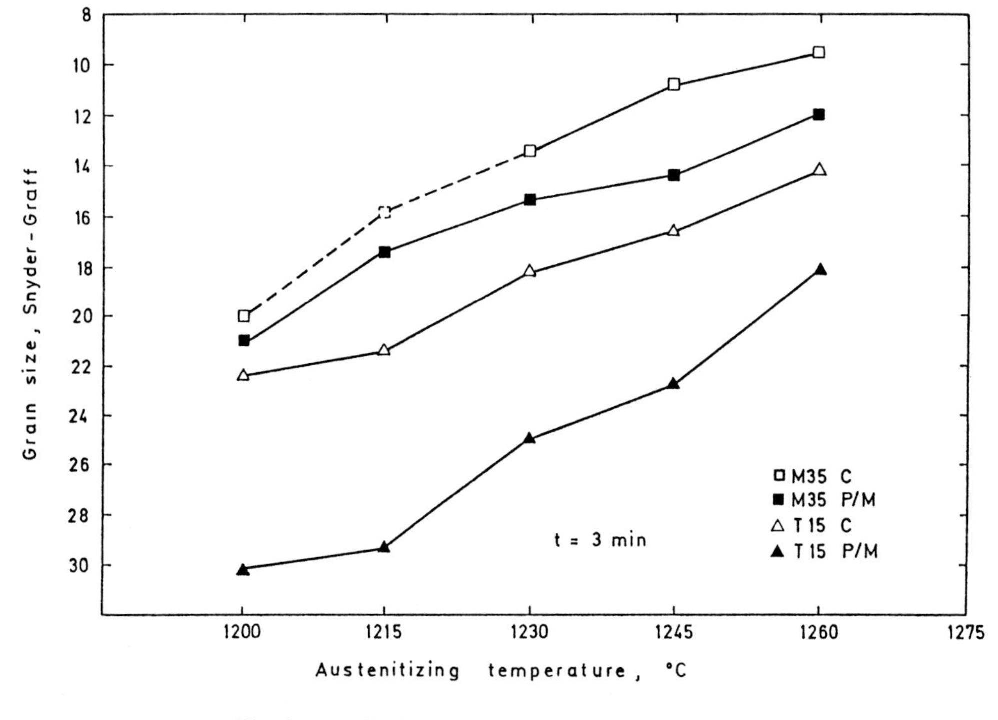

Fig.6 Grain size dependence on austenitizing temperature.

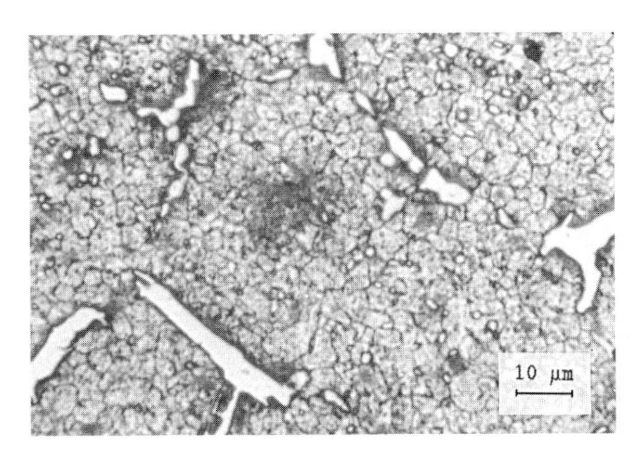

Fig.7 Microstructure of oversintered PM M35 steel sample showing an ultrafine grain size and coarse primary carbides.

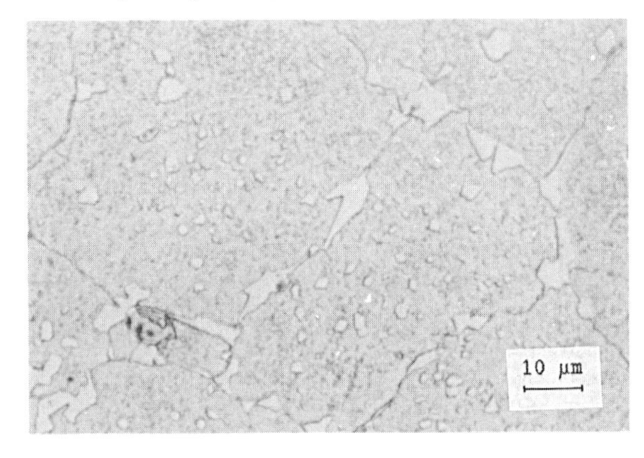

Fig 8 Coarse grained microstructure of hardened and tempered undersintered PM M35 steel specimen.

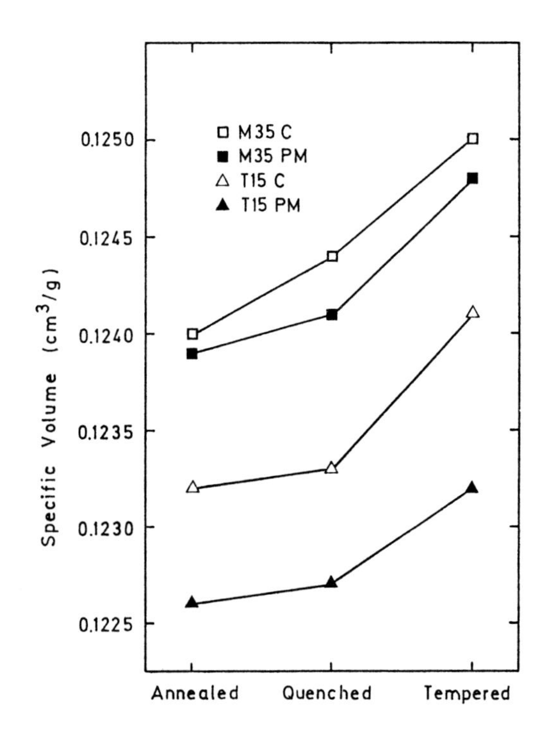

Fig.9 Specific volume changes from annealed to full treated conditions.

Influence of alloying elements in the gas-nitrocarburizing of sintered steels

M ROSSO and G SCAVINO

The authors are in the Dipartimento di Scienza dei Materiali e Ingegneria Chimica at the Politecnico di Torino, Italy.

SYNOPSIS

The study reports results of tests on nitrocarburized sintered steels, alloyed with Cr, Al, Mo and Ni. The samples, previously treated with SO_2 at a temperature of 500 °C for 1h, were subsequently gas nitrocarburized at 630 °C for 1h with an atmosphere constituted by NH_3 and CH_3OH. The surface layers were analysed throughout by means of röntgenographic, microhardness and metallographic techniques, carefully abrading a few tenths of a µm at a time. The influence of the alloying elements on the layers' composition and constitution was determined, demonstrating that the presence of Cr or Al promotes the formation of layers mainly constituted by the ε phase of the Fe-C-N system, whereas Ni-containing samples have surface layers with a larger quantity of the γ'-Fe_4N carbonitride.

INTRODUCTION

The fatigue and wear resistance of ferrous materials are incremented by nitrocarburizing treatments, i.e. thermochemical processes founded on the nitrogen and carbon adsorption at the surface of the treated materials, followed by diffusion in the bulk, with the formation of surface layers constituted by the ε-solid solution of the Fe-C-N system and by the γ'-Fe_4N type carbonitrides.

These treatments can be also applied to sintered steels. The usual technique calls for the use of gaseous treatments. In fact, the wide contact surface, caused by the extensive presence of pores, aids the reaction between the nitrocarburizing atmosphere and the material; however, the interconnected porosity act as preferential penetration paths of the nitrocarburizing chemical agents; consequently, the carbon and nitrogen penetration is generally high but relatively non-uniform, without the formation of a distinct case over the core.

These drawbacks can be attenuated by reducing porosity using various means, for example by adding S or P /1, 2/ to the powder before sintering, or by copper infiltration /3/, or by steam heat-treating /4, 5/. The latter is the most used technique.

To simplify the heat-treatment sequence and to reduce the overall treatment time, we have studied a new cycle /6/, consisting of a pre-oxidation by the action of sulphur dioxide, followed by a nitrocarburizing stage. The execution of the SO_2 atmosphere pretreatment reduces the porosity because a more compact surface layer, mainly constituted by Fe_3O_4, is formed. Significant amounts of FeS are also generated. Therefore the process is interpreted as being a sulphoxidation.

The presence of iron sulphide in the magnetite layer of the sulphoxidated samples is beneficial since it allows the formation of greater quantities of ε-solid solution, with an interstitial elements content (C+N) higher than the one that can be detected in nitrocarburizing specimens steam-oxidated only /6/. The fact was explained hypothesizing that sulphides favour the decomposition of the ammonia adsorbed on the surface, thus promoting the nitrocarburizing action of the atmosphere, if the model suggested by Grabke holds true /7/.

In our previous study /6/ samples containing only 2% of copper and 0.01% of carbon were examined. Tests using the same treatment cycle on sintered steels alloyed with chromium, aluminium, nickel and molybdenum, and with a low to high carbon content have been further performed. On the basis of results already obtained, the pretreatment temperature and time were selected at 500 °C and 1 h, respectively, to avoid excessive thickness and low hardness in the nitrocarburized layers. Nitrocarburizing time was kept at 1 h, since it had been demonstrated that longer times give worse results /6/. Reports on tests at 630 °C are hereby presented. Results at different temperatures are reported elsewhere /8,9/.

EXPERIMENTAL PROCEDURE

Cylindrical samples, 25 mm diameter and 20 mm height, were made in an industrial sintering plant using Höganäs powder mixtures of different compositions. In table I the chemical composition and the volume mass of the samples are reported. The sintering was performed by cold pressing the various mixtures and subsequently heating them at a temperature of 1120 °C in an atmosphere (20% H_2, 80% N_2) of dissociated ammonia with the addition of nitrogen.

The sulphoxidation treatment was carried out in a vertical tube 34 mm (diameter laboratory) furnace, by the action of sulphur dioxide ($1.4\cdot10^{-6}$ Nm^3/h) at 500 °C for 1 h. At the furnace exit, such a flow was adsorbed in a basic solution. The same furnace was used for the nitrocarburizing treatment, performed at 630 °C for 1 hour. The ammonia flow, checked by a flowmeter, was $3\cdot10^{-5}$ Nm^3/h.

Methanol was simultaneously introduced in the furnace by a metering pump set to obtain an atmosphere with a vapor alcohol content of 15% vol. The furnace, provided with a sample release device, let the samples quickly quench in an ethanol bath protected by nitrogen.

After heat-treatment the specimens were examined by X-ray diffraction analysis (radiation Fe-K_α, $\lambda=1.9373\cdot10^{-10}$ m), to determine the surface layer phase constitution and to measure the (C+N) content in ε-phases /10, 11/. To assess the variation of the phase constitution as a function of the distance from the surface, these determinations were repeated after removal of successive layers by polishing with abrasive emery papers and abrasive diamond pastes.

On the polished transverse sections of the specimens, Vickers microhardness tests (load 0.98 N) and optical microscopy were performed.

RESULTS

The phases constituting the surface layers and the inner zones of the samples, detected by röntgenographic analysis, are reported in figures 1÷5. The ε solid solution is always encountered along with, for most of the specimens, the carbonitride γ'-Fe_4N; where present, the amount of γ' rises approaching the inner layer. Below the transformed layer, phases with a fcc and bcc lattice were always detected; whereas the former is clearly austenite, the latter has to be identified as martensite, at least in zones closer to the surface, owing to the hardness peaks there encountered.

The presence of a large quantity of ε phase, after only 1 hour of treatment, emphasizes the high nitrocarburizing potential of the gaseous atmosphere.

As regards the influence of carbon, results from plain carbon steel specimens (samples 1 and 2) show that they have a similar outer layer constitution. In the one with the highest carbon content, slightly larger amounts of austenite are obviously found in the underlayer.

When carbon accompanies chromium, an increase in its concentration (sample 4) favours the formation of ε solids instead of γ' carbonitride. Although some specimens are formed from powders with a high carbon percentage, the röntgenographic analysis has never detected the presence of cementite.

Different nickel amounts in the sintered powder (samples 6, 8 and 10), cause a variation of the relative quantity of the phases: in fact, while increasing the nickel content, the overall amount of ε solid solution in the outer layer decreases and a progressively larger quantity of austenite is detected in the underlayer.

Chromium produces an opposite effect in comparison with nickel. The carbonitride γ' disappears at high C content (sample 4) or its quantity is limited in the absence of C (sample 3); conversely, more ε is present in the outer layer and a lower amount of austenite occurs in the underlayer.

Aluminium additions (sample 5) exalt the effects of chromium: the compound layer is now constituted only by ε solids, even no carbon was present in the original powders (see sample 3 for comparison).

In all the samples, the metallographic analysis, effected on the section of the specimens, show that the carbonitride layer is divided in two zones: the outer with a diffuse porosity, the inner more compact (fig. 6 a). In table II the thicknesses of the compact and total carbonitride layers are reported: the porous layer thickness decreases as the alloy content of the samples increases. Particularly important in this respect is the action of chromium which provokes the formation of the thinnest and most compact layers. Nickel has instead a limited effect. Moreover, in the specimen with the highest nickel content (sample 10), the presence of chromium coupled with the more elevated original volume mass induces the formation of thin layers; in fact, the layer depth depends also on the original volume mass. In the aluminium alloyed specimen (sample 5), the high thickness values may depend on the absence of copper and the slightly lower original volume mass, although no exhaustive explanation can be given now.

The interstitial elements contents of the ε solid solution, determined at different distances from the surface, are reported in figs. 7 and 8. For lack of definition of line peaks it has not always been possible to determine the interstitial content in the most inner zones of the ε solid solution, so that, for some specimens, the interstitial content has been determined only for the first tens of micrometers from the surface.

The higher interstitial elements contents are found in the aluminium or chromium alloyed specimens, whereas the ε solid solutions of the nickel containing samples show a lower (C+N) content. In figs. 1÷5 hardness test results have also been inserted: the hardnesses of the outer layer are dominated by the porosity and are independent of the alloying elements, the values being always about 250 HV 0.1. Therefore, in the surface layers the maximum values have been detected in the compact zone. Here, the influence of the alloying elements is not very evident except as regards carbon: in fact, nickel alloyed specimens have values similar to those of aluminium or chromium alloyed ones. Conversely, the presence of carbon in the sintered powders generally increases the hardness peaks in the samples. In the underlayer the higher values are encountered for aluminium and chromium alloyed samples, considering that here also carbon plays a hardening role, whereas the presence of nickel decreases the values, overcoming any cocurrent inverse effect of chromium and molybdenum.

DISCUSSION

The influence exerted on the layers' constitution by the alloying elements, evident from the examination of the experimental results, can be rationalized as follows.

Chromium and aluminium favour layers with small amounts of γ' carbonitride or completely without it.

To try to explain how these elements affect the layer formation, it is necessary to consider the isothermal section at 630 °C of the metastable Fe-C-N phase diagram (fig. 9) /12/. In fact it is possible to observe that there exists a large immiscibility gap where the austenite is in thermodynamic equilibrium with the ε solid solution only. Thus, it is possible to form ε solids from γ-Fe directly, without following the sequence: γ-Fe / γ'/ ε. It can also be seen that in this case, the nitrogen content of the reacting austenite has to be lower than the maximum value of its solubility in γ-Fe. Therefore, in order that the ε phase can form directly from the austenite, it is necessary that the nitrogen contribution to austenite is not too high in comparison with the carbon one.

Chromium tends to decrease both the nitrogen solubility limits in the austenite /13/, whereas it increases the element solubility in the ferrite /14/. Consequently, the rate of the nitrogen-austenite formation from the ferrite is reduced; then austenite can originate from the reaction between cementite, nucleated from the ferrite, and nitrogen, so that the system may be represented, in this initial step, by a point in the ternary field α-Fe, γ-Fe, Fe_3C and not in the binary field α-Fe, γ-Fe. In this case, the austenite solution has the highest possible carbon content so that further nitrogen intake can easily originate ε crystals. As a proof, it can be seen that the sample containing chromium and carbon, where a large quantity of cementite is present before the nitrocarburizing treatment, has the surface layer constituted by ε solids only. Furthermore, the chromium has a marked tendency to enter the ε phase, replacing the iron atoms, whereas it has a limited possibility of entering the γ' carbonitride /13/.

For the chromium and aluminium alloyed specimen (sample 5), the mechanism of formation of the nitrocarburized layer can be the same.

In the specimens with progressively higher nickel contents, the surface layers show an increase of the γ' carbonitride quantity. Unlike chromium, the presence of nickel in the ferrite decreases the nitrogen solubility in this phase /15/, so that austenite can easily originate from the ferrite. Furthermore, nickel augmenting amounts in the alloy progressively narrow the austenite field /16,17/; in particular, the upper boundary of nitrogen solubility shifts to lower nitrogen content; consequently the austenite has a greater tendency to become nitrogen saturated, thus segregating the γ' carbonitride. In this latter phase nickel can replace a high quantity of iron atoms, and, hence, render the phase more stable. The γ' carbonitride, formed on the specimen surface, can further (partly) react with the nitrogen deriving from the treatment atmosphere and form limited amounts of the ε solid solution, whose lower boundary, as regards nitrogen solubility, is shifted to higher nitrogen content, in presence of nickel.

In the samples without substitutional alloying elements, the austenite that is formed contains carbon and nitrogen in quantities such as to be in equilibrium with the γ' carbonitride and the ε solid solution: three phase zones can be easily detected.

In all the samples, the metallographic analyses have shown the presence of diffuse porosity in the outer layer, where the ε solid solution is prevalent. The cause can be found, owing to the higher temperature, in the more rapid ε layer growth, that can cause crystal mismatch in columnar growth. Observations made on samples nitrocarburized with the same cycle, but at lower temperature /8/, reveal the absence or a settled reduction of porous zone extension, whereas specimens treated in the same manner at higher temperature, and hence, having higher layer growth rate, show a considerable growth of the porous layer /9/. The presence of alloying elements in the samples can reduce the depth of the zone with great porosity because of different reasons. In the chromium alloyed specimens , as previously stated, the presence of chromium in the ferrite increases the nitrogen solubility in this phase and delays the nitrides formation, so that under the same treatment time, the layers' growth rate can be lower, and, consequently, the crystal mismatch in columnar growth is limited. Also the total layer thickness is, of course, reduced, as shown by the metallographic analyses (fig. 6b). Since the porous zone extension coincides with the zone where ε phase is present, except in the case of chromium and aluminium alloyed samples, in the nickel alloyed samples, owing to the limited depth where the ε solution can also be detected, the porous zone extension results were reduced.

The presence of carbon and, hence, cementite, in the specimens before the nitrocarburizing treatment, does not appreciably reduce the zone having a high porosity, as results by comparing data from samples 1 and 2. On the other hand, the presence of carbon can favour the formation of ε solid solution with a high carbon content, deriving from the direct reaction between the cementite and nitrogen atoms apported by the atmosphere and therefore, as noted before /18/, it may decrease the tendency to form porous layers. It has, therefore, to be concluded that, in the obtained experimental conditions, the presence of porosity in the outer layers depends rather on the ε solid rapid growth than on the high nitrogen content of the ε solid solutions. Moreover, the highest thicknesses of the porous layers are encountered either in plain carbon steel samples where the interstitial element concentration is high, or in nickel alloyed samples

where the (C+N) content is lower; what can be taken as a further argument in favour of porosity being increased more by rapid growth than by high nitrogen concentration. Finally, in the case of the presently heat treated sintered steel samples it has to be remembered that the initial presence of iron sulphide and the possibility of this compound increasing the layer growth rate, can favour the pore formation in the outer surface layers.

The interstitial contents present in the ε solid solution, as reported in figs. 7 and 8, depend on the nitrogen affinity of the metal atoms that replace the iron atoms in this phase: the presence of aluminium, which has a high nitrogen affinity, allows the constitution of ε solids with the highest interstitial contents, whereas in the nickel alloyed samples, and above all in the sample 10, the interstitial contents are lower, owing to the lower nitrogen affinity to this element, that overcomes the opposite influence of chromium and molybdenum.

The microhardness tests, effected on the compact zone of the surface layers, have not shown a strong difference of maximum hardness values in samples having different chemical compositions. The compact zone of the layer of chromium and aluminium alloyed specimens is mainly constituted by the ε solid solution. This phase, in the inner, can present a higher C/N ratio, as reported in previous works about the gas nitrocarburizing of plain carbon steels /19/, owing to the decrease of the nitrogen content and rise in the amount of carbon; it has been demonstrated /20/ that upon augmenting the C/N ration the hardness values of the ε solid solutions can rise. It is also possible that the presence of chromium and aluminium in the ε solids contributes to elevate the hardness values owing to chromium and aluminium nitride micro-precipitation phenomena. In the nickel alloyed specimens the outer layer compact zone is constituted quasi-exclusively by the γ' carbonitride, having a hardness characteristic comparable to high carbon ε solids.

The carbon samples present higher hardness values generally, in fact, since the compact layer has, at the interface, the presence of cementite in these samples (fig. 6c), and hence the ε phases of this zone are probably generated by the reaction between cementite and nitrogen, these ε phases have a high carbon content, as can be derived observing the Fe - C - N phase diagram. As noted above, a high amount of carbon in ε solids allows a high level of hardness to be achieved.

The great amount of austenite in the underlayer is due to the nitrogen and carbon contribution in the inner layer of the samples; in the presence of austenite stabilizing elements, like nickel, the γ solution quantity obviously increases. This element, moreover, decreases the thermodynamic metastability of the austenite during the sample quench and, hence, shifts to lower temperatures the austenite transformation into martensite. In fact, in the nickel alloyed specimens, the hardness values in the underlayer are low in comparison with the values of chromium and aluminium alloyed samples.

These latter elements can segregate in the form of nitride or carbide during the cooling and decrease the interstitial elements content of the austenite, favouring the austenite change in martensite or ferrite. Therefore, in these specimens the austenite quantity is limited and higher hardness values occur. In the plain carbon steel samples, it is possible to have γ solution with a high content of this element and, hence, the presence of martensite zones with higher hardness.

Therefore, microhardness measurements, performed on sections, are represented by curves that rise as a function of the depth (figs. 1÷5) because there is an increase of martensite amounts with the depth, justified considering that, at increasing distance from the surface the diffused nitrogen content decreases, thus yielding a more metastable austenite.

CONCLUSION

The effect of the presence of alloying elements (aluminium, chromium, nickel and molybdenum) in steel sintered samples with various carbon amounts has been evaluated after sulphoxidation treatments followed by gas-nitrocarburizing.

It has been observed that nickel favours the constitution of surface layers formed mainly by the γ' carbonitride, with limited amounts of ε solid solution. The effect is more evident as the nickel content in the alloy increases: lower quantities of nickel, together with molybdenum presence in alloy, allow the constitution of appreciable ε solid solution quantities in the surface layers. In the chromium and aluminium alloyed samples, the surface layers are mainly or only constituted by the ε solid solution.

The interstitial content of the ε phases depends on the presence of the alloying elements: upon increasing the nickel amounts in the sintered samples, it is possible to detect a progressive reduction of the interstitial element content, whereas in the chromium and aluminium alloyed samples the ε solid solution have a higher (C+N) content.

The outer layer of the samples show the existence of diffuse porosity, that is, partly reduced by the presence of the chromium and aluminium, whereas the inner layer is compact.

Microhardness measurements performed on transverse sections show that the hardness values in the zone of diffuse porosity are independent of the chemical composition of the samples; nor has a large difference in hardness been ascertained in the compact zone, whereas in the underlayer the presence of chromium, aluminium and molybdenum has a more pronounced influence; i.e. they reduce the amount of the austenite, thus shifting the hardness peak to lesser distances from the surface due to martensite forming under the austenite zone.

The different effects of the alloying elements on the nitrocarburized layer constitution has been rationalized on the basis of the affinity that exists between these elements and the nitrogen (and carbon), together with their more or less pronounced stabilizing action towards all the phases of the Fe-C-N metastable diagram.

ACKNOWLEDGMENTS

The authors wish to thank Dr. E. Galetto and Mr. G. Pesce of the FIPS s.p.a., Torino, for the accurate sample production.

REFERENCES

1) R. M. German, K. A. D'Angelo, Int. Met. Rev., 29 (1984), 249-272.
2) C. Blände, G. Skoglund, J. Tengzelius, Proc. Conf. Powder Metallurgy and Related High Temperature Materials. Bombay (1983), 118-140.
3) H. E. Boyer, Secondary Operations Performed on P/M Parts and Products. Metals Handbook, 7 (1984), 451-462.
4) J. H. Eggleston, R. D. Fisher, Proc. Conf. Progress in Powder Metallurgy. Montreal (1982), 613-619.
5) P. Cavallotti, D. Colombo, C. Mangiacavalli, G. F. Bocchini, A. Mancuso, Metall. Ital., 75 (1983), 703-714.
6) M. Rosso, G. Scavino, Metall. Ital., 79 (1987).
7) J. H. Grabke, Ber. Bunsenges. Phys. Chem., 72, (1968), 553.
8) M. Rosso, G. Scavino, To be presented at XI Conv. Naz. Tratt. Term., AIM 1987.
9) M. Rosso, G. Scavino, Proc. Int. Powder Metallurgy Conf. Düsseldorf (1986), 369- 373
10) D. Firrao, M. Rosso, G. Scavino, Not. Tecn. AMMA, 5 (1984), 5-16.
11) D. Firrao, B. De Benedetti, Metall. Ital., 68 (1976), 4-15.

12) F. K. Naumann, G. Langenscheid, Arch. Eisenhuttenw., 36 (1965), 677-682.
13) D. Firrao, M. Rosso, B. De Benedetti, Atti Acc. Sc. Torino, 114 (1979-80), 383-393.
14) V. G. Permiakov, A. V. Belotskii, R. R. Barabash, Izvest. Vuz Chern. Met., 129 (1972).
15) Y. Imai, T. Masumoto, K. Sakamoto, Sci. Rep. Res. Inst. Tohuku Univ., 20 (1968), 1-13.
16) T. P. Floridis, W. R. Chilcott, Trans. Amer. Soc. Metals, 57 (1964), 360-1.
17) A. J. Heckler, J. A. Peterson, Trans. Met. Soc. AIME, 245 (1969), 2537-41.
18) D. Firrao, M. Rosso, G. Scavino, Metall. Ital., 78 (1986), 285-304.
19) D. Firrao, B. De Benedetti, M. Rosso, Metall. Ital., 73, (1981), 513-522.
20) D. Firrao, M. Rosso, Atti Accad. Sc. Torino, 114, (1980), 171-181.

Table 1: Alloying element contents (wt. percentage) and volume mass of the sintered steel samples.

Sample	Cu	C	Cr	Ni	Mo	Al	Volume mass [Kg/m^3]
1	2	0.01					$6.7\cdot10^3$
2	2	0.55					$6.8.10^3$
3	0.5	0.01	1.5				$7.0.10^3$
4	0.5	0.55	1.5				$7.0.10^3$
5	-	0.01	1.5			1.0	$6.6.10^3$
6	1.5	0.01		1.75	0.50		$6.7.10^3$
7	1.5	0.55		1.75	0.50		$6.7.10^3$
8	1.5	0.01		3.95	0.50		$6.7.10^3$
9	1.5	0.55		3.95	0.50		$6.7.10^3$
10	2	0.50	1.5	7.0	0.50		$7.1.10^3$

Table 2: Compact and total surface layers thicknesses.

Sample	Compact layer thickness [μm]	Total layer thickness [μm]
1	50	180
2	60	170
3	60	120
4	90	120
5	100	250
6	60	160
7	60	170
8	80	170
9	80	180
10	110	130

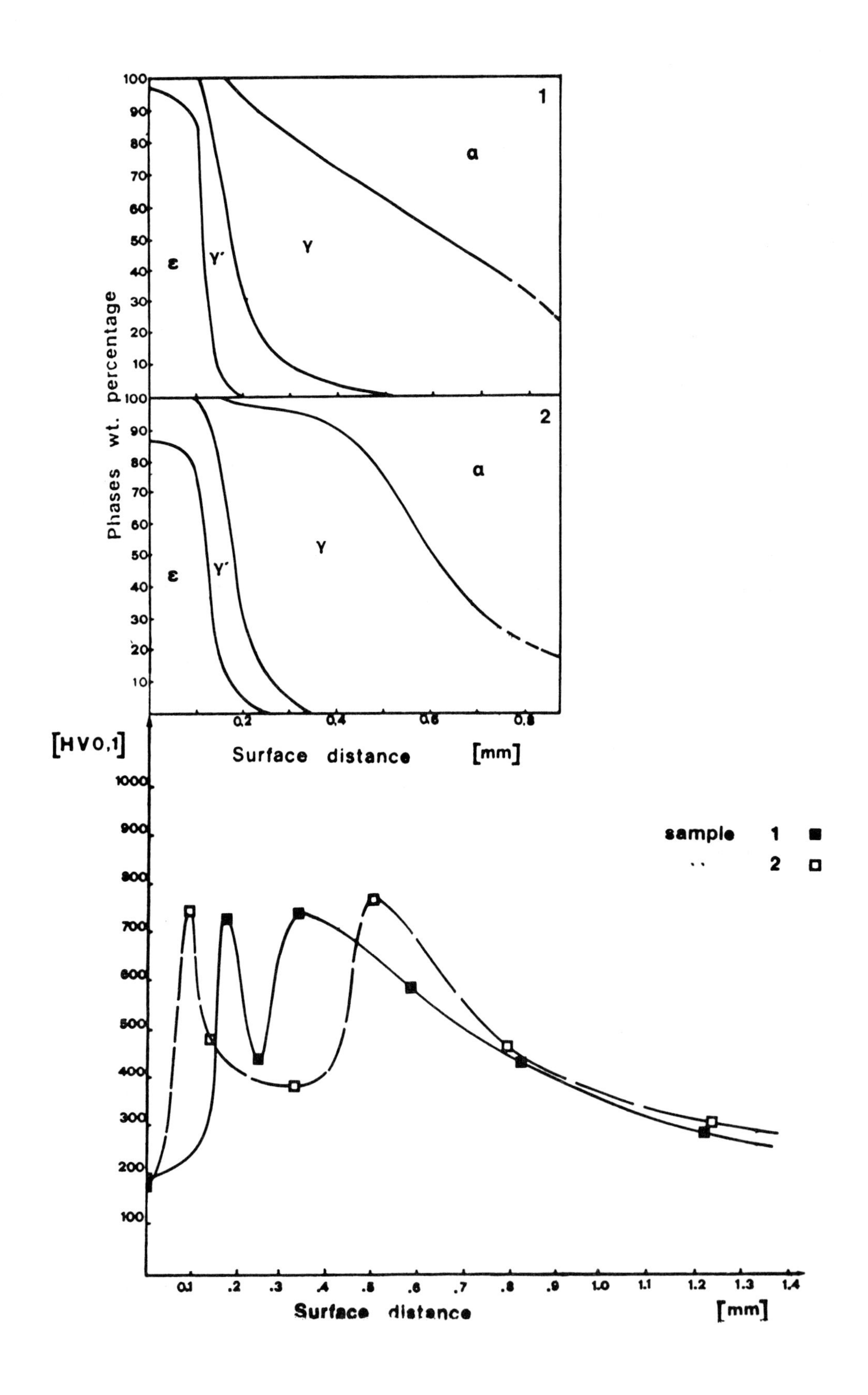

1 - Variation of the phase wt. percentages and microhardness as a function of the surface distance for plain carbon sintered steel samples (specimens 1 and 2, table I).

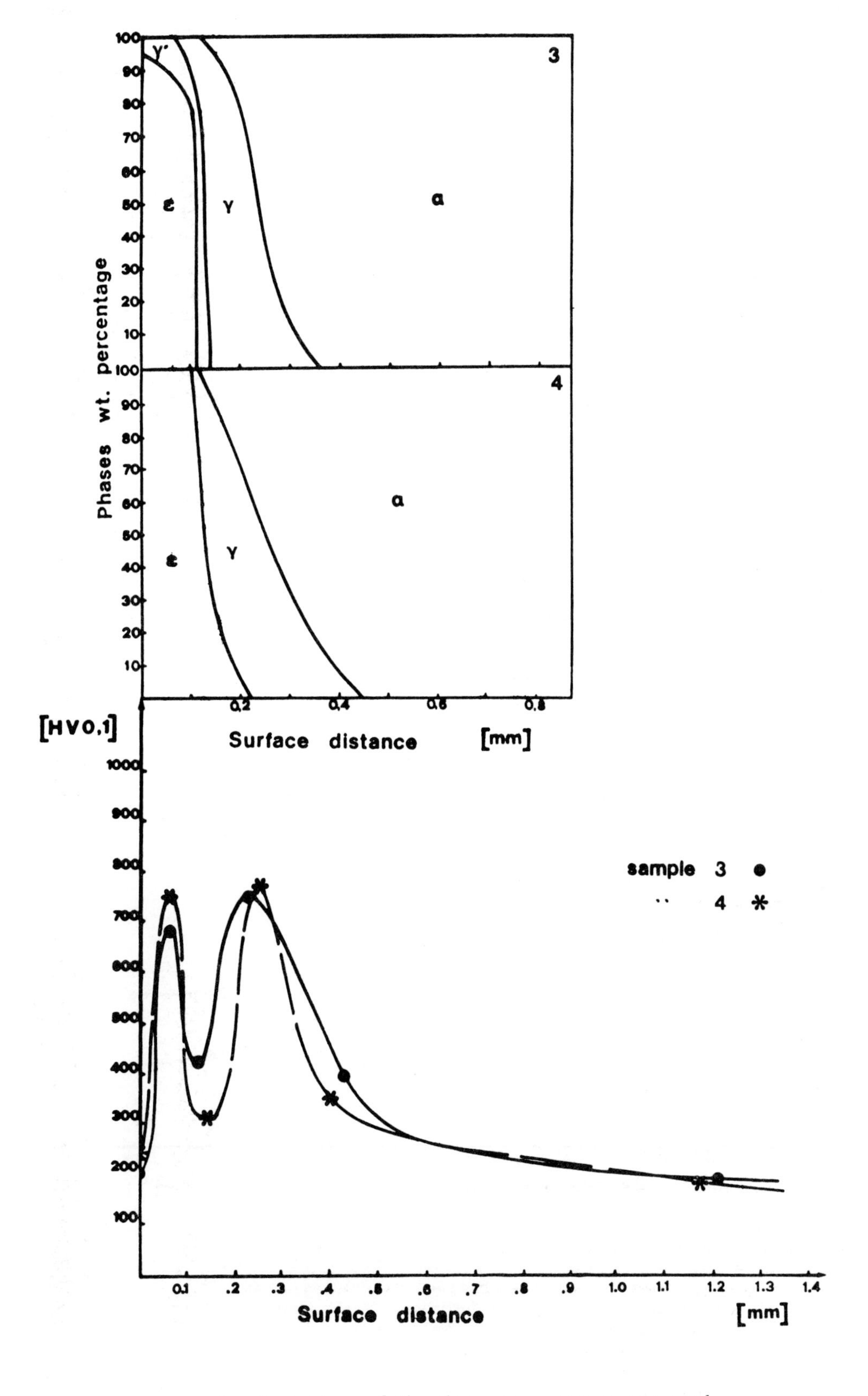

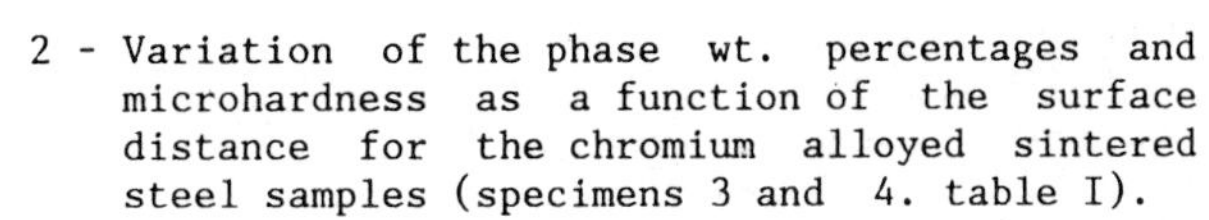

2 - Variation of the phase wt. percentages and microhardness as a function of the surface distance for the chromium alloyed sintered steel samples (specimens 3 and 4. table I).

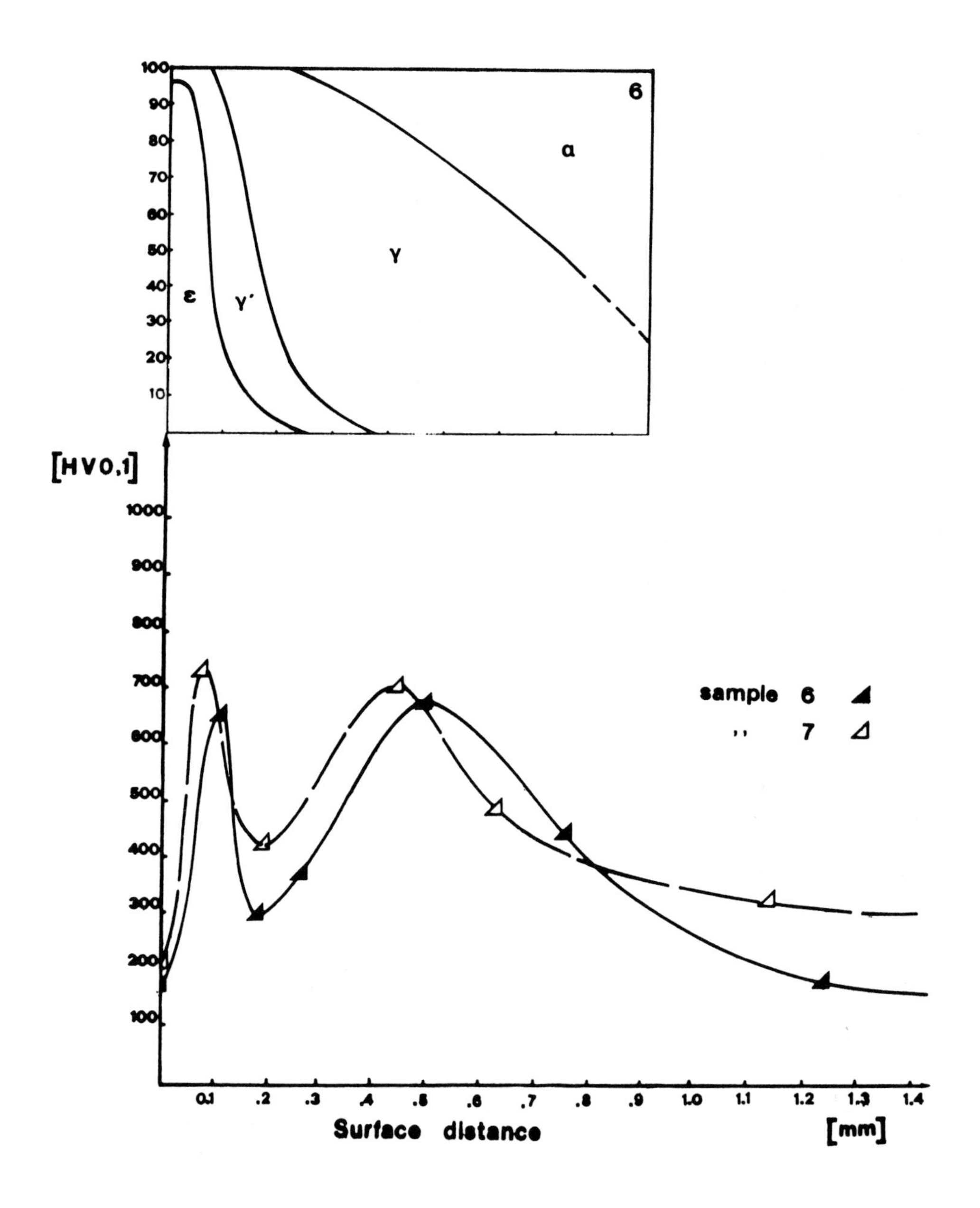

3 - Variation of the phase wt. percentages and microhardness as a function of the surface distance for the nickel and molybdenum sintered steel samples (specimens 6 and 7, table I).

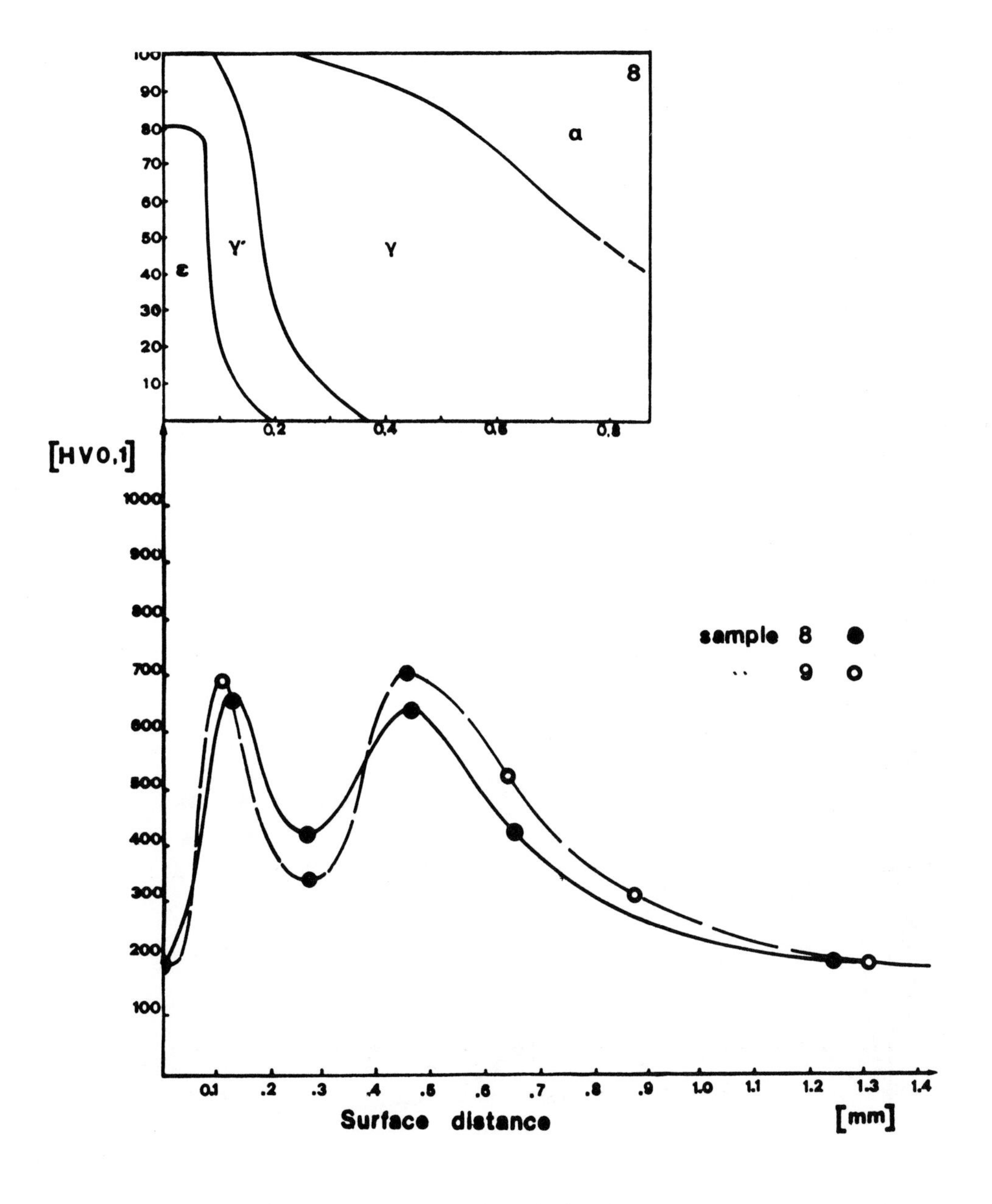

4 - Variation of the phase wt. percentages and microhardness as a function of the surface distance for the nickel and molybdenum sintered steel samples (specimens 8 and 9, table I).

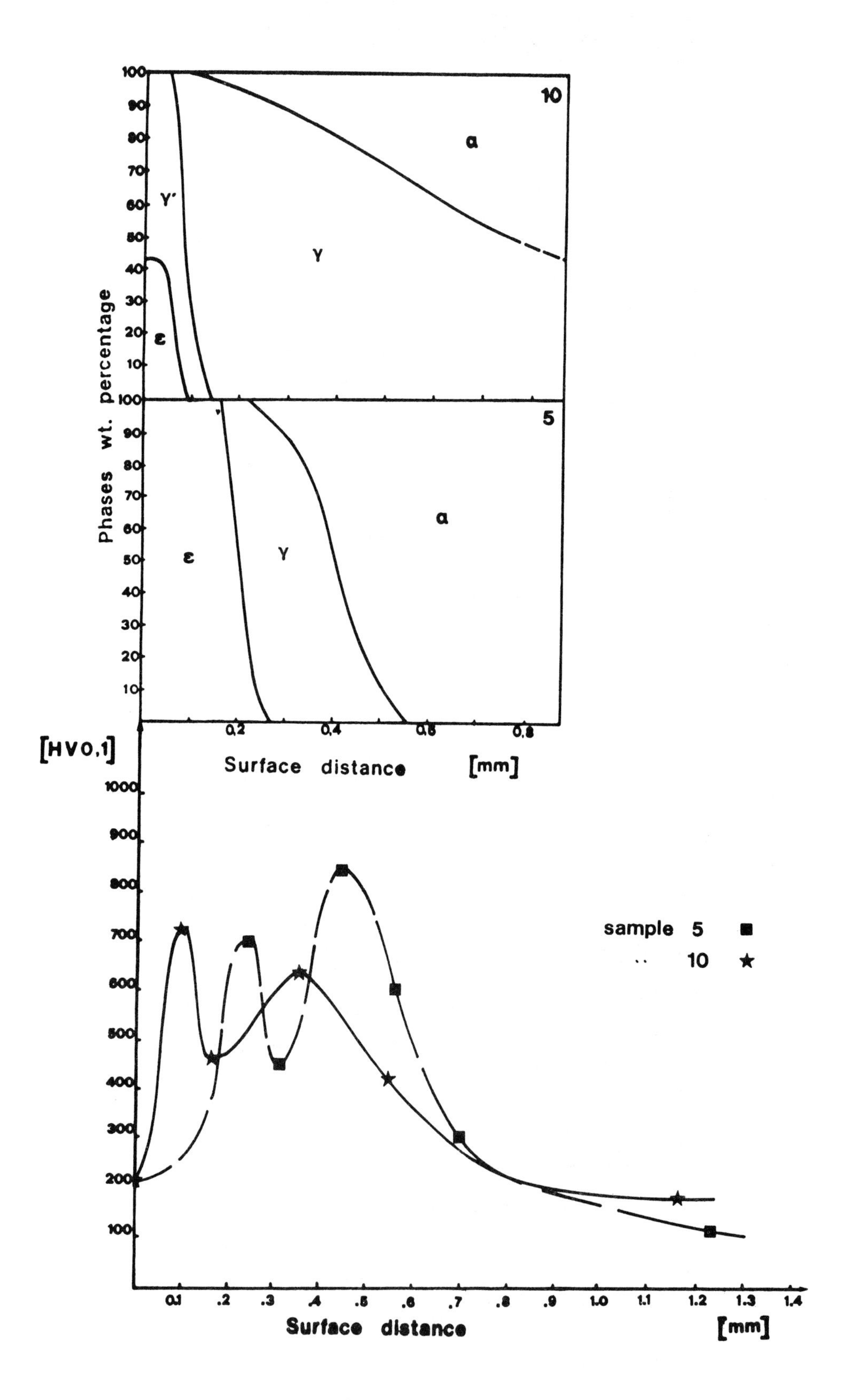

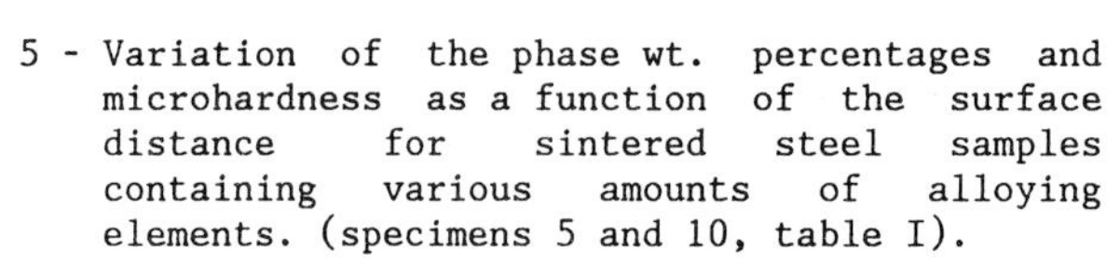
5 - Variation of the phase wt. percentages and microhardness as a function of the surface distance for sintered steel samples containing various amounts of alloying elements. (specimens 5 and 10, table I).

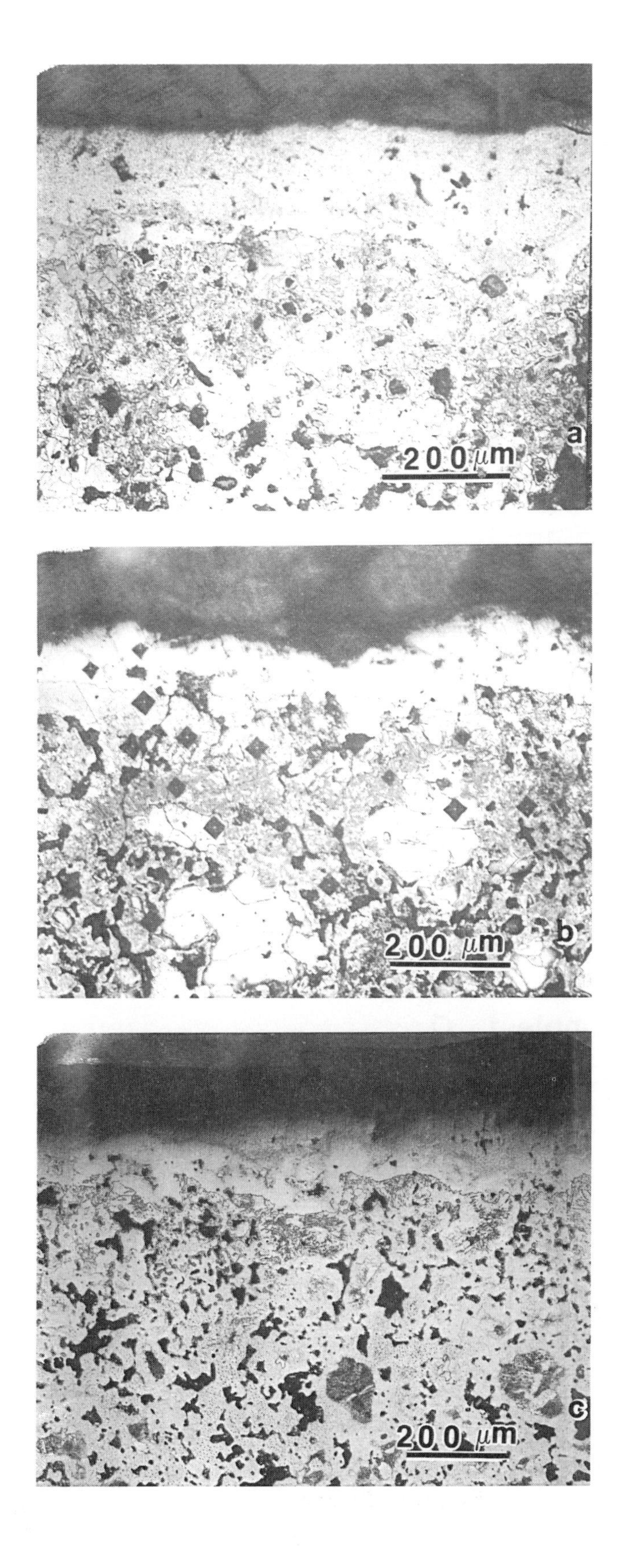

6 - Micrographs of cross section of: a) sample 1, b) sample 4 and c) sample 2 of table I. Nital etch.

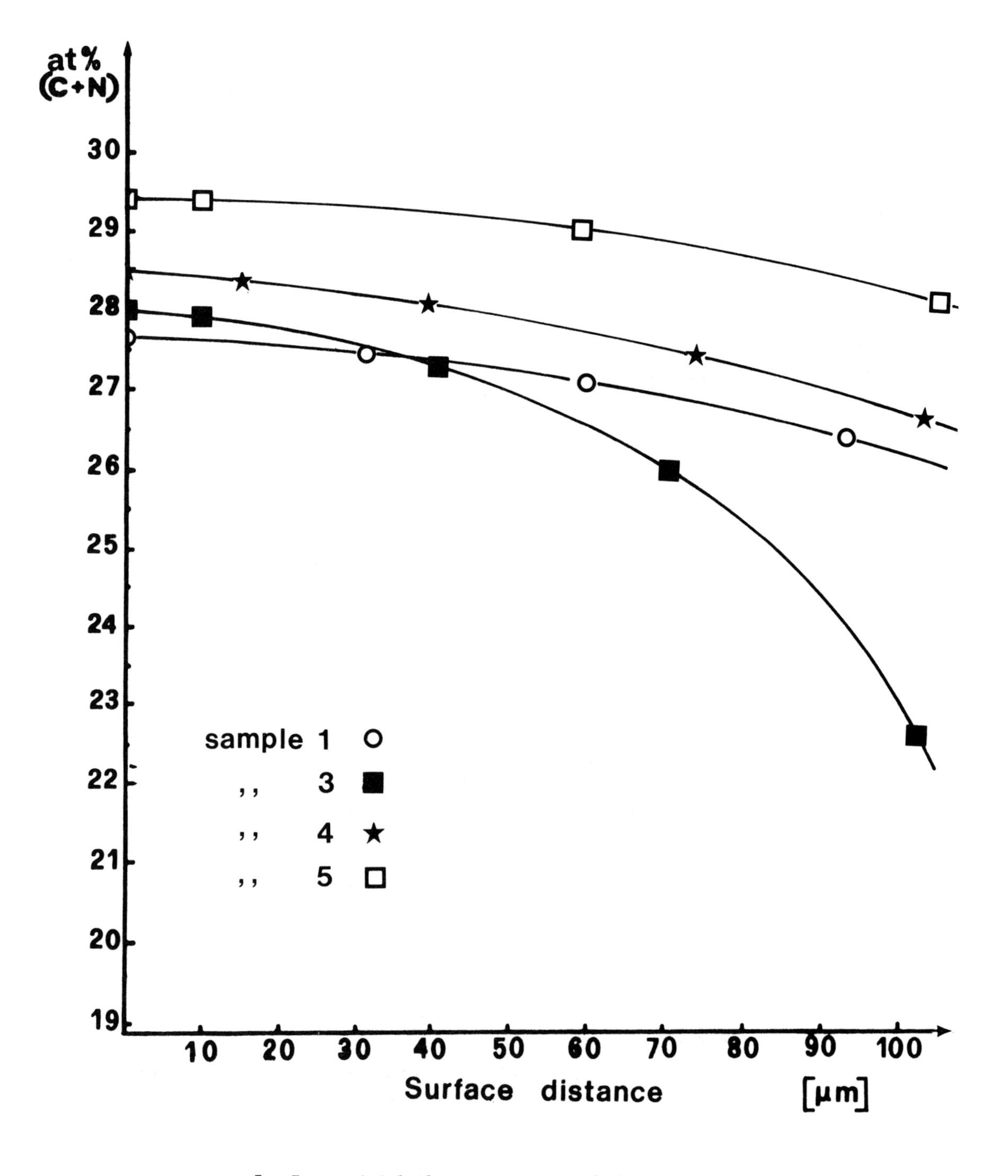

7 - Interstitial elements content of the ε phases (C+N)at.%, at various distance from the surface.

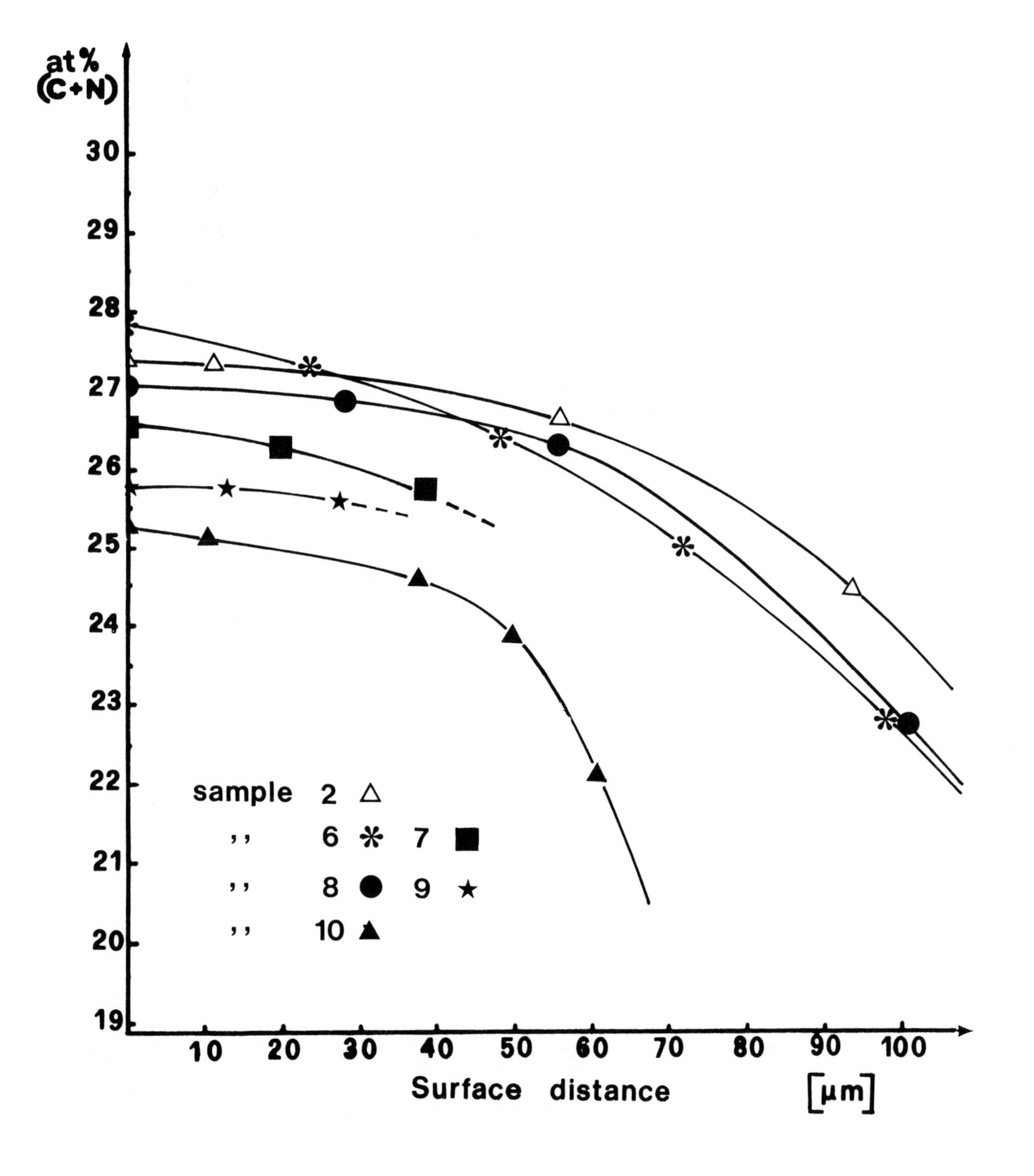

8 - Interstitial elements content of the ε phases (C+N)at.%, at various distance from the surface.

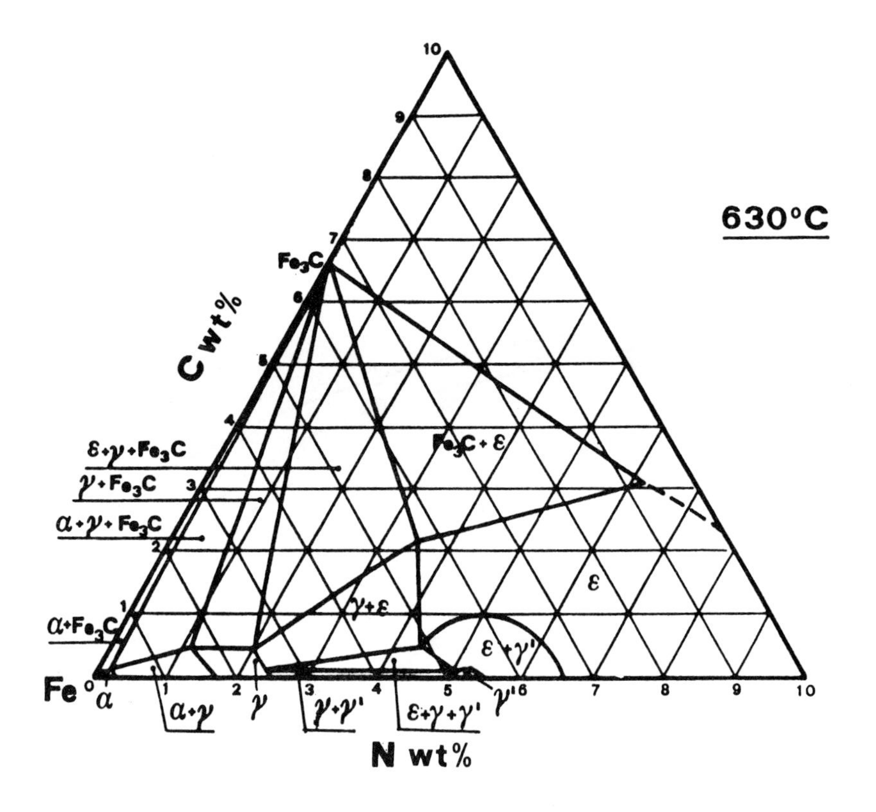

9 - Isothermal section of the metastable Fe-C-N phase diagram at 630 °C.

Laser treatment of thermal spray coatings

E LUGSCHEIDER, A KRAUTWALD and J WILDEN

The authors are with the Aachen University of Technology, Aachen, FRG.

Synopsis

Thermal spray coatings show good resistance against wear and corrosion. These properties can be improved by laser remelting. Due to the absence of cracks and the microstructure of these layers they can be used in cases where thermal spray coatings cannot be applied.

Introduction

Thermally sprayed coatings are widely used in industry where resistance to abrasive wear or corrosion or both is required. Coating a cheap substrate with a suitable alloy allows the production of composite materials with properties that fit the specific necessities of certain industrial applications.

The basic requirements for coatings are good adhesion, low porosity and the absence of cracks. If these requirements are not met gaseous or liquid media may corrode the substrate and cause severe failure. In the case of abrasive wear particles may break out and lead to failure of the coating.

This is the reason why thermal sprayed coatings for special applications are densified by a subsequent surface melting process. The following article demonstrates the laser-remelting of thermally sprayed coatings for protection against wear and corrosion.

Experimental Procedure

Alloys

For laser-remelting of thermally sprayed coatings new developed alloys were selected. A carbo-boridic alloy called A-1 containing carbides of the type M_7C_3 and $M_{23}C_6$ but also Ni_3B and Ni_3Si compounds can be manufactured by flame spraying. Flame spraying is a cheap coating process but it requires a melting point below 1300°C. The melting point of the A-1 alloy lies between 970 °C and 1050°C. Because of the low content of Boron and Silicon a conventional inductive or flame remelting of the surface is impossible. Even in the sprayed condition this alloy shows a very good behaviour against abrasive wear.

Another group of alloys is formed by complex-boridic alloys. These complex-boridic, so-called τ-borides, are only stabilized at the composition $Ni_{21}M_{R2}B_6$. The stabilizing metals are for example the refractory metals Titanium, Tantalum and Niobium. The melting point of such alloys lies in the region from 1080°C up to 1170°C. According to this, τ-boridic alloys stabilized by Titanium, Tantalum or Niobium can be worked by flame spraying.

Table 1 shows the chemical composition, the hardness in the as-cast conditions, the melting region and the particle size of the powders that have been used.

The Coating Process

The coating process selected was flame spraying. The coatings were produced with a Metco 5 P gun. This kind of equipment has the powder container directly connected to the top of the gun. The powder falls into a small chamber, where it is hit by the flame. The powder particles are molten and stinged onto the work piece.

Laser-Remelting

Laser-remelting was carried out using a 2 KW laser with a 10''lens. During the laser process the surface of the coating was centered in the focus of the lens. Using a special handling system it was possible to choose several feed rates just as different distances between the beads.

The power was varied from 400 W to 1500 W, the feed rate from 0.5 m/min up to 4 m/min and the distance between two beads varied between 0.2 mm and 2 mm. To prevent cracks in the laser remolten coating the samples were heated up to 400 °C before processing.

Abrasive Wear Test

The abrasive wear tests were carried out on a pin-on-disk equipment. The coated samples with a diameter of 16 mm are pressed onto a Silicon Carbid paper with a force of 5 N. The paper is rotating with 50 m/min. During testing the weight loss of the samples was measured after every 25 m and the paper with a grit of 600 was changed in every case.

Results and Discussion

τ-Boridic Nickel-Base-Alloy

A typical flame sprayed coating of τ- boridic alloys is shown in fig. 1. Making use of the processing parameters shown in table 2 the layer could be molten completely and a metallurgical connection to the substrate was achieved. An example is given in fig. 2.3, where the remolten layer of a Titanium-τ-boridic-alloy is shown.

In the transition region a heat affected zone is visible containing ferrite, retained austenite and bainite. In the molten layer a fine dendritic microstructure with interdentritic eutectic is visible. This results from the rapid quenching of the melt subsequent to the laser melting process. Slight variations in composition may cause a different etching behaviour of the individual laser tracks. The test cycles done with the Ti-τ-boride alloys show very clearly the influence of laser power on melt depth and layer hardness, (fig. 4).

The melt depth increases nearly linear with the rising laser power, as it was expected. The curve of the layer hardness shows two interesting facts. Firstly the hardness remains nearly constant with increasing laser power. If the laser power produces a melt depth which is exactly the same as the coating thickness, the layer hardness decreases rapidly. This can be explained by the dilution of the coating alloy with iron from the molten substrate and the loss of Titanium due to oxidation during processing. Titanium oxide was found on the surface of the sample. The amount of dilution can be modified by the laser parameters. The hardness cannot take any value below the hardness of the substrate even if the laser power is increased to very high values.

Carbo-boridic-Alloy A-1

The behaviour of this alloy after laser melting is quite different compared to the τ-boride containing samples. The high amount of hard phases and metalloides in the A-1, fig. 5 shows the microstructure in the as-cast condition, leads to a sufficient amount of hard phases after laser melting. The microstructure of a laser melted A-1 layer is presented in fig. 6.

Needles of chromium-carbides and -borides are visible within the nickel matrix. A diffusion zone can be detected between the molten surface layer and the substrate.

The decrease in hardness of 780 HV in the as-cast condition to 770 HV after laser melting is not severe however, the difference in thermal expansion of layer and substrate results in the formation of cracks.

The results of the pin-on-disk experiments are shown in fig. 7. The abrasive wear data of laser remolten A-1 layers are compared with the wear data of an A-1 coating in the as-sprayed condition.

A decrease in wear of about 30% was reached by laser remelting. This can be explained by the microstructure. The carbides and borides are clearly formed in the coating after the laser treatment. This leads to a significant decrease in abrasive wear data. Laser remelting produced an increase of adhesion strength from 300N/mm^2 in the as-sprayed condition to more than 1000N/mm^2 after processing. The reason is the metallurgical connection of the layer to the substrate.

Conclusions

This paper shows the influence of laser treatment of thermal sprayed coatings on the coating microstructure and the coating properties. The advantage of the laser processing can be summarized:

- low heat input and distortion
- dilution of the coating alloy by the substrate material can be adjusted by the laser parameters
- high degree of flexibility.

The coating shows:

- excellent adhesion strength
- the same hardness as in the as-cast condition
- high resistance against abrasive wear
- high corrosion resistance resulting from the absence of pores.

Laser melting of thermal sprayed coatings leads to new applications in surface-technology.

Acknowledgment

Special thanks are directed to B.L. Mordike and H.W. Bergmann as well as to the Fraunhofer Institut für Produktionstechnologie in Aachen for the laser processing tests.

The authors thank H.C. Starck Berlin for the powder production.

Table 1) Physical properties

Alloy	Chemical Composition wt.-%									T_s °C	T_l °C	Hardness HV 50	Particle size µm
	C	B	Si	Cr	Fe	Ti	Nb	Ta	Ni				
A-1	3.0	2.6	3.5	24.0	3.5	-	-	-	bal	961	1024	780	-90 +30
Ti-τ	-	3.0	-	10	-	13.0	-	-	bal	1093	1171	718	-90 +45
Ta-τ	-	2.0	-	8.0	-	7.0	-	24,5	bal	1112	1149	650	-90 +45
Nb-τ	-	3.0	-	20.0	-	-	11.0	-	bal	1079	1143	640	-90 +45

Table 2) Laser processing parameters

Alloy	Power KW	Feed rate m/min	Distance mm	Temperature °C
A-1	0.6	0.5	0.4	400
Ti-τ	0.7	0.5	0.2	400
Ta-τ	0.7	1.0	0.2	400
Nb-τ	0.7	2.0	0.2	400

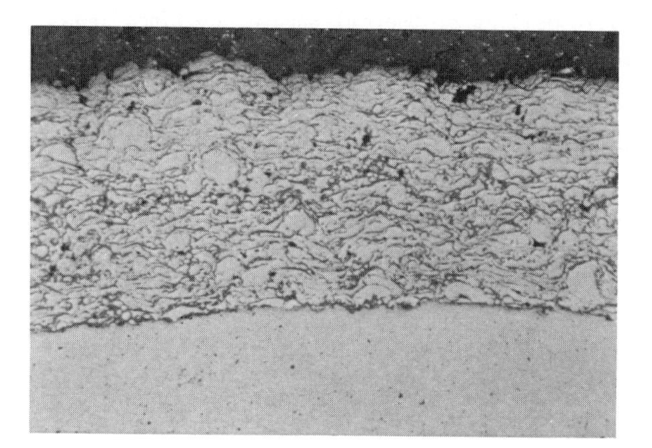

Fig. 1 Flame sprayed coating of τ-boridic alloy

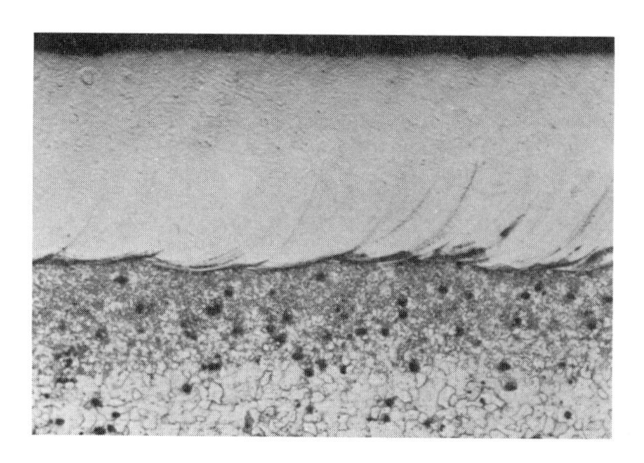

Fig. 2 Laser remolten Titanium -τ-boridic layer

Fig. 3 Laser remolten Titanium -τ-boridic layer

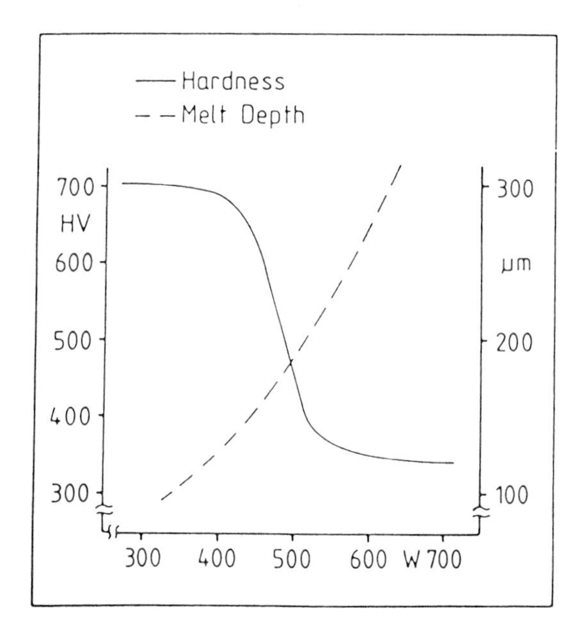

Fig. 4 Influence of laser power on melt depth and layer hardness

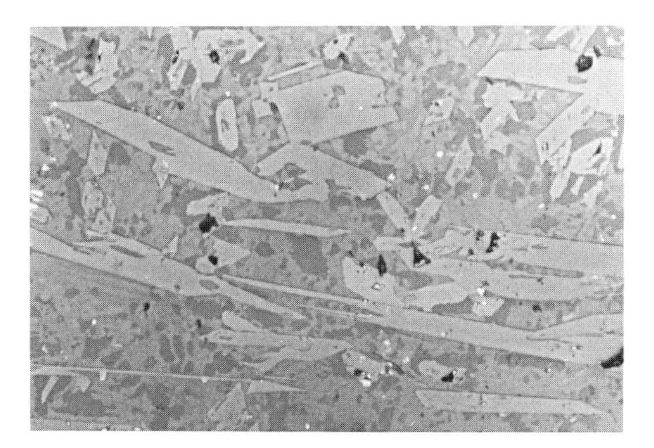

Fig. 5 Structure of A-1 alloy in the as cast conditions

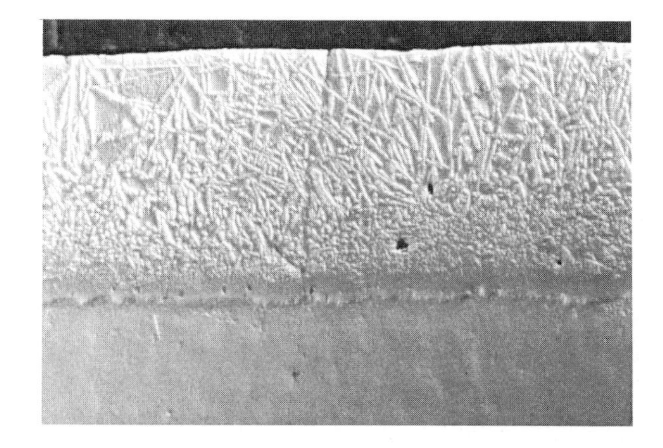

Fig. 6 Structure of laser remolten A-1 layer

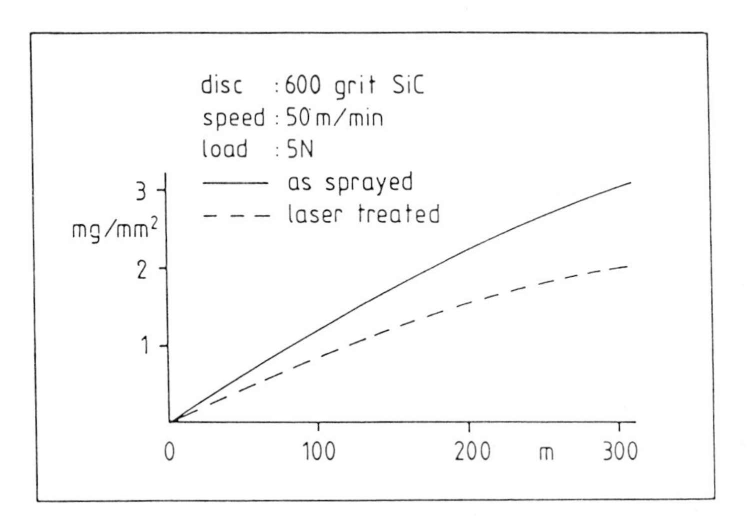

Fig. 7 Abrasive wear data

Laser surface hardening of alloy cast irons

Z D CHEN, D R F WEST, W M STEEN and M CANTELLO

ZDC, DRFW, and WMS are at Imperial College, London and MC is at R.T.M., Turin, Italy.

SYNOPSIS

An investigation is reported of the laser surface melting of a series of alloy cast irons:- Silal, Ni-hard, Ni-resist and a high chromium alloy. Also surface alloying has been investigated using substrates of flake graphite or spheroidal graphite iron. Alloying was achieved by powder feed into the laser melt pool using a stainless steel powder and also chromium powder. In addition, precoating of the substrate with powder was used to produce alloy layers of chromium and nickel, and of chromium and vanadium. Melting and alloying experiments used both a 2 kW and a 15 kW laser. Microstructures and hardness have been determined. The Silal alloy showed the highest degree of surface hardening associated with a fine scale microstructure containing a substantial proportion of carbide. Measurements of wear resistance showed beneficial effects of the laser treatments: the most wear resistant group comprised the nickel-chromium and chromium-vanadium alloyed surfaces and the surface melted Ni-resist alloys.

INTRODUCTION

Laser surface melting and alloying of cast irons enable a wide range of rapidly solidified structures to be produced on the surface of components[1,2]; of particular interest is the potential for property enhancement, especially in terms of hardness, and resistance to wear, erosion and corrosion[3-5]. At a constant laser power, control of beam diameter and traverse speed leads to control of thermal history i.e. "melt time" and temperature, and hence to melt zone width and depth, solidification rate, temperature gradient and cooling rate. Using a 2 kW CW CO_2 laser, melt zone widths up to several mm and depths up to approximately 1 mm are readily produced, with cooling rates of the order of 10^3 - 5.10^4 K/s.

The rapid cooling rates, applied to cast irons, typically produce fine scale white iron structures, high in carbide content and of high hardness. With appropriate choice of alloy composition and thermal conditions cracking during processing can be virtually eliminated. Variation in cooling rate and rate of solidification provide control of dendrite secondary arm spacing and of ledeburite eutectic interphase spacing. In the laser melting of flake and spheroidal graphite cast iron variation of interaction times produces a range of melt times and temperatures which leads to varying degrees of graphite solution and hence to various carbon contents in the liquid alloy[6,7]. Investigation of a 3.8% C 2.9% Si S.G. iron, using a 2 kW laser has shown that at low traverse speeds (relatively long interaction times) hardness levels > ~ 1000 HV were obtained[7]; essentially full solution of graphite occurred during melting, and solidification produced austenite dendrites and ledeburite eutectic. With increasing traverse speeds less solution of graphite occurred and the structures contained increasing proportions of austenite with some martensite and carbide; the hardness decreased to ~ 500 HV. Further increase in speed with a focussed beam resulted in lower carbon contents of the liquid; the hardness increased to ~ 700-800 HV with structures interpreted as predominantly martensitic. A similar effect of traverse speed has been reported in a study of laser processing to improve erosion resistance[5]; S.G. iron showed a greater sensitivity to traverse speed as compared with flake iron, and this is attributable to more rapid solution of flakes because of their higher ratio or surface area/volume.

The literature of laser surface melting covers a wide range of alloy cast irons, containing, for example, silicon, chromium and nickel[8-11]; correlations of dendritic secondary arm spacing were made with cooling rate and melt zone depth.

Laser surface alloying of iron with carbon has been used to produce carbon contents up to the hypereutectic range, enabling studies to be made of the rapid solidification of the white iron eutectic[12]. This eutectic is of the facetted (Fe_3C) and non-facetted (austenite) type showing strongly coupled growth with a skewed coupled zone[13,14]. Observations on laser alloyed specimens demonstrated mechanisms of eutectic formation, previously reported in coarser white iron structures[15]. Thus in hypereutectic alloys nucleation of eutectic on primary Fe_3C plates involved the formation of two-dimensional dendrites on the plate faces; both rod like and plate like eutectic morphologies were observed, the latter being dominant and tending to form parallel to the direction of steep temperature gradients. Interlamellar spacings (λ), in the range ~ 0.2-0.6 μm were obtained, and using estimates of average solidification rate, R, reasonable agreement was obtained between values of $\lambda \sqrt{R}$, for the rapid solidification conditions and those of 'conventional' solidification rates[16]. Transmission electron microscopical examination of rapidly solidified ledeburite showed that the matrix phase was ferrite, rather than austenite[17]: this feature was interpreted as austenite in the eutectic decomposing diffusionally during cooling by rejection of carbon to the Fe_3C plates leading to plate thickening.

The range of structures and properties attainable in cast irons is significantly extended by the addition of substantial proportions of elements such as silicon, nickel, and chromium, and the various industrially used 'alloy cast irons' are designed to match particular applications, e.g. for wear or corrosion resistance. In the present paper results are reviewed of an investigation into the structures produced by laser surface melting to four such alloys i.e. Silal, Ni-hard, Ni-resist and high chromium (Table I)[18] The objective of this research has been to characterise the structures as a function of laser processing conditions and to investigate the benefits achievable in wear resistance.

Additionally, new results are reported of the use of laser surface alloying/cladding to produce surface layers of alloy cast irons on grey flake and S.G. iron substrates. The use of such alloying and cladding techniques opens up a route to the use of substrates providing required combinations of mechanical properties at relatively modest cost, and to develop properties on the surface which may be difficult or expensive to produce in bulk components. The alloying elements used in the present work were nickel, chromium, and vanadium.

EXPERIMENTAL PROCEDURE

Laser processing (surface melting and alloying) was mainly carried out using a 2 kW CW CO_2 laser (Control Laser Ltd.). In the surface melting of the alloy cast iron processing parameters were varied using powers of either 1.8 or 1.0 kW, traverse speeds from 1-140 mm/s and beam diameters 0.4 or 3 mm. Argon was used as a shielding gas to prevent contamination of the melt zone. The specimen surfaces were grit blasted to provide a standard surface finish.

Surface alloying experiments were carried out on and S.G. iron (3.5% C) using continuous feed of powder into the melt pool; a stainless steel powder (100 mesh) was used of composition (wt %) 17.6 Cr, 11.9 Ni, 3.0 Mo, 0.04 C. Typically, a beam diameter of 2.5 mm was used, with a traverse speed of ~ 2 mm/s and with various powder feed rates to control the extent of alloying; other experiments used pure chromium powder. Alloying experiments were also made by precoating a pearlitic flake iron (3.0% C) with a mixture (1/1 by weight) of chromium powder and powder of an alloy containing 81% V, balance Fe: the coatings were applied by spraying a suspension of the mixture onto the substrate surface. The beam diameter was 2.5 mm and traverse speeds ranged from 2-11 mm/s.

A further series of surface alloying experiments was made at R.T.M., Turin, using a 15 kW CO_2 laser with a beam spot of ~ 11 x 11 mm. Alloying was done by precoating the flake iron substrate with powder mixtures of chromium and nickel (3/2 ratio by weight) and also of chromium and vanadium 81%/iron alloy (4/1 ratio by weight). Traverse speeds from ~3-12 mm/s were used, and melt pool depths from ~ 0.5-2 mm were obtained.

Structural investigations were made of laser surface melted specimens using microscopy (mainly optical and scanning electron microscopy). Etching was carried out using 3% nital. Also X-ray diffraction, hardness and microhardness measurements were made. Electron microprobe analysis (EPMA) of alloyed zones was carried out to determine the contents of nickel, chromium and vanadium.

Wear tests were made on samples processed with the 15 kW laser. A pin-on-disc device was used with crystalline silicon (electronic grade) as the disc, whose hardness (~ 1000 HV) was comparable to the highest hardness found in the laser treated materials investigated. The pin (3 x 6 x 10 mm) was held, unlubricated, against the rotating disc. Tests were made at room temperature in air and a maximum temperature of 60°C was obtained during testing. Wear on the disc was determined by profilometric analysis, and total wear (of pin and disc) was also measured.

RESULTS

Laser Surface Melting of Alloy Cast Irons: Microstructure and Hardness

Silal

The initial hardness was 240 HV, associated with a structure consisting of ferrite dendrites with fine graphite flakes. Laser surface melting, for all the processing conditions used, produced high values of surface hardness, in the range ~ 800-1010 HV. The primary dendrites were of austenite (Fig. 1), whose secondary arm spacings decreased from ~ 6 μm to ~1 μm with increase in traverse speed corresponding to increase in cooling rate: (this range of dendrite spacings was typical of the values for all the surface melted alloys studied). Cementite was present in all the Silal samples. At the highest traverse speeds only austenite and cementite were detected, but with decreasing traverse speed some martensite was identified by transmission electron microscopy also some graphite was found at the slowest speeds and possibly some iron silico-carbide is also present. The maximum hardness (~ 1010 HV) corresponded to the intermediate range of speeds (Fig. 2) with dendrite arm spacings of ~ 3 μm.

Ni-hard

The initial structure of austenite-martensite + cementite corresponded to a hardness of ~ 650 HV. Laser surface melting produced structures consisting of fine austenite dendrites + eutectic (formed as austenite + M_3C), and some martensite was present

(Fig. 3). The hardness was reduced by laser surface melting. With 1.8 kW power and a focussed beam the hardness was between ~ 420 and 490 HV (Fig. 2): using a 3 mm beam diameter the hardness ranged from ~ 550 to 615 HV.

Ni-resist

The initial structure consisted of austenite dendrites + eutectic (austenite + graphite) with a hardness of only ~ 160 HV. The laser melted structures consisted of fine austenite + eutectic for the complete range of processing conditions (Fig. 4). The hardening effect of laser melting was relatively small (Fig. 2): the maximum hardness found was ~ 450 HV with the coarsest dendrites, decreasing to ~ 360 HV with the finest dendrites with focussed beam.

The structure of a Ni-resist sample laser surface melted at 6 kW of power and with a beam of ~ 11 x 11 mm is shown in Fig. 5; the smooth profile and freedom from cracking and porosity illustrate the potential for using high power beams.

High Chromium Iron

The initial structure (hardness ~ 230 HV) showed ferrite dendrites and a duplex constituent interpreted as a eutectic of ferrite and M_7C_3 (Fig. 6a) Laser surface melting produced refined structures consisting of ferrite dendrites with refined eutectic (Fig. 6b). X-ray diffraction confirmed the structure as containing ferrite + M_7C_3. The hardening effect was relatively small (Fig. 2); the hardness ranged from 320 HV for the slowest traverse speeds to ~ 400 HV for the highest speeds. Quantitative metallography using back scattered electron imaging (Fig. 6) showed a reduction in the volume fraction of primary dendrites from ~ 74 to 64 vol % as a result of laser melting; correspondingly the proportion of eutectic increased from ~ 26 to 36%; also the proportion of carbide (in the eutectic regions) increased from ~ 13 to 24%.

Surface alloying

Figs. 7 and 8 illustrate the effect on alloy zone profile of increasing feed rate of stainless steel powder. There is an increase in the volume of the zone above the substrate surface level (i.e. thickness of clad), and a decrease in the volume of substrate melted; this gives an increased concentration of alloy content illustrated in Fig. 8. EPMA indicated a good degree of compositional homogeneity. The measured ratios of chromium and nickel agreed with the powder feed composition. Carbon contents were estimated from the dilution effect resulting from the relative volumes of substrate melted and alloy zone: the carbon content decreases with increase in alloy content, reaching a level of ~ 1.4 wt % C at 16% (Cr + Ni).

Fig. 9 shows the increase in chromium and nickel contents up to a value of 16 wt % produced by alloying with stainless steel at increasing feed rates. The microstructures, as viewed by light microscopy, showed dendrites with dark etching interdendritic regions; the proportion of dendrites increased with chromium and nickel content. Alloying resulted in surface hardening, showing a value of ~ 520 HV at an alloy content of ~ 5 wt %, decreasing to ~ 360 HV at ~ 16 wt % alloy content (Fig. 9).

Fig. 10 shows the hardening associated with chromium alloying up to a chromium level of ~ 33 wt %; within the range studied the hardness values lay between 580 and 480 HV. The microstructure showed primary dendrites plus eutectic (Fig. 11a). Austenite, ferrite and M_7C_3 have been detected by X-ray diffraction in an alloyed zone containing 32.5 wt % Cr, and austenite + M_3C in a zone containing 4.4 wt % Cr.

The effect of alloying, using precoating, with chromium and vanadium is also illustrated in Fig. 10. The variation of hardness with alloy content is complex, showing a maximum of ~ 700 HV at ~ 11% alloy content and a minimum of ~ 400 HV at 35 wt % alloy. The microstructures showed dendrites plus dark etching interdendritic regions. (Fig. 11b).

Wear Testing

The wear of the pin with increasing time of testing (i.e. wear distance travelled relative to the disc) tended to follow a very approximate linear relation after an initial period of up to 10 minutes (Fig. 11). Average wear rates of the pin material (i.e. melted and alloyed irons) are shown in Table II.

Of the alloy irons not subjected to laser treatment the Ni-hard showed the best wear resistance. Laser surface melting and alloying produced marked improvements in wear resistance. The surface melted alloy irons specimens investigated, namely Silal and Ni-resist exhibited wear resistance similar to that of untreated white cast iron. The laser surface alloyed samples (Cr + Ni and Cr + V) also showed similarly high wear resistance.

DISCUSSION

Microstructures and hardness of surface melted and surface alloyed material

In the initial as-cast materials, solidification of the Silal and Ni-resist had followed the stable equilibria, producing graphite, although the Ni-resist also showed some carbide eutectic; the Ni-hard and the high chromium alloy formed carbide. The rapid cooling in the laser melted alloys led to solidification according to the metastable carbide equilibria with the exception of some graphite formation in the Silal at the slowest cooling rates, indicative of the strong graphitizing effect of silicon. The exact details of the solidification and solid state cooling paths have not been determined. However, phase diagram data from the literature are available to assist in the interpretation of the laser processed structures and of the associated hardness include particularly the proportions of carbide and of martensite and the refinement of the microstructure.

In the Fe-Si-C metastable system the addition of Si to Fe-C alloys decreases the carbon content of the austenite + Fe_3C eutectic. An invariant reaction:

$$\text{liquid + austenite} \rightarrow \text{ferrite} + Fe_3C$$

is reported at 1145°C (14,19): the solubility of carbon in Fe_3C is very low. There are differences in detail between published phase diagrams[14] but the main features of the Silal solidification in the laser treated material are seen to be primary austenite formation, followed by the austenite + carbide eutectic. In the solid state the phase diagram shows the transformation of austenite to ferrite; however, with rapid cooling, preventing the attainment of the equilibrium solubility of carbon, austenite retention and some martensite is expected. The composition of the Silal (Fe-6Si-2.5C) is such that a small proportion of the iron silico-carbide ternary phase may form.

The Silal alloy is the only one of the series investigated that reached the high hardness levels (up to ~ 1000 HV) which are also attainable by laser melting of flake or S.G. irons. The formation of a large proportion of carbide (consistent with the phase diagram) contributes substantially to this hardness, but martensite formation is involved also. The fact that the Silal hardness varies little with change in laser parameters is attributed to the fine graphite flake size in the initial material; this facilitates the solution of graphite so that the carbon content of the melt zone is essentially that of the original substrate.

In the Ni-hard alloy, the phases found in the original as-cast state are also found after laser melting, namely austenite, martensite and carbide (M_3C), although martensite was only detected for the slower range of cooling rates of melt zones. Notwithstanding the considerable refinement of structure, the laser melted structures have a lower hardness (420-615 HV) than the original material (650 HV). The reduction of martensite content after laser melting is interpreted as due to a greater degree of retention of carbon in supersaturated austenite, causing a lowering of M_s temperature. The present observations agree well with results on rapidly solidified powder particles of a Ni-hard alloy[20] ; most of the particles studied solidified at cooling rates in the range 5.10^4 to 5.10^5 °C/s with secondary dendrite arm spacings of 0.2-0.5 μm. TEM observations showed that the structures consisted of austenite, cementite and martensite; calculations of volume fractions gave typical values of 68% austenite, 25% M_3C, 7% martensite[20] . Microhardness values ranged between ~ 310-410 HV; cooling in liquid nitrogen produced increases in hardness up to ~ 410-610 HV due to martensite formation[20].

In the Ni-resist alloy the transformation from a graphite-containing structure to fine scale structures containing carbide, produced a relatively small hardening effect from ~ 160 up to 400-450 HV. The absence of martensite reflects the very high nickel content and the high degree of carbon supersaturation of the austenite. The slight increase in hardness with decrease in cooling rate may be due to a lower degree of carbon retention in the austenite, and and increase in carbide content.

Phase diagram data for the Fe-Cr-C metastable system[21] show that a 30Cr 1C alloy commences its solidification by the formation of primary ferrite. The precise solidification path is not known but the invariant reaction:

$$\text{Liquid + Ferrite} \rightarrow \text{Austenite} + M_7C_3$$

is involved, followed by the:

$$\text{Liquid} \rightarrow \text{Austenite} + M_7C_3$$

eutectic; γ transforms to α in the solid state at a high temperature. The alloy shows a structure consisting of ferrite + M_7C_3 in both the initial and laser melted states. The observed reduction in the proportion of primary dendrites and the increase in carbide content in the laser melted state is interpreted as a consequence of the coring effects associated with rapid solidification, including a reduction in the extent to which the invariant reactions occur : the increased hardness of the laser melted specimens is mainly attributable to its higher carbide content.

Surface alloying

Further structural information is required to enable the solidification mechanisms to be fully interpreted and the investigation of the nature of the matrix has not yet been completed in all cases. However, some of the features observed can be considered in relation to the Fe-Cr-C and Fe-V-C phase diagrams[21,22]. In the case of alloying with stainless steel, bearing in mind the austenite stabilising effect of nickel, the primary dendrites in the composition range up to ~ 16 wt % Cr + Ni form as austenite. The eutectic stage of solidification is expected to be γ + M_7C_3, except at the lowest levels of Cr + Ni when γ + M_3C is expected. The increased proportion of γ dendrites with increase in Cr + Ni content is explicable on the basis of carbon dilution, and the reduced hardness follows from the lower volume fraction of carbide-containing eutectic. When alloying involves only chromium, the primary phase will be austenite until the Cr level increases to above ~25-30wt%.[21] The presence of M_7C_3 in a 32.5 wt % Cr alloy and M_3C in a 4.4 wt % Cr alloy is consistent with the phase diagram.

The alloy carbides in the vanadium-containing alloys are expected to be VC, V_2C and a ternary carbide[22]. At a carbon content of ~ 3% the proportion of VC is expected to increase at the expense of M_3C with increase in V content: in the higher range of V covered in the present work V_2C and the ternary carbide are expected. At the highest levels of Cr + V the primary phase is expected to be α . The carbon dilution effect resulting from increase in Cr + V content leads to a decrease in carbide content.

Wear resistance

Further investigation is required to elucidate the wear mechanisms involved in the present work. but some comments can be made on microstructures and hardness in relation to wear. The untreated graphite-containing alloys (flake graphite, Silal, and Ni-resist) of low hardness, showed poor wear resistance. The Ni-hard untreated, showed a good wear resistance, accompanying its hardness of 650 HV associated with the austenite + martensite + carbide structure. Also, the untreated white iron, with its ferrite matrix and high carbide content showed a very good wear resistance.

The surface melted Silal (HV ~ 880) and Ni-resist (HV ~ 450) showed similar, very good resistance to wear inspite of the hardness difference. The wear resistance of the Silal is attributed largely to its high carbide content, while in the Ni-resist an important factor may be a high work hardening capacity of the austenite, including the possibility of deformation-induced martensite. The very good wear resistance of the surface alloyed layers is associated with hardness levels of significantly less than that of the laser melted Silal - 510 HV (Cr +Ni) and 620 HV (Cr + V). The compositions of these layers has not yet been determined but from the observations such as those in Fig. 11 carbide-containing eutectics will be an important factor in controlling hardness.

In practical terms the laser melting of Silal offers a means of achieving a very good wear resistance, while presumably achieving also the beneficial effects of silicon on corrosion resistance. The laser surface alloying technique offers considerable interest in achieving a wear resistant surface on a relatively cheap substrate such as flake iron, with good mechanical properties.

SUMMARY AND CONCLUSIONS

1) Changes in hardness resulting from the laser surface melting of alloy cast irons vary considerably depending on the major alloy elements present, and their concentrations. In the case of Silal laser melting changes the initial grey iron structure to a refined white iron structure containing substantial propertions of carbide and martensite: surface hardness values of ~ 880-1010 HV were obtained as compared with the initial value of ~240 HV. Ni-resist, after laser melting, consisted of primary austenite dendrites with a fine (austenite + M_3C) eutectic: the initial hardness of ~ 160 HV was increased to ~ 400-450 HV. In a high chromium alloy the initial ferrite + M_7C_3 carbide structure (~ 230 HV) was refined by laser melting and contained an increased carbide content, giving hardness values of ~ 320-400 HV. In a Ni-hard alloy the initial hardness of ~ 650 HV associated with a structure of austenite-martensite + carbide, was reduced by laser melting; this is attributed to a reduction in martensite content by greater retention of carbon in solution in austenite. Of the four alloys studied, the Silal is of particular interest in terms of laser surface melting, as providing a combination of high hardness and corrosion resistance.

2) Effective surface hardening of S.G. or flake iron substrates can be achieved by laser surface alloying. The use of combinations of Cr + Ni and Cr + V have produced hardness levels ranging up to ~ 700 HV, resulting from the formation of fine scale structures consisting of primary austenite dendrites together with carbide-containing eutectics.

3) Laser surface melting produces considerable improvements in wear resistance in Silal and Ni-resist: similarly laser surface alloying of grey iron with Cr and Ni or Cr and V is very effective in achieving wear resistance.

ACKNOWLEDGEMENTS

Acknowledgements are made to the Chinese Government for financial support to one of the authors (Z.D.C.), to B.C.I.R.A. for supplying the cast irons and to Dr. K. Scrivener for her contribution to the quantitative metallography.

REFERENCES

1) H.W. BERGMANN. Surface Engineering 1985, 1,2, 137.

2) D.N.H. TRAFFORD, T. BELL, J.H.P.C. MEGAW and A.S. BRANSDEN. Met. Tech. 1983, 10, 69,

3) G. RICCIARDI, F. PASQUINI and S. RUDILASSO. Proc. Conf. Lasers in Manufacturing Technology, IFS Nov 1983

4) A. BLARASIN, S. CORCENITO, A. BELMONDO and D. BACCI. Wear, 1983, 86, 315.

5) C.H. CHEN, C.J. ALSTETTER and J.M. RIGSBEE Met. Trans. 1984, 15A, 719

6) I.C. HAWKES, W.M. STEEN and D.R.F. WEST, Electroheat for Metals Conference Proc., Sept. 1982, 4.4.

7) I.C. HAWKES, H.M. WALKER, L. LUNDBERG, W.M. STEEN and D.R.F. WEST. 3rd Int. Symp. The Physical Metallurgy of Cast Iron Eds. H. Fredriksson and M. Hillert. Mat. Res. Soc. Stockholm, Elsevier Science Publ., 1984, p447

8) B.L. MORDIKE. Z. Werkstofftech, 1983, 14, 221

9) H.W. BERGMANN and B.L. MORDIKE. ibid, 1983, 14, 228.

10) H.W. BERGMANN, B.L. MORDIKE and H.V. FRITSCH ibid, 1983, 14, 237

11) H.W BERGMANN, G. BARTON and J. METZ. ibid, 1983, 14, 244

12) A.M. WALKER, D.R.F. WEST and W.M. STEEN. Met. Tech. 1984, 11, 399.

13) H. JONES and W. KURZ. Met. Trans. 1980, 11A, 1265

14) I. MINKOFF. The Physical Metallurgy of Cast Iron. John Wiley and Sons 1983.

15) M. HILLERT and H. STEINHAUSER, Jernkont Ann. 1960, 144, 520

16) I.C. HAWKES, A.M. WALKER, W.M. STEEN and D.R.F. WEST. Lasers in Metallurgy 2, Eds. K. Mukerjee and J. Mazumder, ASM 1984.

17) H.M. FLOWER, A.M. WALKER and D.R.F. WEST. J. Mat . Sc., 1985, 20, 989.

18) Z.D. CHEN, D.R.F. WEST and W.M. STEEN. ICALEO, Boston 1986.

19) J.E. HILLIARD and W.S. OWEN JISI, 1952, 268

20) G. FROMMEYER. Z. Metallk, 1985, 76, 10, 662.

21) W.R. THORPE and B. CHICCO. Met. Trans. 1985, 15A, 1541

22) C.E. DREMANN. Metals Handbook A.S.M., 8th Edition, 1973, p415

TABLE I

SPECIFICATION COMPOSITION (wt %)

Alloy	*C*	*Si*	*Mn*	*Ni*	*Cr*	*Initial Structure*
Silal	2.5	6.0	0.65			Ferrite + graphite
Ni-hard	3.2	0.5	<0.5	3.5	1.5	Austenite+martensite + cementite
Ni-resist	3.0	1.8	1.25	21	2.0	Austenite + eutectic (γ + M_3C) + graphite
High-Cr	1.0	0.5	0.5		30	Ferrite + carbide

TABLE II

WEAR TEST DATA

Material	*Wear rate of pin µm/min*	*Hardness of pin material HV*
Silal	1.45	240
Flake graphite grey iron	1.02	180
Ni-resist	0.62	160
Ni-hard	0.37	630
Silal (Surface melted)	0.29	980
Ni-resist (Surface melted)	0.24	460
White cast iron	0.24	670
Flake graphite grey iron surface alloyed with:		
a) Cr and Ni	0.23	510
b) Cr and V	0.22	620

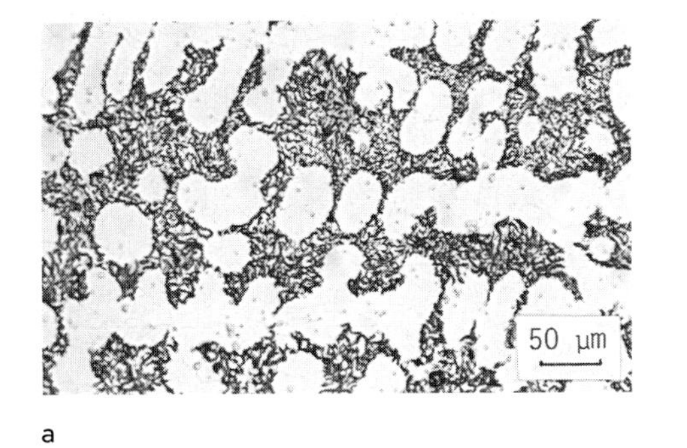

a

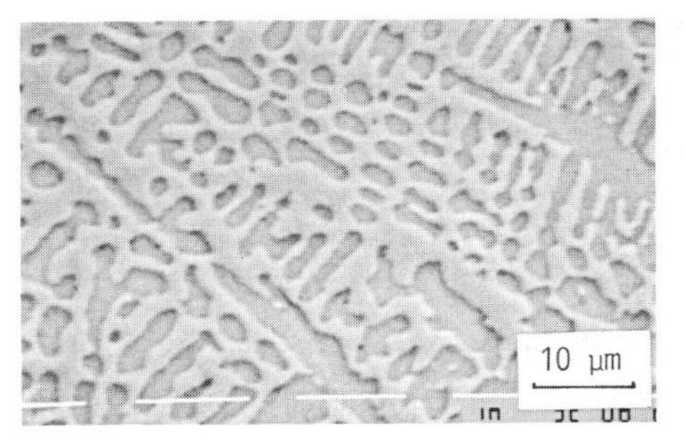

b

FIGURE 1

Silal a) substrate structure. Ferrite + graphite-containing eutectic. b) Laser melt zone structure Austenite + carbide containing eutectic. 1kW, 3mm 10 mm/s.

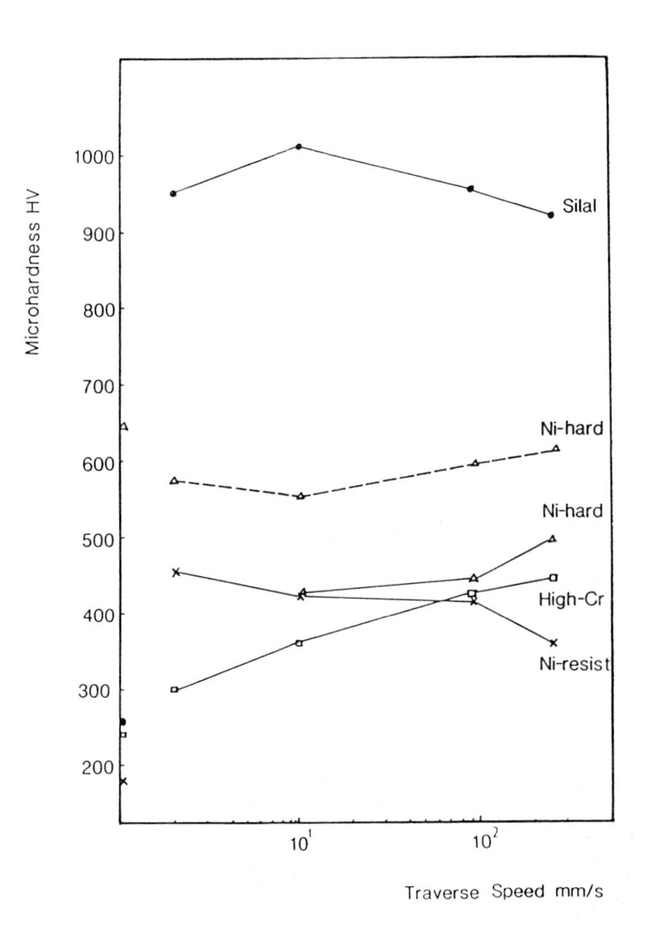

FIGURE 2

Alloy cast irons. Laser melt zone hardness vs traverse speed. Substrate values shown on the hardness axis 1.8 kW, 0.4 mm. beam diam. ——; 3mm - - -.

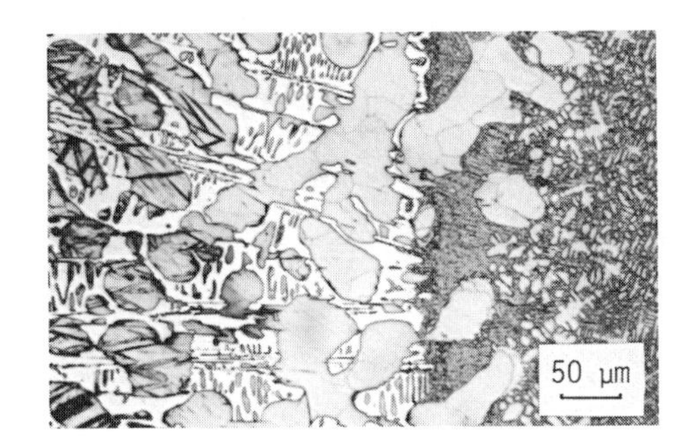

FIGURE 3

Ni-hard. Part of substrate (austenite/martensite + M_3C) and laser melt zone: refined structure of austenite + eutectic 1.8 kW, 0.4 mm, 10 mm/s.

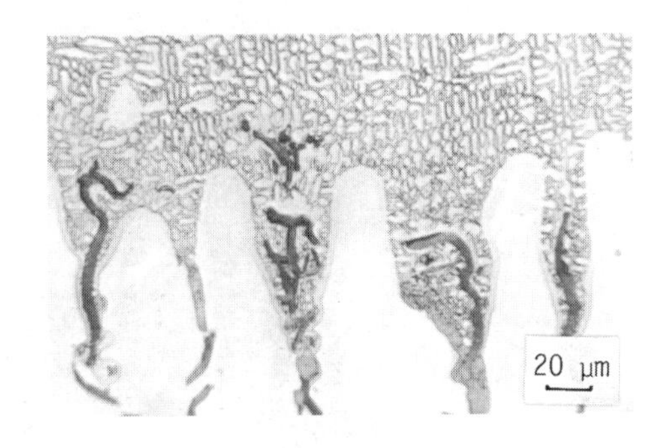

FIGURE 4

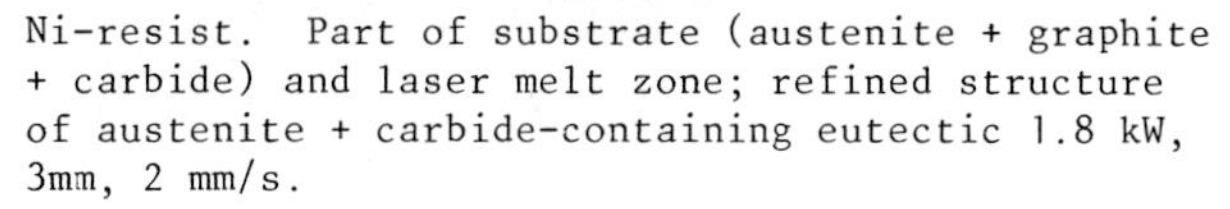

Ni-resist. Part of substrate (austenite + graphite + carbide) and laser melt zone; refined structure of austenite + carbide-containing eutectic 1.8 kW, 3mm, 2 mm/s.

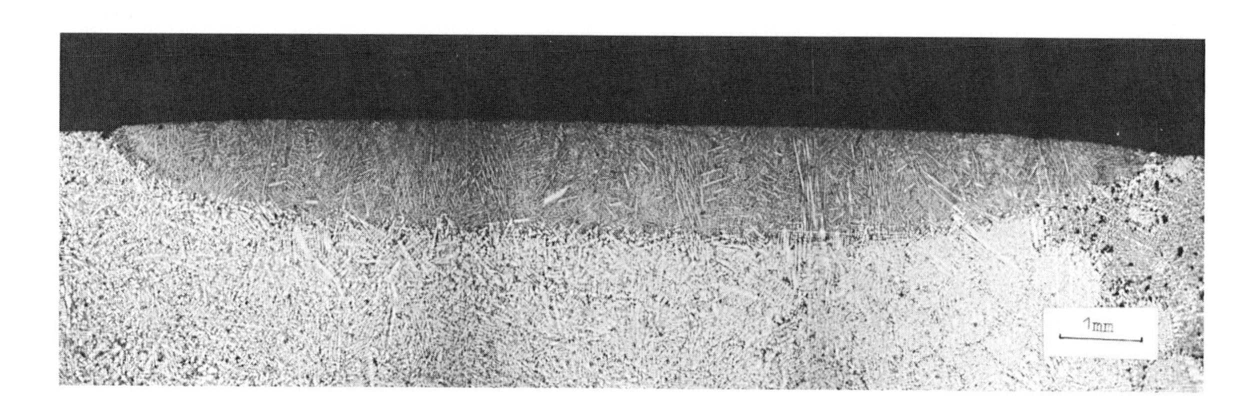

FIGURE 5
Ni-resist. Laser melt zone: austenite + eutectic. 6kW, beam spot 11 x 11 mm^2

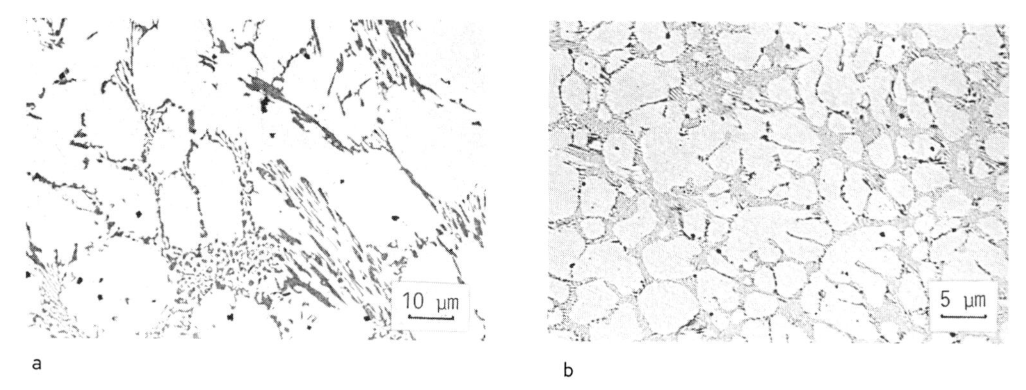

FIGURE 6
High chromium iron a) substrate structure ferrite + eutectic containing M_7C_3 b) laser melt zone structure: refined structure of ferrite + eutectic containing M_7C_3 . Back scattered electron image 1.8 kW, 0.4 mm, 1 mm/s.

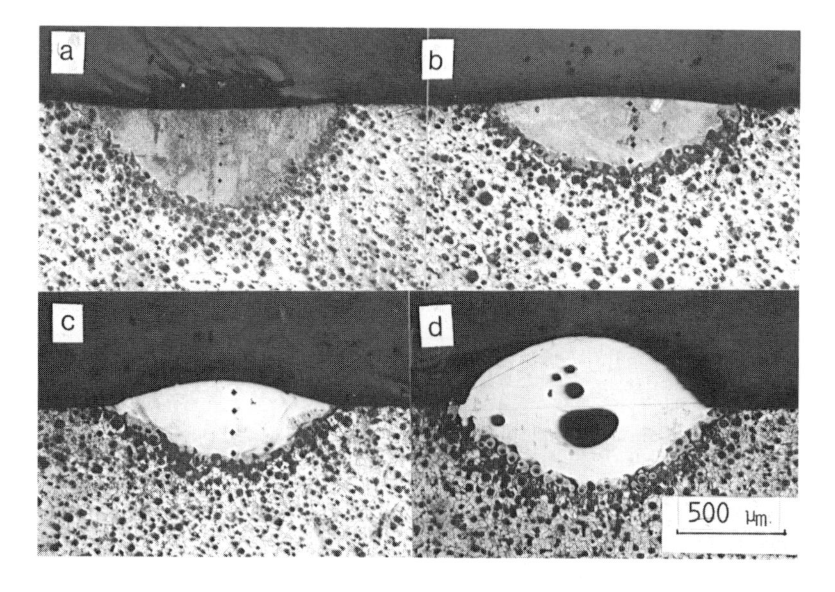

FIGURE 7
S.G. iron surface alloyed, using stainless steel powder (~ 18Cr, 12 Ni). a) surface melted - no powder feed. (b-d) showing effect of increasing powder feed speed (b-10, c 20, d 60; arbitary units) on profile and dimensions of melt zone 1.5 kW, 2.5 mm, 2.3mm/s.

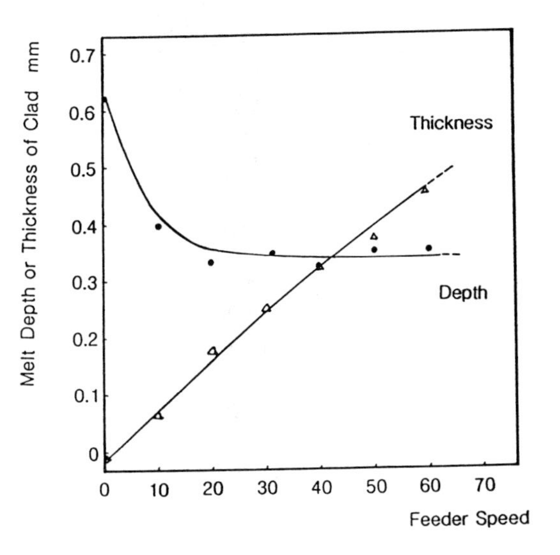

FIGURE 8

S.G. iron alloyed as in Fig. 7. Variation of melt zone depth and thickness of 'clad' layer (i.e. above level of substrate surface) with powder feed speed (arbitrary units).

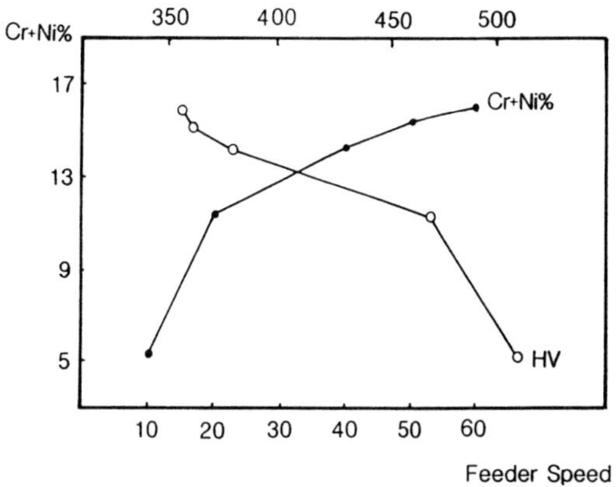

FIGURE 9

S.G. iron alloyed as in Fig. 7. Variation of Cr + Ni content and hardness of melt zone with powder feed speed.

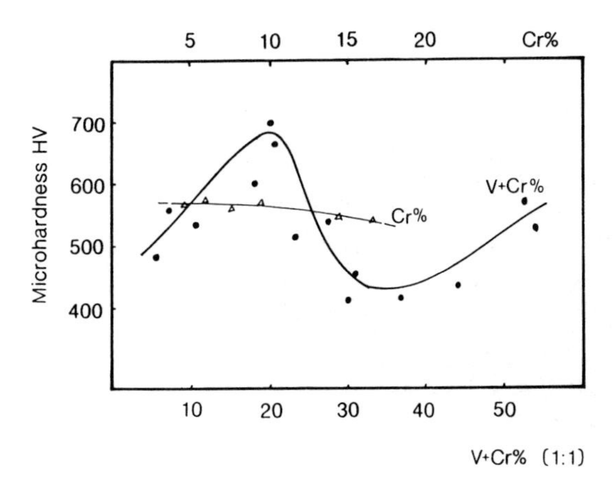

FIGURE 10

Flake graphite iron, surface alloyed by continuous powder feed of Cr or by precoating with Cr + V, powder Variation of melt zone hardness with alloy content. Cr 1.5 kW, 1.5 mm, various powder feed speeds, Cr + V. 1.7 kW, 2 min, 10-35 mm/s

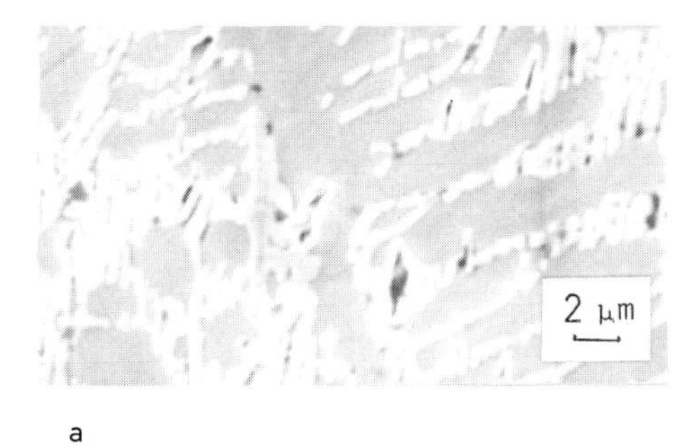

a

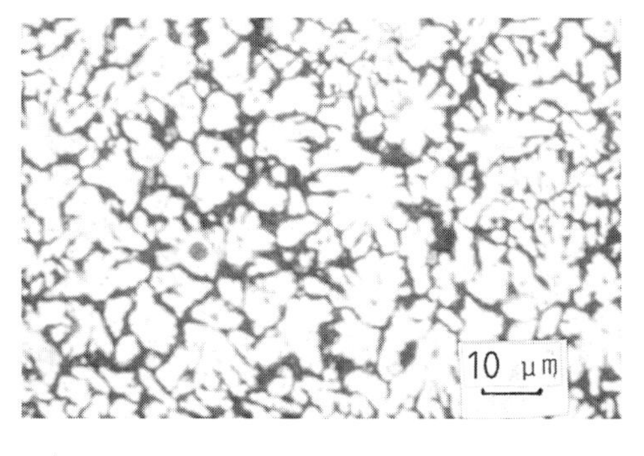

b

FIGURE 11
As in Fig. 10. Laser surface alloyed structure. a) 32.5% Cr b) 15.8% Cr, 14.8% V, 1.7 kW, 2mm, 7.8 mm/s. Structures showing primary dendrites + carbide - containing eutectics.

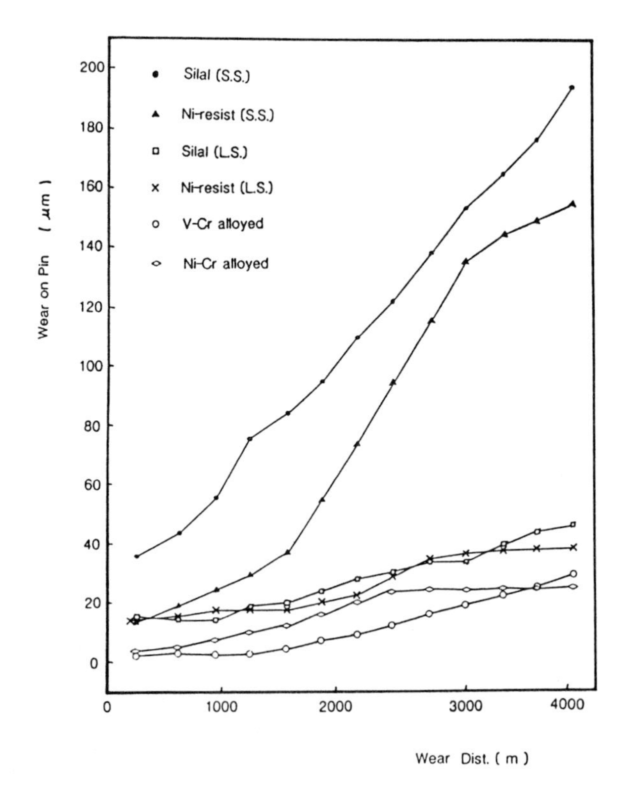

FIGURE 12
Pin-on-disc wear test data for surface melted and surface alloyed irons. Wear on pin vs distance travelled by disc relative to pin. SS = substrate LS = laser surface melted (Silal 12 kW, Ni-resist 10 kW) Alloyed 15 kW

Microstructure of high-speed steel by electron beam surface alloying and post heat treatment and its effect on wear resistance

HAIOU ZHAO, YANWU FENG, CHUANFANG TANG and YUANPU ZHU

The authors are in the Department of Heat Treatment at the Beijing Research Institute of Mechanical and Electrical Technology, MIC, China.

SYNOPSIS

An electron beam surface alloying technique on W18Cr4V (AISI T1) high-speed steel with TiC/WC-Co powder has been studied. A microstructure survey and electron diffraction analyses for the alloying layer have been carried out and the relationship among alloying parameters, microstructures and wear resistance was discussed. The effect of post heat treatment on the microstructures and properties of the alloying layer was investigated. It was found that the alloying layer in W18Cr4V HSS with prior annealing structure consisted only of alloying zone and transition zone while no transformation region existed. There are optimum parameters for the electron beam surface alloying process. The results of wear resistance tests showed that EB alloying can significantly improve wear resistance, which is 3.1 times as great as that of conventional heat treatment in the static load condition. Under shock loading conditions, on the contrary, the alloying layer exhibits some decrease in wear resistance but can be improved by post heat treatment.

1 INTRODUCTION

Electron beam heat treatment, a new process for modifying the surface by applying a high energy density heat source to ferrous materials, has recently been developed. A significant feature of the process is its high-rate heating and cooling. These result in extra refinement of the microstructure, extension of solubility levels and dissolution of high melting point carbides. In this respect, much work has been carried out on materials such as carbon steel, cast etc.[1–4]

However, we are interested particularly in high-speed tool steels. In these alloys one expects that a significant improvement in machining properties and tool life will result from an extremely small grain size and a homogeneous dispersion of fine carbide particles. The previous work on electron beam and/or laser transformation hardening and melting of high-speed steel (HSS) showed that[5–8] these processes have great potential for improving the wear resistance of HSS. However, only electron beam surface alloying (EBSA) used on HSS is under study - the process of modifying the surfaces of relatively inexpensive substrates by adding small amounts of alloying elements to the metal surface by a high energy electron beam heat source. Thus this process can provide an alloying layer of special properties according to application requirements.

This paper emphasizes the study of the process, the microstructures and the phase structure of EBSA for W18Cr4V HSS (AISI T1).

Firstly, the influence of the EBSA process parameters on the microhardness of the alloying layer was investigated. By means of a metalloscope, SEM and TEM, microstructures and phase structures of the alloying layer were studied.

Secondly, the effect of post heat treatment (oil quenching at 1260°C, tempering at 560°C for 1 h, thrice repeated) on the microstructures and properties of the alloying layer is discussed.

Finally, laboratory wear resistance tests were carried out under various friction conditions.

2 EXPERIMENTAL METHOD

A commercial W18Cr4V HSS, received as annealed plate, was used for the substrate material. The chemical composition of this steel is (wt-%):

0.7-0.74 C 17.64-18.05 W 3.88-4.15 Cr and 1.02-1.11 V.

A mixed powder (25 wt-% TiC + 75 wt-% WC-Co) was coated onto the surface of the substrate by brushing. Then an EB process was performed on the surface of the specimen.

After EBSA, the specimens were cut transversely and observed by optical and electron microscopes. A Nital reagent (4-10% NHO_3, ethyl alcohol solution) was used for revealing microstructural features. The microstructures of specimens were studied mostly by the SEM (X-650 scanning electron microscope). The alloying elements of the layer were measured by electron probe microanalysis. Thin-foil transmission electron observations (H-800 TEM, operating voltage 200 kV) were carried out on specimen sections in the upper alloying region.

Static load sliding wear tests, in the form of dry friction, were conducted in an MM-200 wear machine. The wearing surfaces of upper samples were treated by the EBSA process and the lower samples by conventional hardening and tempering. The rotating speed of the lower samples was 200 rev/min. The upper samples moved back and forth parallel to the rotating axis at 35 times per minute, in order to make good contact with the wearing surfaces. Under constant rotation (total 30 000 rev) the loss of weight was a measure of the wear resistance. The worn surfaces were examined by SEM. The shock load sliding wear tests were carried out in an MLD-10 type wear machine. The upper samples processed by EBSA moved up and down at 150 times per minute to produce a shock force. The lower samples of 45 steel obtained by conventional heat treatment (heating at 840°C and water-quenching, tempering at 180°C for 1.5 h), rotated at 200 rev/min. The loss of weight after 3 h testing was a measure of the wear resistance.

3 RESULTS AND DISCUSSION

3.1 Effect of EBSA on composition and hardness

Figure 1 shows the composition and the hardness in the alloying layer on W18Cr4 HSS treated by EBSA. The alloying layer can be divided into three zones: alloying zone, transition zone and base material. The results of electron probe analysis show that after EBSA the alloying element content of the alloying layer is obviously higher than that of the substrate, and its distribution is not uniform. The alloying element content, highest at the surface, gradually decreases from surface to substrate. This will affect the microhardness curves of the alloying layer in which there are significant differences in hardness (Fig. 1b). The microhardnesses shown are $HV_{0.1}$ 1000 in the alloying zone and approximately $HV_{0.1}$ 900 in the transition zone; the microhardness thereafter suddenly decreases to that of the annealed steel substrate.

It is found by experiment that the alloying layer in W18Cr4V HSS with prior annealing structure consists of alloying and transition zones only; no transformation region exists. However, both a transformation region and a high-tempering region can be produced beneath the transition zone when EBSA is used in the prior quench-tempering structure. These observations had been made in EB transformation hardening of W18Cr4V HSS. Research [9] has shown that EB transformation hardening cannot be carried out in W18Cr4V HSS with prior annealing structure. This may be explained by the different dispersivity of the prior structure. In the rapid heating process such critical points as Ac_1, Ac_3 and Ac_m will move up to higher temperature with higher rate of heating [10–11] and the transformation points can also be raised with the decrease in dispersivity of the prior structure.[12]

At the same time, the effect of alloying elements on austenitization should be considered. We may point out that in the EB rapid heating process the austenite transformation temperature of W18Cr4V HSS with prior annealing structure may approach the melting point on account of the poor dispersivity; this makes the steel melt only, and no transformation hardening can be carried out. However, the prior quench-tempering structure is better in dispersivity. Its critical point rises less and does not come near to the melting point. Accordingly, a transformation hardening zone can be produced in EB heating. It can be seen that there is a limit for the steel species and its original structure in EB transformation hardening. For high-alloy steels containing more carbides which are more difficult to dissolve, EB melting or alloying are suitable techniques.

In our experiments the factors influencing alloying layer composition are qualitatively investigated as follows.

3.1.1 Control of thickness of alloy powder coating

The coating thickness affects not only the alloy content in the layer but also the roughness of the treated surface. In order to obtain a higher alloy content of alloying layer, one may expect to produce a thicker coating; however, the difficulty in obtaining a smooth surface on a thicker coating sample is increased, and is liable to cause powder splashing. So the coating brushed onto the surface should not be too thick. Experiments show that powder with large particles and heavy density is suitable for use in EBSA. When pure TiC powder is used in alloying the splashing is serious and makes it difficult to form a smooth layer because of its light density. However, a mixed powder (25 wt% TiC + 75 wt% WC-Co), in which the density of the WC-Co powder is heavy, should be used in alloying. The TiC powders adhere to the WC-Co powders and splashing can therefore be controlled.

3.1.2 Alloying parameters control

The melting depth of the substrate can be controlled by changing the EB alloying parameters. Figure 2 shows the effect of the parameters on the microhardness of the alloying layers. For a given thickness of coating, too low a beam power can melt only part of the powder coating, so that the alloying element content in the layer is lower and the increase in microhardness is small. The alloying powder coating can be fully melted with an increase of beam power, with little melting of the substrate. Thus the alloying element content in the alloying layer is increased and maximum hardness can be attained. When beam power is raised further still, there is a greater 'dilution' in alloying composition because of an increase in melting depth of the substrate. This decreases the alloying content of the layer, so that microhardness falls (Fig. 2). Otherwise powder splashing will be increased with beam power and will result in a lowering of the alloying content in the layer.

3.2 Microstructures in alloying layer

Figure 3 shows the SEM morphology of dendrite microstructures for EBSA. There are black particles containing Ti (as found by electron probe analysis) in the dendrite boundaries. From their form and distribution, they are TiC carbide particles which are not fully dissolved during the alloying process. The undissolved particles will be retained finally and distributed in the dendrite boundaries during solidification. However, all the WC carbide particles have been dissolved in alloying.

Investigation of the microstructures in the alloying layer by TEM show that the SEM dendrite structures are, in fact, martensites (Fig. 4). The morphology of these martensites is a mixed structure of lath-dislocation and plate-twin martensites. There are thin-film retained austenites at the lath boundaries and locally mass-retained austenites. Because the contents of alloy elements and carbon are higher in the layer, the plate martensites are more than laths. The electron diffraction results demonstrate that the phase of the dendrite boundary is a eutectic structure of α-ferrite/M_6C carbide with a fish-bone morphology (Fig. 5). Thus the microstructures in the alloying zone are martensite, retained austenite and a eutectic structure of α-ferrite/M_6C carbide.

3.3 Effect of post heat treatment on alloying microstructures

As mentioned above, the microstructure of EBSA is dendritic, which is not suitable for increasing surface strength. In order to eliminate the dendrite and refine the structure, post heat treatment can be used. After post heat treatment (oil quenching at 1260°C and tempering at 560°C for 1 hour, three times) the microstructure of the alloying layer is as shown in Fig. 6. It is found that post heat treatment can entirely eliminate original dendritic structures from the layer. Very fine and well-distributed carbide particles are precipitated in a layer saturated with alloying elements. TEM observations show that long particles in the networks along the grain boundary are MC-type carbides (Fig. 7) with fcc lattice; lattice parameter a = 4.16Å. Electron probe analysis shows that

$MC = Fe_{0.5}(W,Ti,Cr,V)_{0.5}C$.

However, there are two types of carbide particle precipitated within the grains (Fig. 8), in which the square particles are M_6C carbides (fcc, a = 11.08Å) and the polygonal particles are TiC carbides with fcc lattice; lattice parameter a = 4.32Å. The precipitation of these hard carbide particles contributes to improving the wear resistance of the alloying layer.

3.4 Wear-resistance properties

In order to study the wear resistance of the alloying processes under different friction conditions, both static load and shock load wear tests were carried out in the laboratory.

The results of tests under static load sliding friction conditions show that the EBSA process can significantly improve the wear resistance of W18Cr4V HSS, which is 3.1 times as great as that of conventional heat treatment (Fig. 9a). Post heat treatment makes little contribution to enhancing wear resistance of the alloying layer, which may even fall somewhat; however, wear resistance of the layer after post heat treatment is still higher than that of conventional heat treatment. Observation of the wearing surface topography shows that oxidized wear occurs mostly under this friction condition (Fig. 10). Abrasive dust found by X-ray diffraction analysis is Fe_2O_3 oxide, and there is no metallic matrix being peeled off. However, there are scratches of varying degrees of severity on the different wearing surfaces. Scraping of the conventionally heat-treated surface is very serious (Fig. 10a). The scuffing is coarse and originates at sites where a substantial quantity of large carbides are gathered. Some large carbides often remain in HSS after conventional heat treatment and these are liable to be peeled off the metallic matrix and to play an abrasive role in decreasing wear resistance. After EBSA, the large carbides can be fully dissolved and banded structures of carbides in the original materials can also be eliminated. Thus the topography of the scuffing on the alloying surface is not obvious (Fig. 10b) and anti-friction of the alloying layer is significantly improved.

SEM micrographs of the wearing surface show that very fine eutectic carbide networks will help to increase the anti-wear properties of the alloying layer. The fine networks appear to reinforce the matrix, so restraining the plastic yielding produced in the matrix. From this it can be seen that wear resistance of the alloying surface decreases to some extent after post heat treatment. Because the eutectic networks were eliminated and substituted for the carbide particles along the grain boundary by post heat treatment, the effect of the particles upon resistance to matrix deformation is much weakened. However, after post heat treatment in the alloying layer, there are MC-type carbides and precipitations of many M_6C-type and TiC carbide particles, and these well-distributed hard particles will contribute greatly to the improvement of alloying layer wear resistance, which is thus far higher in post heat treatment than in conventional heat treatment.

On the other hand, the results of the shock load wear tests show that (Fig. 9b) under shock load sliding friction conditions the EB alloying layer exhibits some decrease in wear resistance. The shock load wear resistance can be improved by post heat treatment, but it remains poor in comparison with conventional heat treatment. In our tests, however, there was no peeling off and separating from the surfaces, which shows that the binding of layer with substrate is metallurgical in nature. In the shock load condition the toughness of materials is much more important than their hardness.

The eutectic network on the EB alloying layer is obviously deficient in shock load wear resistance. Post heat treatment can eliminate the eutectic network and help improve wear resistance. Distribution of carbide particles along the grain boundary is still harmful to further improvement of the shock load wear resistance. So post heat treatment techniques will need to be investigated. Otherwise, decreased wear resistance of the alloying layer in the shock load condition is related to extra alloying elements which result in an increase in embrittlement.

4 CONCLUSIONS

1. The EB alloying layer in W18Cr4V HSS with prior annealing structure consists only of an alloying zone and a transition zone; no transformation region exists. The microhardness is $HV_{0.1}$ ∿1000 in the alloying zone, and approximately $HV_{0.1}$ 900 in the transition zone.

2. Alloying element distribution is not uniform in the alloying layer. The alloying element content will be highest at the surface, gradually decreasing from surface to substrate. Thus there is an optimum processing parameter in the EBSA technique, in which high alloying element content and maximum hardness in the alloying zone can be reached, with a smoother alloying surface.

3. TEM results show that the microstructures in the alloying zone are martensite, retained austenite and a eutectic structure of α-ferrite/M_6C carbide. After post heat treatment, there are three types of carbide particle precipitated in the tempered martensite matrix of the alloying layer: $Fe_{0.5}(W,Cr,Ti,V)_{0.5}C$ carbide networks along the grain boundary; square M_6C carbide particles within the grains; and a small amount of polygonal TiC carbide particles.

4. The results of wear resistance tests in the static load sliding friction condition show that the EBSA process can significantly improve the wear resistance of W18Cr4V HSS to 3.1 times that of conventional heat treatment. However, post heat treatment contributes little to the enhancement of wear resistance of the alloying layer. Under shock load sliding friction conditions, on the other hand, the alloying layer shows some decrease in wear resistance, which can be improved by post heat treatment although it remains poor compared with that by conventional heat treatment.

So EBSA processing is suitable for application to tools and dies worked at low shock loads; service life may be expected to increase to a great extent.

REFERENCES

1. J A Knapp and D M Follstaedt: 'Laser and electron beam interactions with solids', ed. B R Appleton and G K Celler, 1982, p407
2. B L Mordick and H W Bergmann: 'Rapidly solidified amorphous and crystalline alloys', ed. B H Kear , B C Giessen and M Cohen, 1982, p463
3. K J A Mawella and R W K Honeycombe: J. Mater. Sci., 19 (1984), p3760
4. 李光霞等，华中工学院学报，10(1)，1982，P.57.
5. P R Strutt and H Nowotny: Mater. Sci. Eng., 36 (1978), p217
6. B G Lewis et al.: in Proc. 2nd Int. Conf. on Rapid Solidification Processing, 1980, p221
7. P R Strutt: Mater. Sci. Eng., 44 (1980), p239
8. 杨可传等，'激光热处理对W18Cr4V高速钢组织与性能的影响'，全国激光热处理学术会议'86(论文摘要集)，1986.4.上海，P.24.
9. 赵海鸥，硕士论文，机械科学研究院,北京机电研究所，1986.
10. 川寄 一博等，热处理(日文)，20卷，6号，1980.10，P.281.
11. 横田 清义，铁钢 急速加热变态 研究(日文)，发行者:中岛宏，株式会社产报，1971，P.10.
12. B.C. 柯瓦林科等著[苏]，郭仁东，胡隆庆译，零件的激光强化，国防工业出版社，1985，PP.9,18,99.

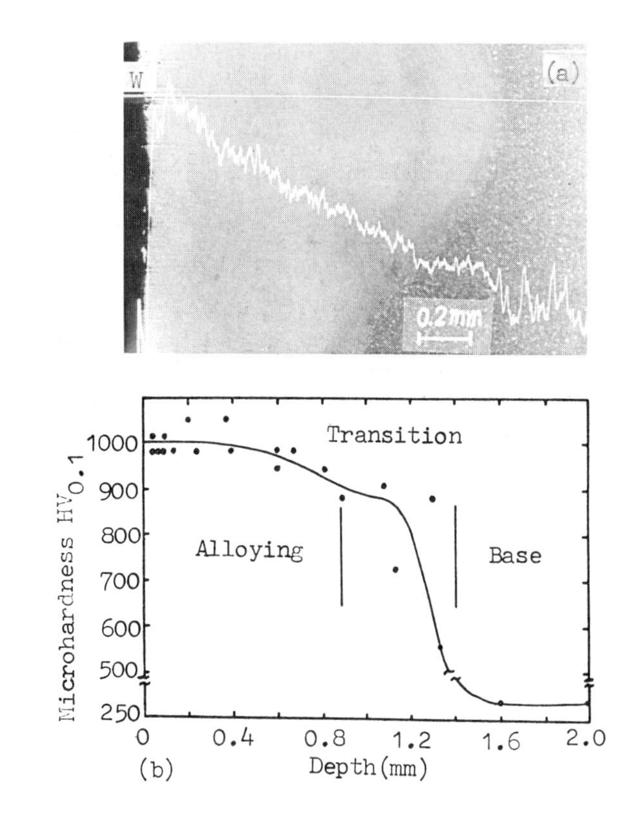

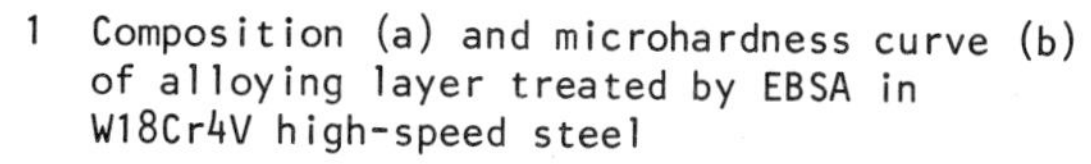
1 Composition (a) and microhardness curve (b) of alloying layer treated by EBSA in W18Cr4V high-speed steel

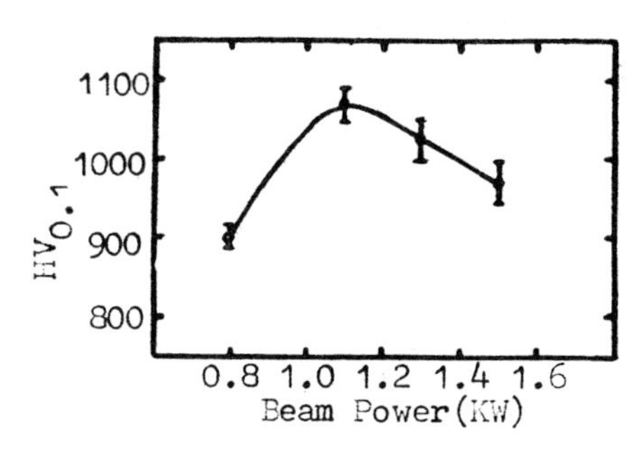

2 Effect of parameters on microhardness of alloying layers; workpiece speed 5 mm/s

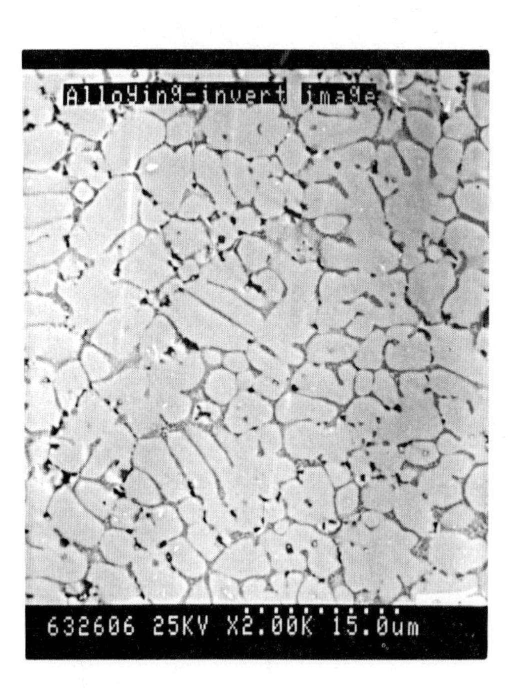

3 SEM morphology of microstructures for EBSA

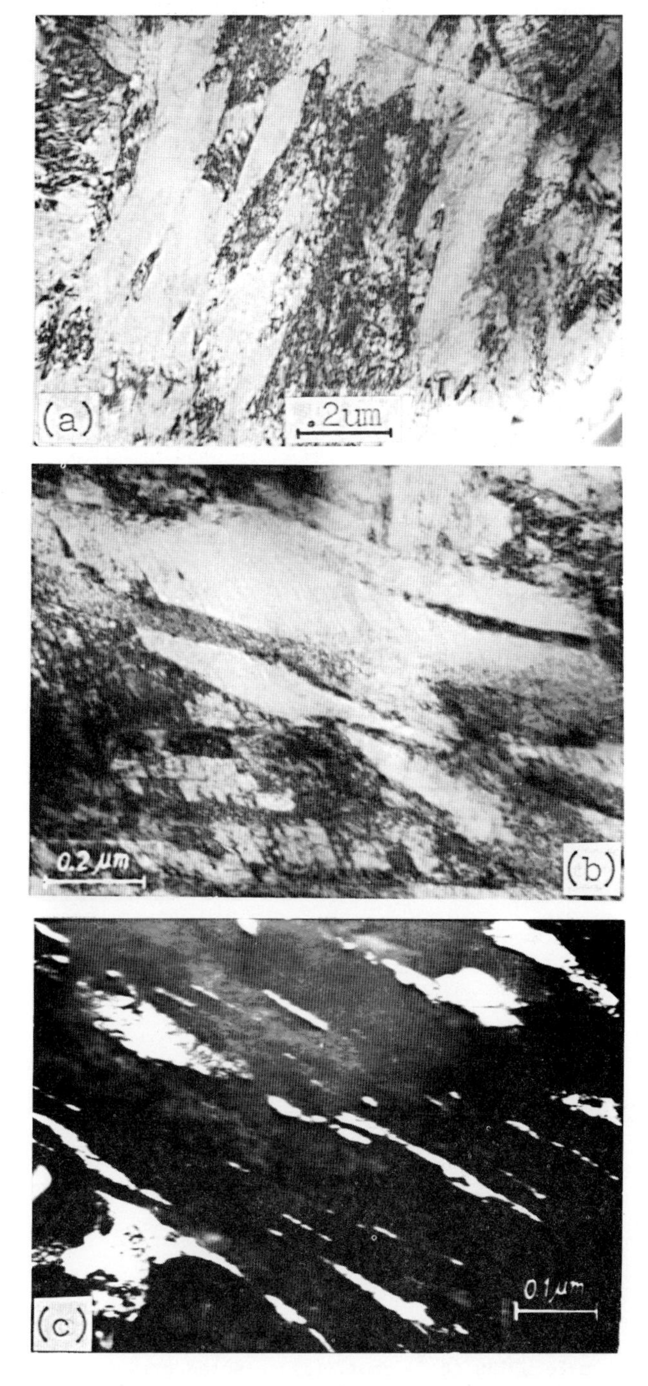

4 Morphology of martensites in alloying layer by TEM: (a) lath martensite; (b) plate martensite; (c) retained austenite (dark field)

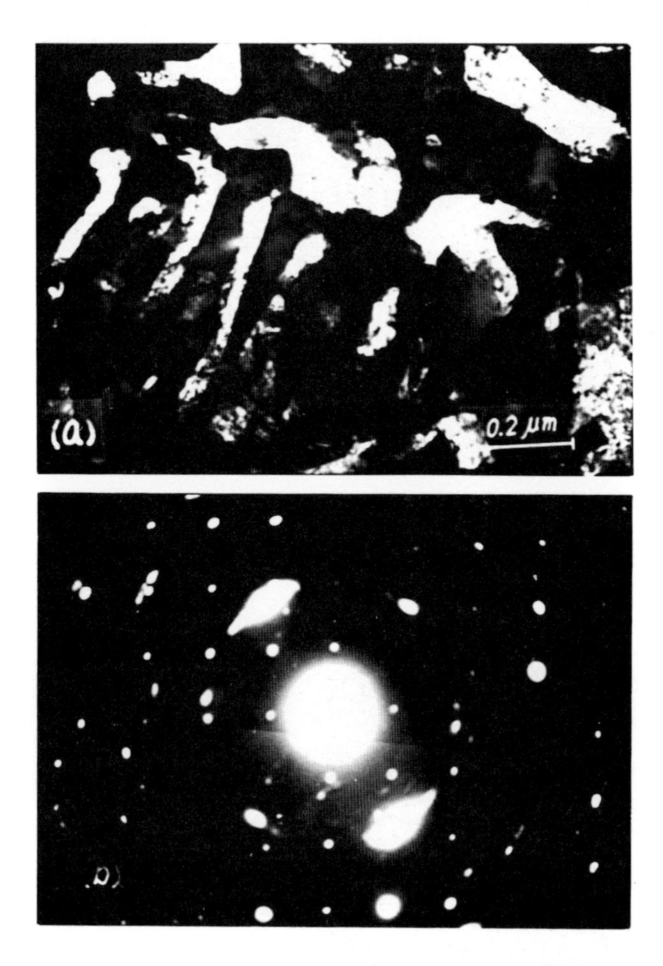

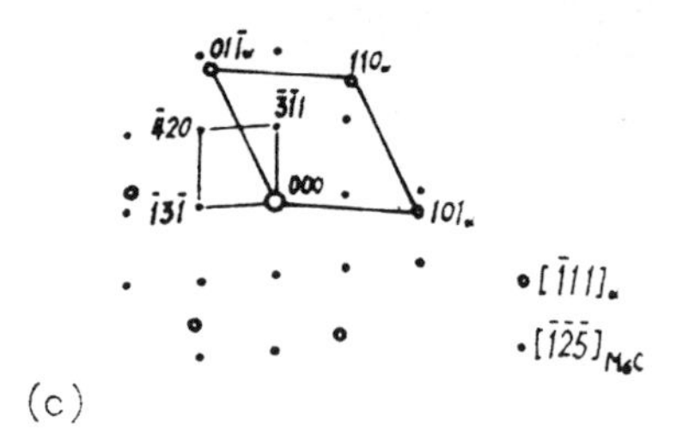

(c)

5 Eutectic structure of α-Fe/M_6C carbide in dendrite boundaries; (a) dark field showing α-Fe using (01$\bar{1}$) spot; (b) selected area diffraction pattern; (c) schematic of SAD

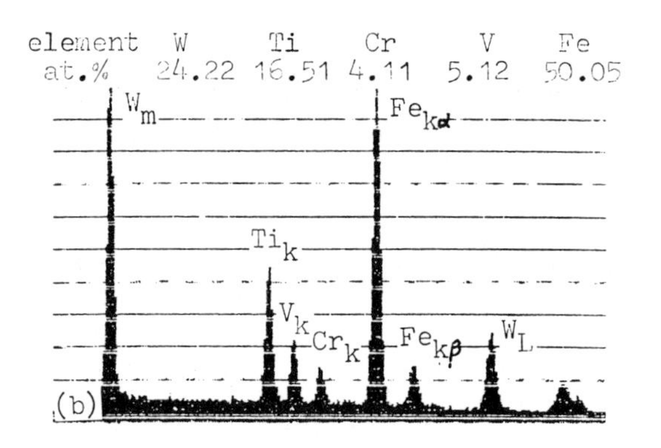

6 After post heat treatment: (a) microstructure of alloying layer (SEM); (b) composition of long particles in networks by EPA

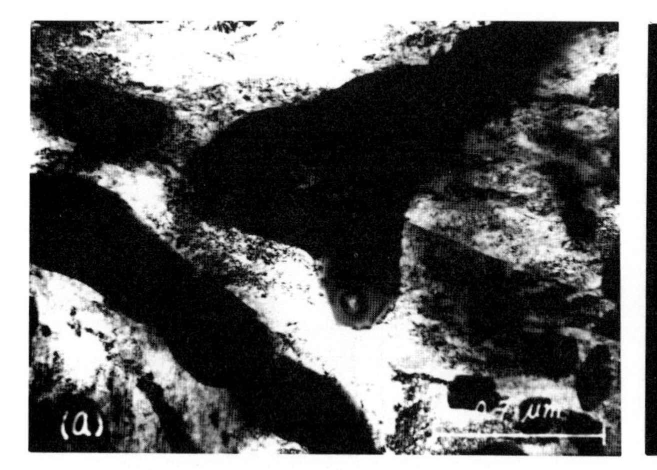

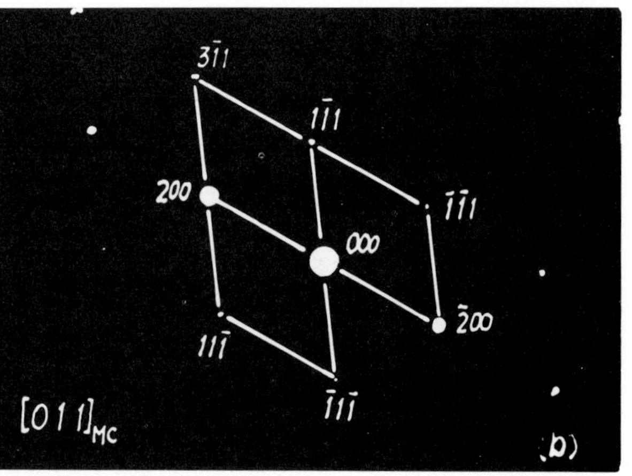

7 TEM of MC-type carbides: (a) bright field; (b) SAD pattern and schematic

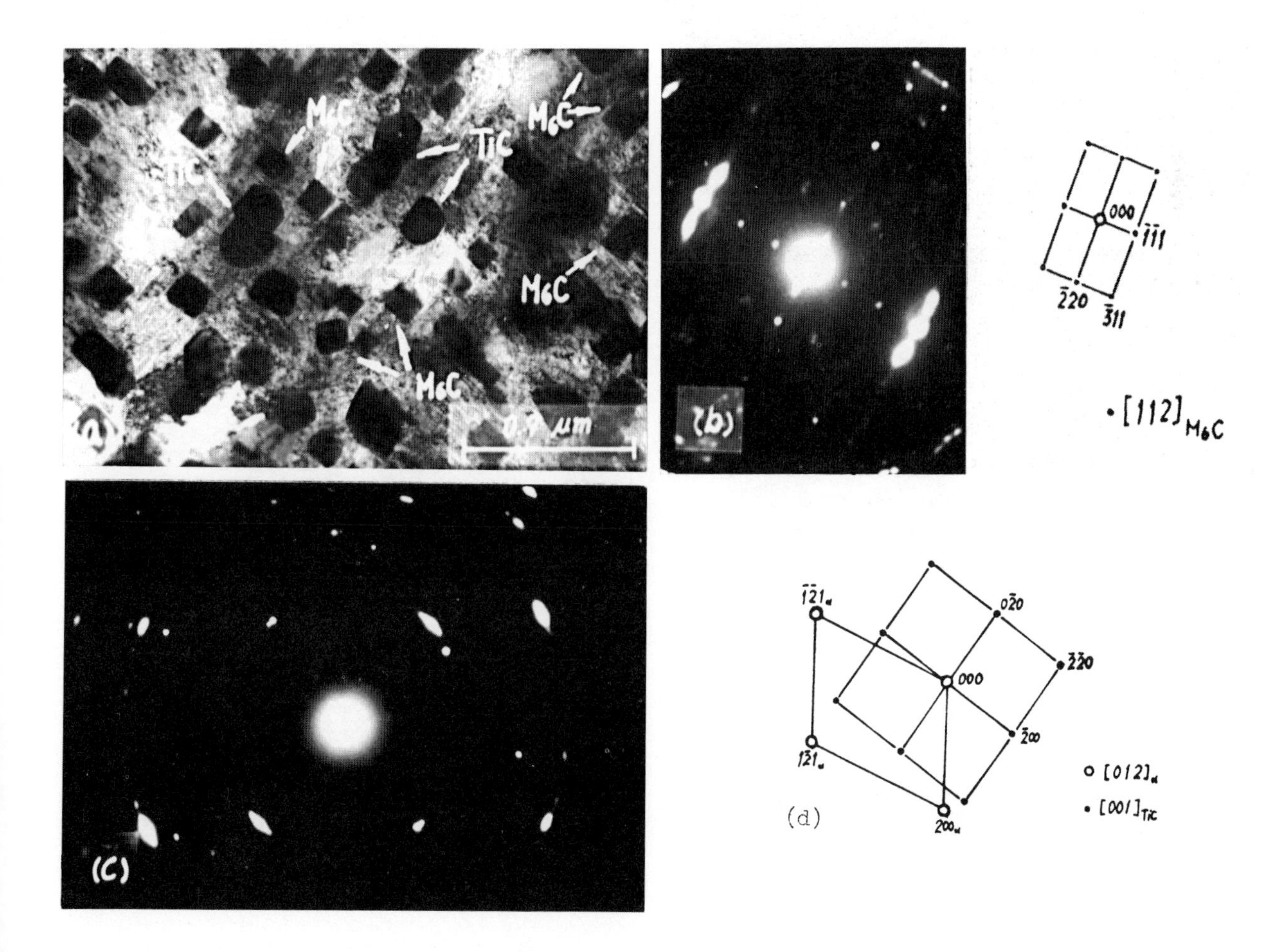

8 TEM of TiC and M_6C-type carbides precipitated in grains: (a) bright field; (b) SAD pattern and schematic of M_6C-type carbides; (c) SAD pattern of TiC carbides; (d) schematic of SAD in (c)

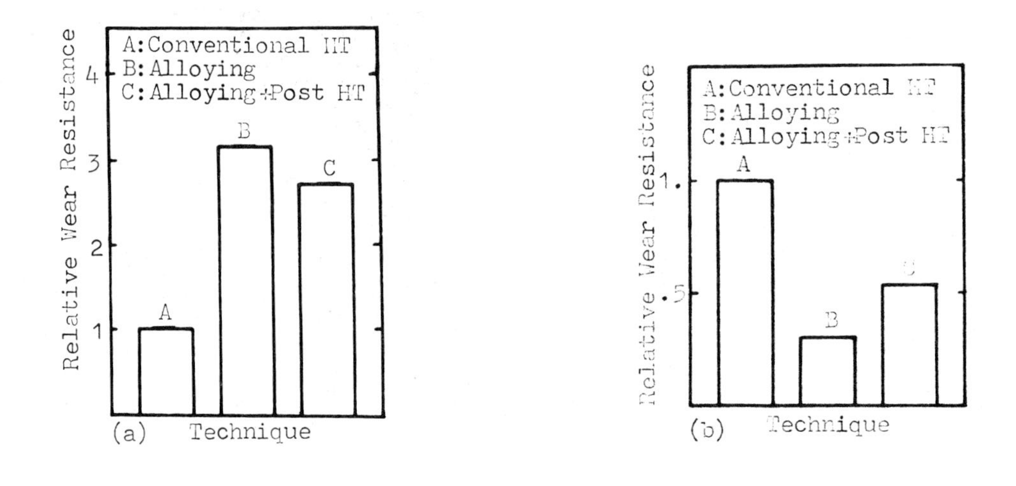

9 Results of wear resistance test in different friction conditions: (a) static load wear test, load 5 kg; (b) shock load wear test, shock energy 0.11 kg.m

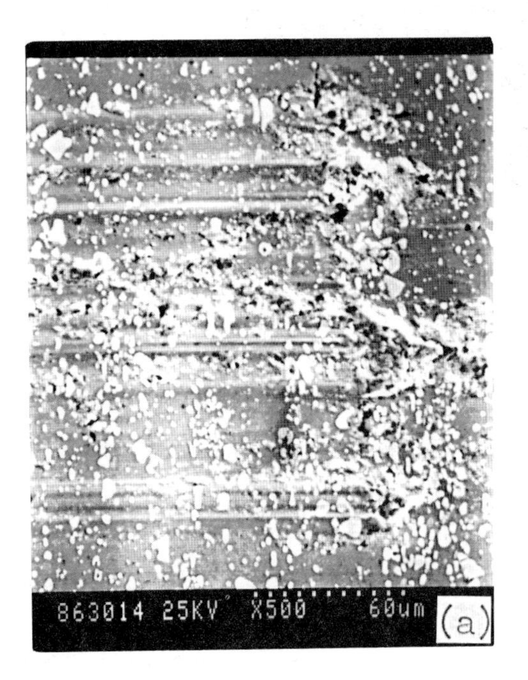

10 SEM topography of worn surfaces: (a) conventional heat treatment; (b) alloying layer treated by post heat treatment

Laser modification of high speed steel

I M HANCOCK, A BLOYCE, A S BRANSDEN
and T BELL

IMH, AB and TB are with the Wolfson Institute for Surface Engineering, Department of Metallurgy and Materials, University of Birmingham; ASB is in the Laser Applications Group, UKAEA Culham Laboratories, Abingdon, Oxon.

SYNOPSIS

This paper describes the laser surface melting of M42 HSS and the resultant modification of the secondary hardening response of this material by rapid solidification. Semi-orthogonal cutting trials have demonstrated improved results for this material. In addition to laser melting wrought M42, as cast M42 has been laser melted in order to compare the cutting performance of the two materials. Laser alloying has shown how a comparable cutting performance may be achieved by an alloy lean substrate.

INTRODUCTION

Although the development of laser processing of materials is now finding application in industrial situations, a relatively early paper by Schmidt(1) demonstrated laser alloying of high speed steel (HSS) tool tips with tungsten carbide, improving cutting performance by 2 - 4 times. The idea was restricted at this time by the absence of suitable high powered lasers. More recently, in studies of rapid solidification processing (RSP), specifically splat quenching, Rayment and Cantor(2), observed a significant modification of secondary hardening characteristics in a range of high speed steels. This both increased the maximum hardness achieved and the tempering temperature at which the peak occurred. The increase in maximum hardness was attributed to an extension of solid solubility and a refinement in structure. Possible explanations for the displacement of the peak to higher temperatures were thought to be a low vacancy concentration, restricting diffusion, or the required austenite -> martensite transformation delaying the onset of carbide precipitation.

The scope of this project was to improve the cutting performance of M42 HSS by laser surface melting, refining the structure and extending solid solubility to produce modified secondary hardening behaviour. A further aim was to utilise one of the major features of laser surface engineering: the ability to treat discrete areas of a component. This involved laser alloying of an alloy lean substrate to create a composition close to M42 at the cutting edge.

EXPERIMENTAL PROCEDURE

a) Material

The composition of M42 HSS is given in Table 1. It is a high performance, high hardness high speed steel used generally in turning operations. Before laser treatment, the surface was roughened by vapour blasting to enhance beam/workpiece coupling.

b) Laser Treatments

The lasers used were high powered, continuous wave, carbon dioxide lasers in operation at UKAEA Culham Laboratory. Both focused and defocused beams were employed with incident power densities ranging from 1×10^5 - $7.2 \times 10^6 Wcm^{-2}$. Tracks were created by use of moving optics or X-Y tables. Traverse speeds ranged from 0.01 - 10 ms^{-1}. Helium gas shields were used to prevent oxidation during both the melting and subsequent solidification processes.

c) Heat Treatments

Where conventional hardening was necessary, the sample was immersed in a fused salt bath at 1200°C 5 minutes and step quenched. Tempering treatments were carried out in neutral salt baths between 500 and 650°C.

d) Microstructural and Compositional Investigation.

Standard metallographic techniques employing optical and scanning electron microscopy, and x-ray diffraction techniques have been used to identify the phases present after laser treatment. EDX was employed to compare variations in composition in different regions of the melt and between melts in surface alloying.

e) Mechanical Performance

The performance of the laser treated materials has been evaluated in terms of hardness tests and semi-orthogonal cutting trials. The latter involved the cutting of a solid bar by a straight-edged tool, normal to the direction of cutting and normal to the feed direction. Figure 1 a and b shows the relative geometries of workpiece and tool material together with cutting tool terminology for semi-orthogonal cutting.

RESULTS AND DISCUSSION

Microstructure

As can be seen from Figure 2, the growth morphologies achieved after laser surface melting were cellular / dendritic. The refinement of the structure becomes obvious when this is compared with Figure 3, a micrograph of chill-cast M42. One immediately noticeable feature of Figure 2 is the absence of primary carbides. These have been dissolved in the liquid phase during laser melting. Partial melting of the primary carbides can be seen in the substrate below the melt track (Figure 4). Preferential melting has occurred around the carbide due to a reverse eutectic reaction.

The fact that the primary carbides are absent from the melt zone means that the carbon and other alloying elements have been segregated to the interdendritic material and solid solubility in the austenite dendrites has been increased due to the rapid solidification. There are two basic mechanisms by which solid solubility may be increased. Firstly, if one equilibrium phase is prevented from nucleation, a metastable equilibrium extension to solid solubility may result(3). Secondly, if the solidification rate exceeds that necessary for the onset of solute trapping(4), the equilibrium partition coefficient will increase from the equilibrium value. The most obvious evidence for increased solid solubility would be the achievement of planar solidification. Traverse speeds of up to $10ms^{-1}$ did not produce this, but a very fine cellular structure was obtained (Figure 5). The combination of microexamination, x-ray diffraction and tempering response indicates the cells/dendrites to be austenite and the interdendritic material to be a eutectic of austenite and carbides. Under slower cooling conditions, this can be seen to be a lamellar structure (Figure 6).

Tempering Treatments

The changes in the secondary hardening peak for two laser treatments after a single, double and triple 30 minute tempers are shown in Table 2.

The greatest increase in peak hardness over conventionally hardened material arises from treatments at lower power densities and lower traverse speeds (100 - 150 HV.3 increase). The shift in the peak was independent of the laser treatment parameters used and found to be 15K higher than that of the conventionally treated material at 565^oC.

A second temper (2 x 30 mins) reduced the differences found in maximum hardness after single tempering due to different laser treatments, but increased the actual maximum hardness by around 130 HV.3. The shift in the peak was reduced to around 5^oC. This reduction in the tempering temperature at which peak hardness occurs with multiple tempering is typical of that observed in the conventionally hardened and tempered material(5). On triple tempering (3 x 30 mins), the differences in peak temperature in the laser treated material over conventionally hardened material were negated, but the maximum hardness increased still further, but only with the same increment over the conventionally treated material. This tempering behaviour is described in Figures 7 and 8.

The increase in maximum hardness of the laser melted material over the coventionally treated material is attributed to the refinement in structure giving a finer precipitate distribution. The apparent shift in the hardness peak to higher temperatures may be explained by the increase in solid solubility of alloying elements in the austenite, associated with laser surface melting producing a slower response to the ageing treatment. This point is further illustrated by the differing response from the high and low power density treatments. Higher power density treatments produce greater cooling rates and hence greater solid solubility. On the second and third tempering treatments, this difference is removed. It is possible that previous work(1) compared rapidly solidified and single tempered material with conventionally hardened and triple tempered steel. The other point of interest from the splat-quenching experiments is the high hardness of the material processed in a nitrogen atmosphere. It was assumed by the investigators that this was due to nitrogen pick up whilst the material was molten, producing a nitriding effect(1). On laser surface melting, treatments carried out in a nitrogen atmosphere show no improvement over those utilising argon and helium. This underlines a basic difference between these two RS processes. Splat quenching requires a relatively long time with the alloy in the liquid state, whereas laser surface melting is a rapid heating and cooling process, allowing the material only a very short time as a liquid.

It is interesting to note that helium, when used in the oxidation protection gas jet, consistantly gives a higher maximum hardness than argon after laser surface melting and tempering, for all power densities. This may be due to the higher specific heat of the helium providing more efficient additional quenching than the argon.

Cutting Trials

Having ascertained a set of suitable laser treatment parameters, 4.7mm x 6.25mm M42 bar was surface melted along one edge, triple tempered and then ground back to produce single point cutting tools in which the laser track formed the cutting edge. The tool was given a nose radius of 0.75mm and the combined geometries of tool and tool holder gave a positive rake angle of 6° and a 6° clearance angle. The samples produced were run in semi-orthogonal cutting trials (akin to normal turning pracitise(6)) on 080A40 (EN8) and 870M40 (EN24) work material. Results suggested that the laser treated and triple tempered material performed significantly better than the conventionally hardened and triple tempered material (Table 3), where the performance is quantified by the maximum rate at which metal may be removed from the workpiece. This is governed by the ability of the tool to withstand the temperature and stress at the cutting edge.

Samples of 'as cast' M42 bar were also laser surface melted: subsequent cutting trials of this material showed the cutting performance to be comparable to the conventionally wrought and heat treated material.

An alloy lean substrate material with the minimum alloying element requirements for producing the secondary hardening peak was laser surface alloyed to produce a discrete band of M42 material at the cutting edge. Table 4 shows laser alloyed tool tip cutting trials. It can be seen that they perform as well as the laser treated wrought material.

Laser Alloying

A variety of pre-placement techniques were employed and the variation in composition of the resultant tool tips are presently being evaluated by EDX (7). These techniques were as follows:

a) thermal spraying of a mixed powder of specific composition,

b) foil, pre-cast to suitable composition, and

c) powder cored strip, in which a powder of specific composition is placed in a thin walled tube which is reduced to the required dimensions by a combination of drawing and rolling.

The alloying additions from techniques b) and c) were subsequently spot-welded into position prior to laser melting.

Difficulties were encountered in (a) and (c) due to volatilization of the powder at the high power densities involved. This resulted in an unstable plasma above the melt pool, producing poor surface finish, an uneven melt depth, cracking and porosity. These commonly encountered problems were not present laser melting of wrought and cast substrates, nor foil alloyed material.

All three techniques required the calculation of composition of the alloying addition, which on melting and subsequent dilution would produce the desired surface alloy. With reference to the result for sprayed powder alloying in Table 6, the inferior cutting performance was caused by over dilution due to the unstable plasma giving uneven melt depth.

CONCLUSIONS

(1) M42 HSS may be laser surface melted with a good surface finish and without cracking or porosity.

(2) Laser melted material exhibits a significantly higher peak hardness on triple tempering than does conventionally hardened material.

(3) There is only a marginal shift in the temperature at which the peak hardness occurs is apparent with laser surface melted material, after triple tempering.

(4) Laser melting, as opposed to conventional hardening, demonstrates a superior performance in semi-orthogonal cutting trials after triple tempering.

(5) As-cast M42 gives a satisfactory cutting performance after laser surface melting and tempering. This suggests that the forging operation may be avoided in the production of M42 tool tips.

(6) Laser alloyed surfaces demonstrate excellent cutting properties, suggesting a bulk saving of strategic materials such as cobalt, molybdenum and tungsten if desired.

ACKNOWLEDGEMENTS

The authors would like to thank Professor R. E. Smallman for provision of laboratory facilities at the University of Birmingham. I.M.H. would like to thank UKAEA Culham Laboratories for the provision of laser facilities and the funding of this project.

REFERENCES

1) Schmidt A.O., J. Eng. for Ind., Aug, 1969.

2) Rayment J.J. & Cantor B., Met. Sci., 1978, 12, p156-165.

3) Mehrabian R., Int. Met. Rev., 1982, 27, (4), p185-208.

4) Baker J.C. & Cahn J.W., 'Thermodynamics of Solidification', 1971, p23-58, ASM Metals Park.

5) Fletcher S.G. & Wendell C.R., ASM Met. Eng. Qtly., 1966, 6, (1), p1-9.

6) Trent E.M., Metal Cutting, 2nd Edn., 1982, p15, Butterworths.

7) Hancock I.M., Phd. Thesis, to be submitted.

Tables

1. Composition of M42 high speed steel.

ELEMENT	STANDARD SPECIFICATION Min %	Max %	EXPERIMENTAL ANALYSIS %
Carbon	1.00	1.10	1.08
Manganese	—	0.40	0.23
Silicon	—	0.40	0.25
Cobalt	7.50	8.50	8.00
Tungsten	1.00	2.00	1.61
Vanadium	1.00	1.30	1.15
Chromium	3.50	4.25	3.75
Molybdenum	9.00	10.00	9.25

3. Comparison of semi-orthogonal cutting data for conventional and laser treated M42.

Work Material: 080A40 (En8)

Depth of Cut: 2.0mm

Feed Rate: 0.25mm/rev

Duration of Test: 30s

SAMPLE	CUTTING SPEED m min^{-1}	CHIP CHANGE s	CRATER DEPTH (A) mm	CONTACT LENGTH (B) mm	START OF CRATER (C) mm	FLANK WEAR (D) mm
A1	30	-	NIL	-	-	NIL
A2	40	-	0.04	1.58	0.47	0.07
A3	50	23	0.10	1.72	0.26	0.24
A4	55	7	FAILED AFTER 24 SECONDS			
B1	30	-	NIL	-	-	NIL
B2	40	-	0.03	1.42	0.60	NIL
B3	50	-	0.10	1.74	0.30	0.24
B4	55	22	0.11	1.86	0.29	0.22
B5	60	-	FAILED AFTER 18 SECONDS			

A - Conventionally Hardened and Tempered Wrought M42 Tool

B - Laser Surface Melted Wrought M42 Tool

2. Secondary peak hardness and peak tempering temperature for multiple 1/2 Hr tempering of laser surface melted M42.

Substrate: Wrought M42

Condition: Hardened and Triple (1/2hr) Tempered at 560°C

Shielding Gas: Helium

LASER PROCESSING PARAMETERS			HARDNESS		TEMPERING TEMPERATURE	
TREATMENT NO.	FOCUS (mm.)	TRAVERSE SPEED (mms^{-1})	PEAK ($HV_{0.3}$)	INCREASE OVER CONVENTIONALLY TREATED MATERIAL	OCCURENCE OF PEAK (°C)	INCREASE OVER CONVENTIONALLY TREATED MATERIAL
SINGLE 1/2 HR TEMPER						
1	+9	100	1170	115	580	15
2	+0	500	1120	65	580	15
DOUBLE 1/2 HR TEMPER						
1	+9	100	1230	150	560	5
2	+0	500	1205	125	560	5
TRIPLE 1/2 HR TEMPER						
1	+9	100	1260	160	560	5
2	+0	500	1240	130	555	0

Work Material: 080A40 (En8)

Depth of Cut: 2.0mm

Feed Rate: 0.25mm/rev

Duration of Test: 30s

SAMPLE	CUTTING SPEED m min^{-1}	CHIP CHANGE s	CRATER DEPTH (A) mm	CONTACT LENGTH (B) mm	START OF CRATER (C) mm	FLANK WEAR (D) mm
H1	40	-	0.01	1.36	0.77	0.19*
H2	50	-	0.01	1.58[2.17]	0.65	0.24
H3	55	-	0.04	1.60	0.55	0.30
H4	60	19	FAILED AFTER 21 SECONDS			
I1	40	-	0.03	1.38	0.71	NIL
I2	50	-	0.07	1.65	0.38	NIL
I3	55	11	0.10	1.79	0.38	*
I4	60	6	FAILED AFTER 26 SECONDS			
J1	40	21	0.11	1.79	0.58	0.35
J2	50	11	0.13	1.89	0.22	0.23
J3	55	6	FAILED AFTER 20 SECONDS			

* Cracked

H - Laser Alloyed Foil

I - Laser Alloyed Powder Cored Strip

J - Laser Alloyed Thermal Spray

4. Semi-orthogonal cutting data for laser surface alloyed material.

Figures

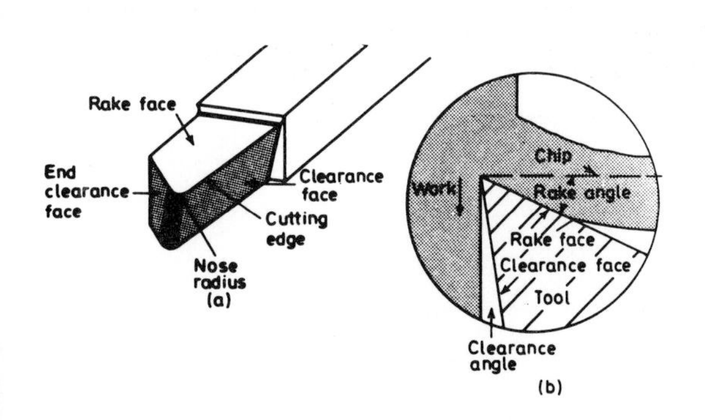

1. Semi-orthogonal cutting terminology.

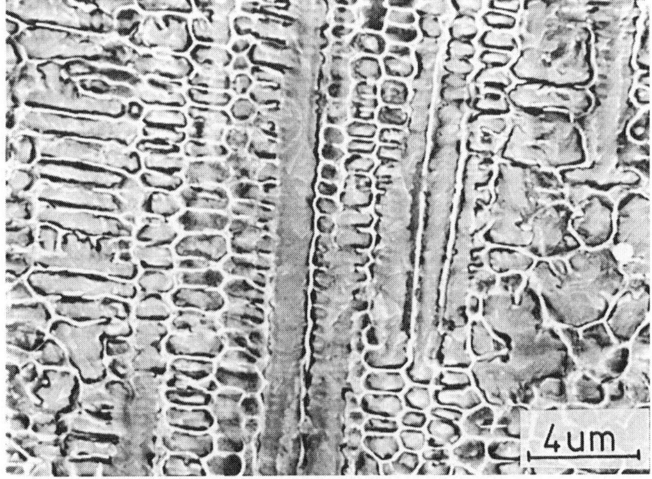

2. Optical micrograph of laser surface melted material exhibiting a dendritic growth morphology. Incident power density 0.89×10^6 Wcm^{-2}, Traverse speed $100mms^{-1}$.

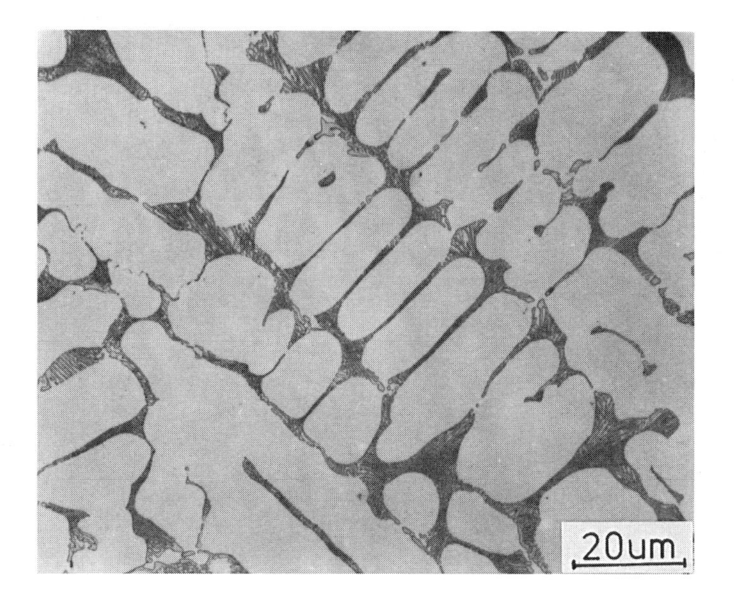

3. Optical micrograph of chill cast M42. Estimated cooling rate 3.9×10^{2} °Cs^{-1}.

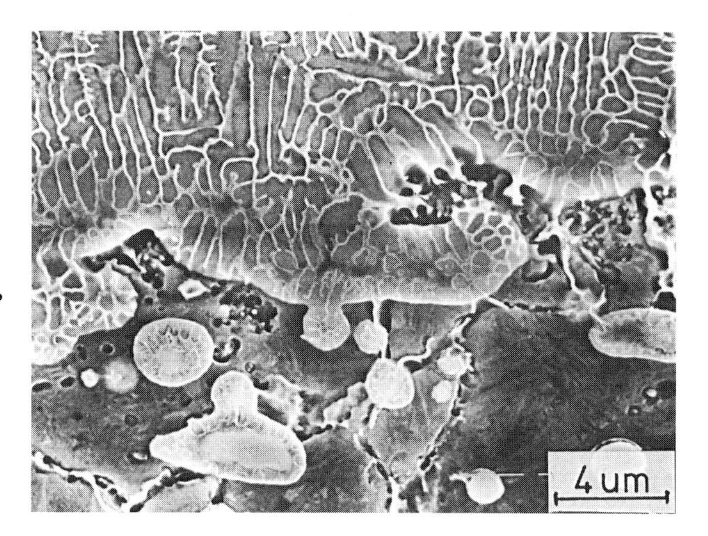

4. SEM micrograph of laser surface melted M42 showing partial melting of carbides at melt boundary. Incident power density 2.8×10^{5} Wcm^{-2}, Traverse speed $100mms^{-1}$.

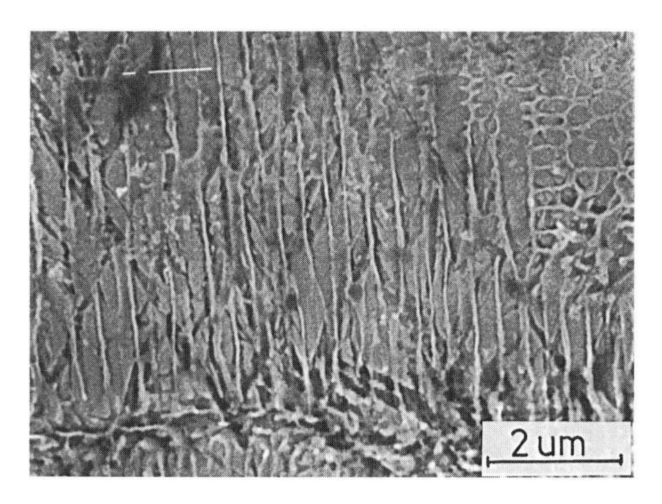

5. Optical micrograph of laser surface melted M42 exhibiting a cellular growth morphology. Incident power density 3.3×10^{6} Wcm^{-2}, Traverse speed $5{,}000mms^{-1}$.

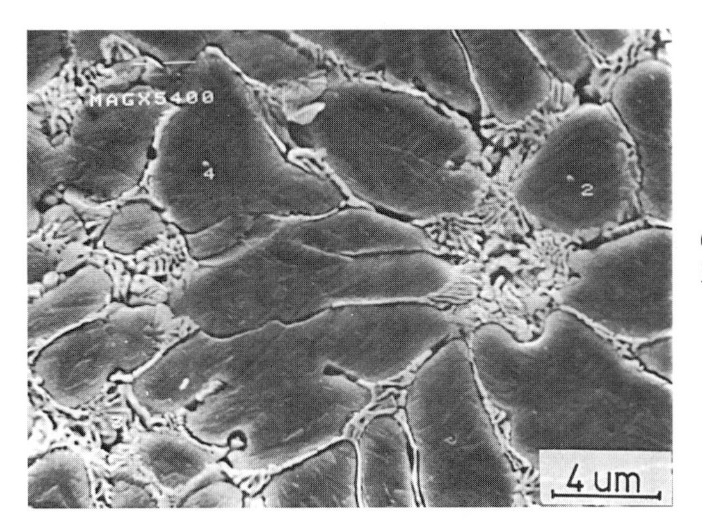

6. SEM micrograph of inter-cellular lamellar structure. Incident power density 1.0×10^5 Wcm^{-2}, Traverse speed $20mms^{-1}$.

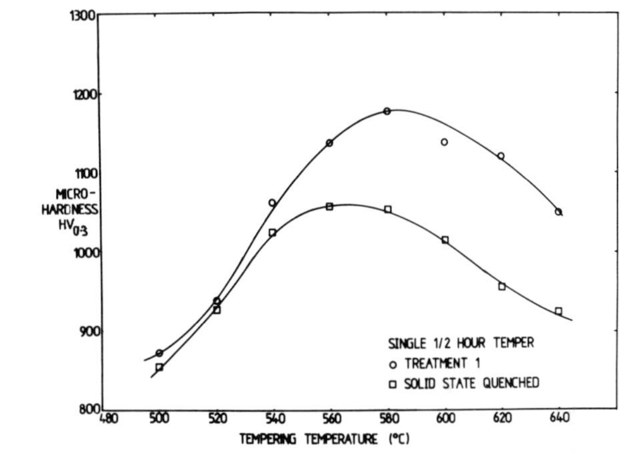

7. Secondary hardening peak of conventional and laser surface melted material after a single 1/2 hour temper.

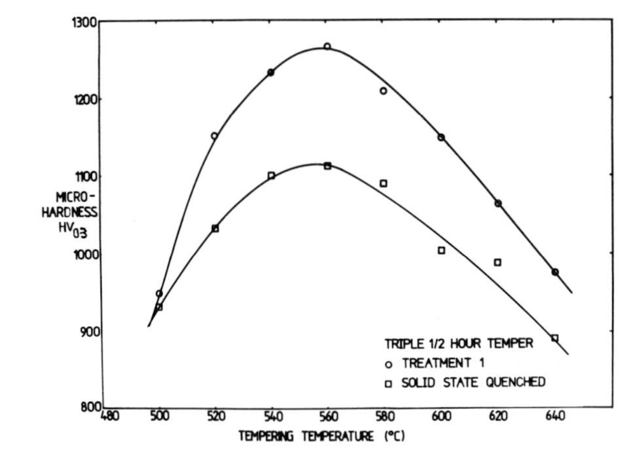

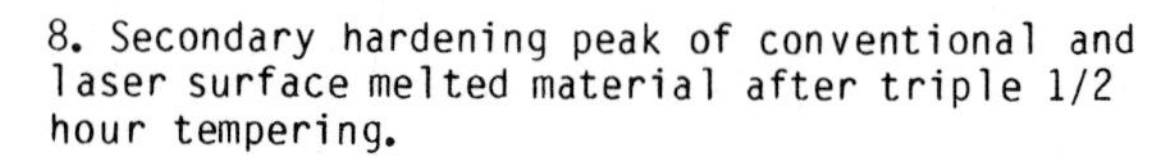

8. Secondary hardening peak of conventional and laser surface melted material after triple 1/2 hour tempering.

Development of ε carbonitride at compound layer/substrate interface during nitrocarburizing: production of compound layers deficient in γ' carbonitride

M A J SOMERS and E J MITTEMEIJER

The authors are in the Laboratory of Metallurgy, Delft University of Technology, The Netherlands.

Synopsis

During nitrocarburizing pure iron in an ammonia/hydrogen gas mixture containing a small amount of carbon monoxide, nucleation of ε-$Fe_2(N,C)_{1-x}$ at the γ'-Fe_4(N,C)/α-Fe interface was observed in an advanced stage of the treatment. The amounts of nitrogen and carbon in this substrate-adjacent ε indicated that the nitrogen concentration in the ε phase can be smaller than in the γ' phase, while the total interstitial content (N+C) in ε is larger.
The present investigation provides an understanding for the development of compound layers deficient in γ' phase on commercial nitrocarburizing.

1. *Introduction*

The compound layer produced on the surface of iron and steels by a nitrocarburizing treatment is usually predominantly composed of iron carbonitrides: ε-$Fe_2(N,C)_{1-x}$ and γ'-Fe_4(N,C) [1,2]. Good anti-corrosion and tribological properties can be ascribed to specific compound layers. For good tribological properties monophase ε layers are preferred [1,3].

According to published Fe-N [4] and Fe-N-C [5] phase diagrams thermodynamical "equilibrium" is not possible between adjacent ε-(carbo)nitride and ferrite at usual treatment temperatures and pressures. Instead, the γ' phase serves as an intermediate between ε and α. However, on commercial nitro-carburizing the occurrence of ε/α interfaces has frequently been reported [i.e. 1,3,6].

During an investigation of the kinetics of compound-layer development on nitrocarburizing pure iron, ε was observed to form at the γ'/α interface in an advanced stage of nitrocarburizing [7]. On the basis of these experiments, an explanation is presented for the formation of ε/α interfaces in commercial nitro-carburizing.

2. *Model experiments for compound-layer formation on iron*

The increase of mass (Δm) of a pure iron specimen per unit of initial surface area (A_0) on nitrocarburizing at 843 K in a gas mixture comprising 53.1 vol.-% NH_3, 3.0 vol.-% CO and 43.9 vol.-% H_2 is shown in Fig. 1 (for experimental details, see [7,8]). Different stages of nitrocarburizing have been indicated. A change in mass-increase rate can have various origins such as the occurrence of another rate-determining step in the nitrocarburizing kinetics or an increase/decrease of contact area between nitrocarburizing medium and specimen. Only for stages II, V and VIII a linear relationship exists between the increase of mass per unit of the initial surface area and the square root of treatment time (t).

A detailed description of stages I-VI has been given in Refs. 7 and 8. The different rate-determining mechanisms indicated for these stages in relation to the microstructural evolution of the compound layer are briefly summarized below.

I, II: Compound-layer formation starts with nucleation of individual γ' nuclei; ε develops on top of γ' nuclei. Growth and coalescence of the dual-phase nuclei (stage I) leads to an isolating ε (top)/γ' (bottom) compound layer. Then a parabolic rate law for the increase of mass occurs (stage II).

III: Development of pores and, by pore coalescence, of channels along ε-grain boundaries is caused by the recombination of dissolved nitrogen atoms into precipitated nitrogen-gas molecules. At some distance from the outer surface of the compound layer *preferential carbon uptake occurs via channel walls,* which is ascribed to nitrocarburizing conditions in the channels different from those at the outer surface. Within the channels the dissociation of NH_3 is counteracted by the presence of N_2 molecules resulting from recombination in the ε phase. This stage of nitrocarburizing is characterized by a positive deviation from the parabolic rate law for mass increase prevailing in stage II as a result of the increase of total contact area between nitrocarburizing gas and compound layer by channel development.

IV, V: The process of formation of a continuous carbon-rich zone within the ε sublayer by coalescence of the laterally expanding carbon-rich regions enveloping the channels along ε-grain boundaries is denoted as stage IV. After the carbon-rich band within ε has appeared, a parabolic rate law for mass increase results, suggesting a constant effective diffusional cross section (stage V).

VI: Within the carbon-rich band in ε, cementite (θ) nucleates at the channel walls. This process of $\varepsilon \rightarrow \theta$ conversion is accompanied by severe pore (N_2) development, as a result of the low solubility of nitrogen in θ. Eventually, a continuous, finely porous, θ sublayer, sandwiched by ε sublayers (porous ε above; "massive" ε below) results (see Fig. 2). Since diffusional fluxes of carbon and nitrogen through θ are much smaller than those possible through ε, stage VI is characterized by a negative deviation from the parabolic rate law for mass increase prevailing in stage V.

The transition of stage VI into stage VII is taken at a point of inflexion of the curve of $\Delta m/A_0$ vs. $\sqrt{t}$ (Fig. 1). At this stage VII, surprisingly, nucleation of ε phase at the γ'/α interface occurred (Murakami-stained regions in Figs. 3 and 4; the Murakami etchant allows sensitive discrimination between nitrides and carbonitrides (also carbides), since it stains the latter phases [9,10]).

The occurrence of very thin ε regions bounded by parallel lines in γ' (denoted by arrows in Fig. 4) suggests that ε nucleates at twin boundaries in the γ'-sublayer. Twinning in γ' (f.c.c. iron sublattice) occurs along (111) planes and is associated with the introduction of stacking faults [11]; at the twin boundaries a local h.c.p. stacking can be discerned. Therefore twin boundaries can be favourable nucleation sites for ε (h.c.p. iron sublattice) in γ' (cf. [12]).

Point concentration measurements by electron probe microanalysis (EPMA) at the compound layer/ substrate interface of an oblique section of a specimen (after 30 h nitrocarburizing) indicated that the ε nuclei contain 4.4. wt% N and 1.0 wt% C, whereas the interstitial weight contents of adjacent γ' are 5.0 wt% N and 0.3 wt% C. Hence, conversion of γ'-carbonitride into ε-$Fe_2(N,C)_{1-x}$ is accomplished by nitrogen depletion and carbon enrichment. Nitrogen and carbon contents in the massive ε sublayer (about 5.0 wt% N and about 1.5 wt% C; see Fig. 2b) as well as in the ε grains developing at the γ'/α interface are within the two-phase ($\varepsilon + \gamma'$) region as obtained at 843 K by interpolation between the published [5] Fe-C-N phase diagrams at 838 K and 848 K, which in turn are derived [5] by interpolation between measured data at 823 K and 873 K. Thus, these results provide another indication that, compared to the proposal made in Ref. 5, the ε-phase field at 843 K may be extended towards smaller nitrogen contents (see also Refs. 13 and 10).

The following explanation is proposed for the development of ε at the compound layer/substrate interface.

As was discussed in [7], the continuous carbon-enriched zone in the ε sublayer, which developed in stage IV of nitrocarburizing, leads to reduced nitrogen uptake from the nitrocarburizing medium, because a positive nitrogen-activity gradient, defined with reference to the distance to the surface, is formed (carbon enhances the activity of nitrogen [14]). Further, a closed cementite sublayer, obtained in stage VI, forms a very effective obstacle for nitrogen penetration. Accordingly, continued growth of the massive part of the compound layer ($\varepsilon + \gamma'$) is accompanied with redistribution of the nitrogen atoms already present, implying that the average nitrogen concentration in the ε sublayer decreases on prolonged nitrocarburizing. EPMA indeed confirms this decrease of the nitrogen concentration in the massive region of the compound layer (cf. data for nitrocarburizing times of 4.5, 15 and 30 h in Fig. 5a; a comparison of average N contents on an absolute scale is hindered by the ragged nature of θ/ε and γ'/α interfaces). Therefore, it can be concluded that supply of carbon to the massive part of the compound layer can lead to layer growth by increasing the amount of ε phase.

The homogeneity range of the γ' phase with regard to the nitrogen concentration is small and only a small amount of carbon can be incorporated in γ' [4,5]. From the data discussed above, it follows that ε can exist at nitrogen concentrations smaller than in γ', provided a certain amount of carbon is present. At a certain stage the nitrogen-activity in the ε sublayer ("massive") has become so small (owing to N redistribution) that a nitrogen activity gradient can not be maintained in the γ' sublayer. Hence, a further shift of the γ'/α interface into the substrate can not occur. The nucleation of ε phase at the γ'/α interface could now be due to carbon segregated there (see also section 3). In particular, if the substrate is not saturated with nitrogen, this ε nucleation could occur as a result of "dissolution" of the γ' sublayer in order to supply nitrogen to the substrate.

Growth of ε nuclei from the γ'/α interface towards the massive ε sublayer can be realized by (i) diffusion of carbon through the γ' sublayer (this kind of upward growth of ε nuclei suggests carbon diffusion through γ' is rate-limiting, cf. Ref. 15), and (ii) nitrogen supply by consumption of the γ' sublayer.

The carbon transported from the ε sublayer above the γ' layer can be supplemented by carbon absorption through channel walls present in this ε sublayer. Originally, this ε sublayer was denoted as "massive", but at this stage of nitrocarburizing appreciable porosity arises there too [7]. Then, an enhanced carbon uptake will occur, which corresponds with the observed net increase of the rate of mass increase (see Fig. 1; stage VII). The absorption of a considerable amount of carbon by the "massive" part of the compound layer in stage VII of nitro-carburizing can be deduced from the carbon concentration-depth profiles shown in Fig. 5b.

As a result of upward growth (in the direction of the carbon source) of the ε nuclei, ε regions develop, bridging the distance between the "massive" ε and the substrate (Fig. 6). In an ε "bridge" a distinct carbon-concentration gradient exists as evidenced qualitatively by Murakami staining differences (Fig. 6) and quantitatively by EPMA (Fig. 7). Consequently, an increasing contact area between the "massive" ε sublayer and the ε "bridges" will lead to enhanced carbon diffusion. At the moment when the contact area between ε "bridges" and "massive" ε is so large that all the carbon that can be transported through "massive" ε is absorbed by the ε "bridges", carbon transport through "massive" ε becomes rate determining. Then a linear relationship between mass-increase per unit area and the square root of time is expected (stage VIII).

3. *Compound-layer formation during commercial nitro-carburizing.*

The presence of ε phase at the compound-layer/ substrate interface has frequently been reported in the literature as the result of a commercial nitrocarburizing process (e.g. [16,17]). In such cases it was assumed, or suggested on interpreting experimental data, that the γ' phase was not present in the compound layer. Its absence was ascribed to a relatively rapid growth of the ε phase (due to the possibly large concentration gradient in this phase) with the consequence that the γ' phase could be absorbed during nitrocarburizing [16], or even that nucleation of the γ' phase was prevented [17]. It should be realized that in all these cases a *steel* substrate was employed. Considering a possible nucleation of ε phase on a steel surface, distinction is made between nucleation on cementite and on ferrite.

A *direct* nucleation of ε phase on a *steel* surface *is* possible, because cementite (θ) phase adjacent to the surface can be transformed into ε phase under conditions where only γ' nitride should develop on ferrite [10] (note the small carbon solubility in γ' [5] and the crystallographic similarity of ε and θ [18]).

In our experiments a direct nucleation of ε phase on an α-iron (ferrite) surface at the start of compound-layer formation was not observed. However, the results presented in section 2 demonstrate that ε nucleation at the interface with α iron is possible in an advanced stage of nitrocarburizing, already when only a minor fraction of carbon supplying agents (3.0 vol.-% CO) is added to the nitriding medium. On commercial gaseous nitrocarburizing about 10-20 vol.-% CO is present.

In a separate experiment, armco iron was nitrocarburized commercially at 843 K in a gas mixture of 50 vol.-% NH_3 + 50 vol.-% endogas. Murakami etching reveals carbon enrichment in two regions within the compound layer (Fig. 8):

- carbon has been absorbed through channel walls;
- carbon-containing ε nuclei, apparently nucleated at the compound-layer/interface, have grown out and have established ε "bridges".

It is emphasized that in this experiment ε nucleation at the compound-layer/substrate interface occurred before a cementite sublayer had developed in the upper half of the compound layer, as was the case with the model experiments reported above. The occurrence of two carbon-enriched regions within the compound layer separated by a zone that is relatively poor in carbon (see Fig. 8), implies that the carbon enrichment adjacent to the substrate did not result from carbon uptake via channel walls. Instead, the data hint at a relatively strong carbon absorption in an early stage of nitrocarburizing. A relatively strong carbon absorption in the initial stage of nitrocarburizing was also observed in the present model experiments [7,8] as well as on nitrocarburizing pure iron powder [19].

It is recalled that the nitrogen concentration at the surface of α iron in NH_3/H_2 gas mixtures at usual nitriding temperatures approaches relatively slowly the equilibrium value, which leads to an incubation time for γ' nucleation at α iron surfaces [20]. Then, recognizing that ε carbonitride can exist at nitrogen concentrations smaller than that required for the development of γ' (see section 2), the relatively strong carbon absorption in an initial stage of commercial nitrocarburizing treatments could bring about ε nucleation at the ferrite surface in an early stage of nitrocarburizing already (this implies a modification of our earlier interpretation of literature data [7]).

In our model experiments, the maximum carbon content at the γ'/α interface in an early stage of nitrocarburizing was only 0.12 wt.-% [7,8], which indeed is within the homogeneity range of γ' [5].

During salt bath nitrocarburizing, the development of $\varepsilon/\gamma'/\varepsilon$ compound layers can be observed, as is demonstrated by a separate experiment: pure iron was nitrocarburized for 0.5 h according to the TF1 process at Degussa, Hanau, FRG (Fig. 9 and see also [21]). The existence of an $\varepsilon/\gamma'/\varepsilon$ sequence within the compound layer is clearly revealed by the application of positive phase contrast light microscopy (γ' phase appears dark [9]; see Fig. 9a). Murakami staining of ε in the bottom region of the compound layer (Fig. 9b) points at an initially strong carbon absorption similar to that observed on commercial gaseous nitrocarburizing (Fig. 8). The absence of an appreciable Murakami staining in ε at the outer surface (Fig. 9b) indicates that carbon can not easily be absorbed from the salt bath by ε. On gaseous nitrocarburizing, carbon absorption by ε occurs preferably via channel walls. An analogous mechanism does not seem likely for salt bath nitrocarburizing, because the liquid salts would not deeply penetrate into the channels ([21]; see also discussions in [7,8]).

The compound layers presented in Figs. 8 and 9 show the occurrence of ε phase bridging a γ' layer. When lateral growth of ε bridges and, accordingly, the "consumption" of γ' (see section 2) has not been completed, some residual γ' phase may be present in the bottom region of the compound layer (Fig. 10).

The above discussion provides an interpretation of the pronounced carbon enrichment observed in the bottom region of compound layers (deficient in γ' phase) on nitrocarburizing pure iron and steels [2,9,10,22,23]. However, in addition to the initially strong carbon absorption from the nitrocarburizing medium, in the case of a steel substrate, a minor part of the carbon enrichment near the layer/matrix interface appears to stem from the substrate (see in particular Fig. 8 in [10]).

4. *Conclusions*

- If a relatively small amount of carbon supplying agents is present in the nitrocarburizing medium, ε nuclei develop at the γ'/α interface in an advanced stage of nitrocarburizing. The nitrogen content of this phase is smaller than that of the adjacent γ' phase, whereas the total interstitial content (nitrogen + carbon) is larger.

- A relatively strong carbon absorption in an initial stage of commercial nitrocarburizing could bring about nucleation of ε phase on ferrite.

- Lateral growth of ε bridges across a γ' sublayer and, consequently, consumption of γ' leads to the production of a compound layer deficient in γ' phase.

Acknowledgements

We are indebted to Professor B.M. Korevaar for constructive criticism and discussions and to Messrs. P.F Colijn and P.J. van der Schaaf for skilful experimental assistance. We thank Ir. D. Schalkoord and Mr. D.P. Nelemans for their EPMA work. Financial support of the Foundation for Fundamental Research of Matter (FOM) is gratefully acknowledged.

References

[1] T. Bell - Heat Treat. Met., 2, 1975, p. 39-49.
[2] E.J. Mittemeijer, H.C.F. Rozendaal, P.F. Colijn, P.J. van der Schaaf, R.Th. Furnée - Proc. Heat Treatment '81, The Metals Society, London, 1983, p. 107-115.
[3] C. Dawes, D.F. Tranter, C.G. Smith - Metals Techn., 6, 1975, p. 345-353.
[4] O . Kubachewski - Iron Binary Phase Diagrams,1982, Springer-Verlag.
[5] F.K. Naumann, G. Langenscheid - Arch. Eisenhüttenwes., 36, 1965, p. 583-590.
[6] "Source book on nitriding", 1977, Metals Park, Ohio, American Society for Metals.
[7] M.A.J. Somers, E.J. Mittemeijer - Surf. Eng., 3, 1987, in press.
[8] M.A.J. Somers, E.J. Mittemeijer - Härterei-Techn. Mitt., 42, 1987, in press.
[9] P.F. Colijn, E.J. Mittemeijer, H.C.F. Rozendaal - Z. Metallk., 74, 1983, p. 620-627.
[10] H.C.F. Rozendaal, P.F. Colijn, E.J. Mittemeijer - Surface Eng., 1, 1985, p. 30-42. Also published in: Proc. Conf. Heat Treatment '84, The Metals Society, London, 1985, p. 31.1-31.16.
[11] B. Gérardin, H. Michel, J.P. Morniroli, M. Gantois - Mém. Sci. Rev. Metallurgie, 74,1977, p. 457-467.

[12] W. Pitsch - Arch. Eisenhüttenwes., 29, 1958, p. 125-128.
[13] A. Wells, T. Bell - Heat Treat. Met., 10, 1983, p. 39-44.
[14] W. Schröter - Wiss. Z. Techn. Hochsch. Karl-Marx-Stadt, 24, 1982, p. 795-809.
[15] R.A. Rapp, A. Ezis, G.J. Yurek - Metall. Trans. A, 4A, 1973, p. 1283-1292.
[16] B. Prenosil - Härt.-Techn. Mitt., 20, 1965, p. 41-49.
[17] K. Sachs, D.B. Clayton - Heat Treat. Met., 6, 1979, p. 29-34.
[18] E.J. Mittemeijer, W.T.M. Straver, P.J. van der Schaaf, J.A. van der Hoeven - Scr. Metall., 14, 1980, p. 1189-1192.
[19] F.K. Naumann, G. Langenscheid - Arch. Eisenhüttenwes., 36, 1965, p. 677-682.
[20] H.C.F. Rozendaal, E.J. Mittemeijer, P.F. Colijn, P.J. van der Schaaf - Metall. Trans. A, 14A, 1983, p. 395-399.
[21] J. Matauschek, H. Trenkler - Härt. Techn. Mitt., 31, 1976, p. 177-181.
[22] F. Hoffmann - Härt.-Techn. Mitt., 36, 1981, p. 255-257.
[23] J. Zysk - Härt.-Techn. Mitt., 31, 1976, p. 319-324.

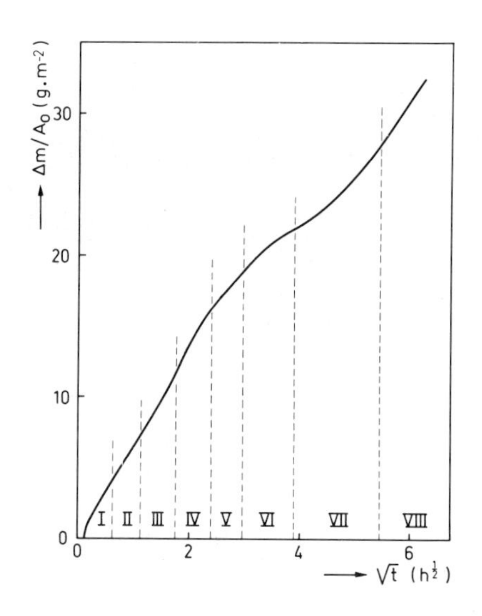

1. Mass increase, Δm, per unit area, A_0, as a function of the square root of nitrocarburizing time, t, (T = 843 K); 53.1 vol.-% NH_3/43.9 vol.-% H_2/3.0 vol.-% CO; specimen thickness 0.78 mm; the same conditions hold for Figs. 2-7.

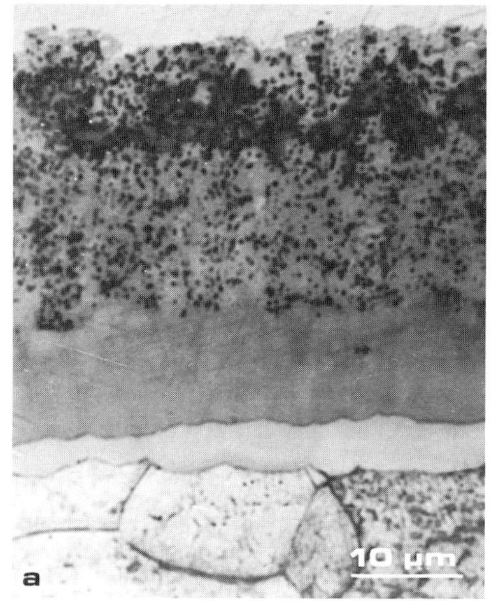

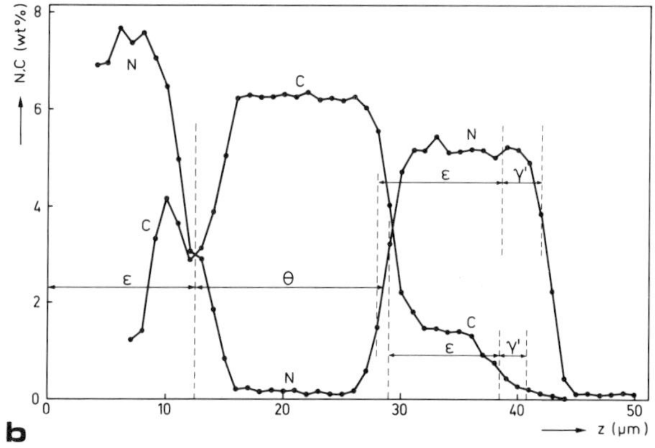

2.a. Cross section of specimen nitrocarburized for 15 h, showing a continuous cementite sublayer sandwiched between an ε sublayer with a larger amount of carbon above, and a smaller amount of carbon beneath.

b. Nitrogen- and carbon-concentration depth profiles of the ε/θ/ε/γ' compound layer shown in Fig. 2.a.

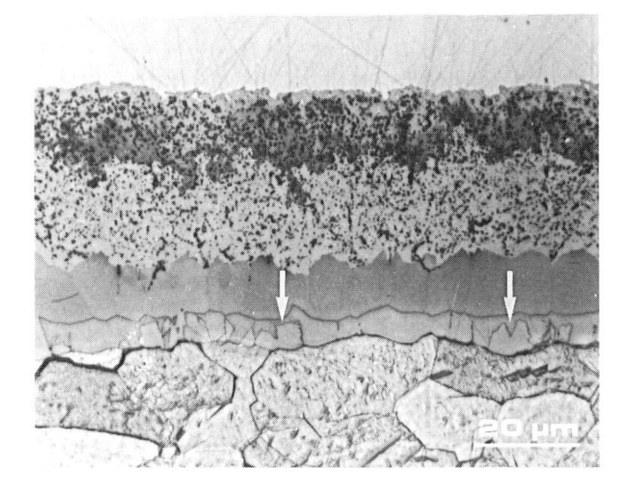

3. Cross section of a specimen nitrocarburized for 24 h, showing nucleaton of ε grains (arrowed) at the γ'/substrate interface of an ε/θ/ε/γ' compound layer.

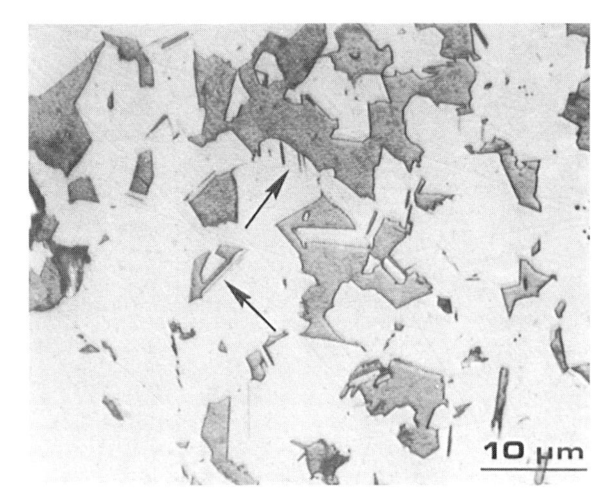

4. Oblique section of a specimen nitrocarburized for 24 h (quenched in brine). Very thin regions of ε bounded by parallel lines (arrowed), suggesting ε nucleation along twin boundaries in γ' (depth 56 µm).

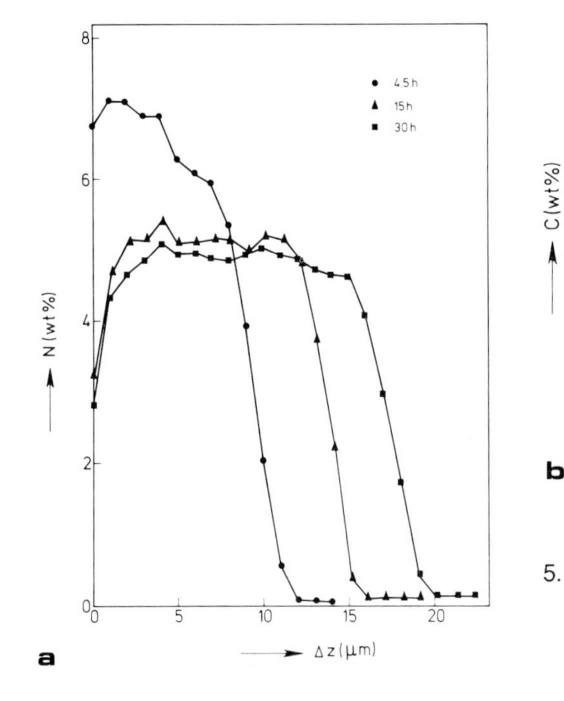

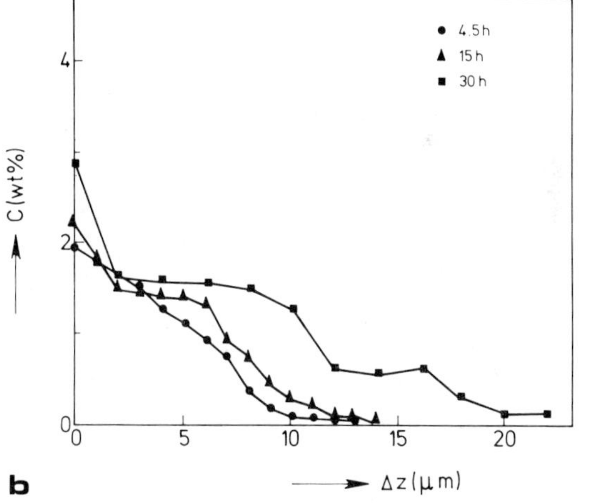

5. Nitrogen- (a) and carbon- (b) concentration depth profiles in the massive region of the compound layer at different nitrocarburizing times (i.e. 4.5 h, 15 h and 30 h).

6. Cross section of a specimen nitrocarburized for 30 h. Growth of ε grains nucleated at the γ'/α interface leads to ε "bridges" across the γ' sublayer. Double strength Murakami etching reveals a carbon gradient in ε "bridges".

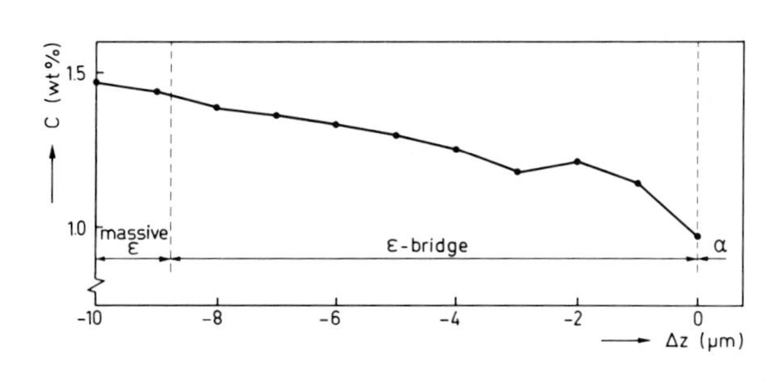

7. Carbon concentration depth profile in an ε "bridge" of the specimen shown in Fig. 6.

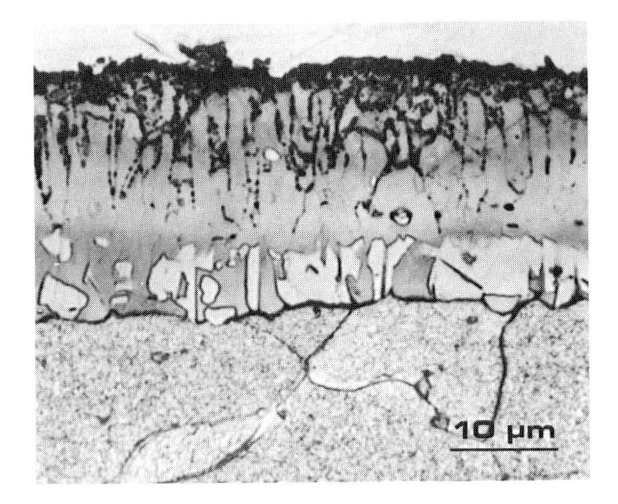

8. Cross section of an armco iron specimen nitrocarburized for 3 h in a 50 vol.-% NH_3/50 vol.-% endogas mixture at 843 K. Murakami staining in both the porous zone as well as in the ε bridges can be observed.

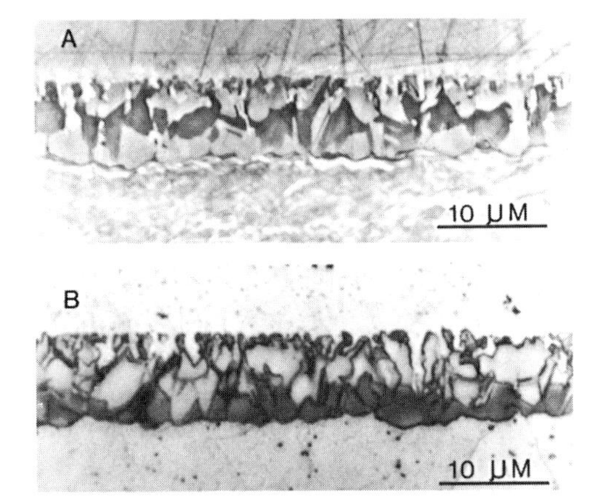

9. Cross section of a pure iron specimen nitrocarburized in a salt bath (TF1 process) for 0.5 h.
   a. Positive phase contrast microscopy showing a ε/γ' /ε phase sequence in the compound layer.
   b. Murakami staining indicates the presence of carbon in ε at the compound layer/matrix interface.

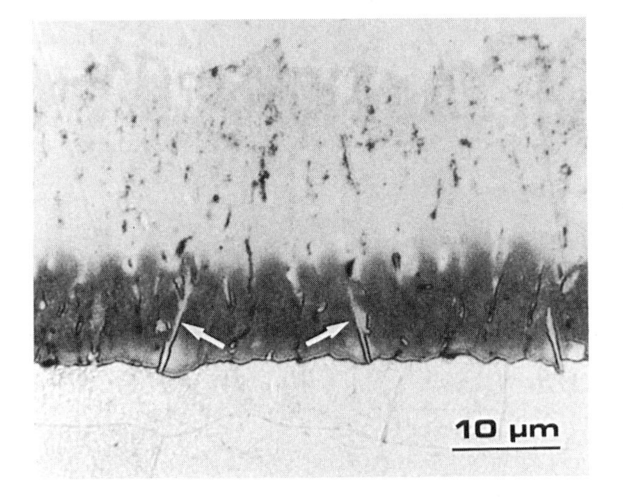

10. Cross section of a pure iron specimen nitrocarburized in a salt bath (Sursulf process; high sulfur variant) for 1.5 h, showing a pronounced carbon enrichment at the compound layer/matrix interface and retained γ' (arrowed) [10].

□

Boronized steels and their properties

D N TSIPAS, J RUS and H NOGUERRA

DNT is with MIRTEC S.A., Volos, Greece; JR and HN are in the Institute of Engineering, Caracas, Venezuela.

ABSTRACT

The thermal diffusion of boron into the surface of steel generally results in the formation of boride layers with improved corrosion-erosion properties. The formation of Fe_2B and/or FeB boride layers depends on the boron chemical activity, alloy composition, temperature and time of treatment, as well as the boronizing method. In this work we present the results of boronizing various commercial steels, 1020, 5Cr - 1Mo, 304 and hadfield steel, by the pack cementation method, with two different powder mixtures, at 815 and 900°C, and for periods varying between 6 and 12 hours. The obtained boride layers were characterized using optical and electron microscopy and the role played during boronizing, by the major alloying elements was analysed.

Finally an evaluation of the corrosion resistance of boronized 1020 steel in Naphthenic acid and during high temperature oxidation will be presented.

1. INTRODUCTION

The diffusion of boron into the outer layers of metals by thermochemical methods generally results in the formation of boride layers.

These layers are increasingly used in various branches of engineering mostly for protecting metal pieces against abrasive and/or adhesive wear. It has also been reported that boride layers are quite resistant to attack by molten metals alkali and acid media, but the full potential of boronizing for protecting against corrosion in different media still remains to be evaluated [1-17].

The type of steel under treatment, the boronizing method and composition of boronizing media, temperature and time of treatment, play an important role in deciding the quality and disposition of obtained boride layers.
Also the alloying elements during boronizing of alloy steels can preferentially enter into the iron borides, or form stable compounds with boron, depending on the base metal/alloying element reactivity towards boron and compatability of their boride structures.

The formation of a single Fe_2B tooth type layer is desirable for many industrial applications, because of its lower degree of brittleness compared to the FeB layer.

Solid boronizing agents for pack cementation generally contain a boron-releasing substance, e.g. elemental boron, B_4C, or ferroboron, an activator in the form of a halogen containing compound, such as chloride or flouride, which promotes the release of boron and its eventual diffusion into the metal surface, and an inert dilution compound, e.g. Al_2O_3 or SiC, which due to its refractory property remains inert at the elevated temperature of treatment.

In this paper we present the results of boronizing various commercial steels by pack cementation and we attempt to clarify the effects of some alloying elements during boronizing.
We also present the results of corrosion resistance of boronized 1020 steel in naphthenic acid and during high temperature oxidation.

2. EXPERIMENTAL PROCEDURE

The boronizing treatment was carried out in specially designed cylindrical retorts where the steels to be coated were placed with the intimately mixed powder compounds, ensuring that a layer of the compound mixture at least one centimeter thick was present around each sample. Two different mixtures of powders

$B/NaF/Al_2O_3$ [mixture No 1] and $B_4C/SiC/KBF_4$ [mixture No 2] were used at temperatures of 815 and 900°C respectively and for periods varying between 6 and 12 hours.

The treated steels were: A low carbon SAE 1020, Hadfield steel grade A-128, AISI steel 304, and a 5% Cr - 1% Mo ferritic steel. Characterization of the resulting boride layers was done with optical microscopy, X-Rays and Scanning Electron Microscopy. Corrosion testing in naphthenic acids was carried out under static conditions 1. in the liquid phase at 210°C, using a JOBO Venezuelan crude oil distillation fraction of the temperature range 375-450°C, neutralization No 1.10 (neutralization No is expressed as the milligrams of KOH required to neutralize the acids in a 1 gram sample) which was adjusted to 18.6 by the addition of 6% by weight of concentrated naphthenic acid; 2. in both liquid and vapour phase at 250°C using concentrated naphthenic acid.

High temperature oxidation was carried out in air at 650°C.

3. RESULTS AND DISCUSSION

The boride layers obtained with powder mixture No 2 were greater and more uniform than mixture No 1 [see tables I and II, and Figure 1]. Both powders were of high boron chemical activity producing FeB and Fe_2B layers, FeB being the external layer. The 1020 steel presented a tooth shaped morphology with both powder mixtures, and the Hadfield steel in all cases presented an almost flat boride-matrix interface. The 5 Cr - 1 Mo steel, when treated with mixture No 1 presented a flat interface, while when treated with mixture No 2 a tooth-shaped interface was observed. Finally in the case of 304 steel and with both powder mixtures a flat boride layer-matrix interface was observed.

In the case of Cr and Ni as alloying elements in Fe, acting independently or in combination with each other in industrial steels, it could be deduced that:

1. Cr either enters iron borides or concentrates at the interface between the boride layer [19-21] and the substrate and in certain cases also forms boride layers entirely composed of CrB, figure 2.

2. Ni has been found to concentrate beneath the boride coating, enter the Fe_2B layer and finally, in some cases, to precipitate as Ni_3B out of the FeB layer [19-22].

3. Both Cr and Ni reduce boride layer thickness and flatten out the tooth shaped morphology generally observed on low carbon steel.

4. Mn as an alloying element in Fe, has been found to concentrate during boronizing within the boride coating generally the Fe_2B layer, [23]. Similarly to other alloying elements, e.g. Cr and Ni, Mn was also found to reduce the boride layer thickness and flatten out the tooth shaped morphology observed in low carbon steels. The preferential entry of Mn into the boride layer was also observed in the present study, figure 3.

Amongst the factors influencing alloying element behaviour during boronizing we could mention:

1. The effect of alloying element on the base metal reactivity towards boron.

2. The ability of the alloying element to preferentially enter iron borides.

3. The ability of the alloying element to react and form stable compounds with boron.

4. Boronizing time and temperature.

5. Boron activity of the powder mixture.

In the case of Cr and Ni their role could be better understood considering some of the processes believed to take place during power boronizing, which are as follows:

a. On heating the powder mixture, gaseous boron halide, BX_3, is formed.

b. BX_3 interacts with the surface of the base material to be coated, forming FeB or Fe_2B on it, depending on the activity of our boronizing mixture.

c. The activity of boron is lower as we go from the gas phase BX_3 to the FeB layer and the FeB/metallic matrix interface, thus in the case of high activity boronizing mixture conducing to the formation of an Fe_2B inner layer.

For low activity boronizing powder mixture, Fe_2B will be the only layer formed from the early stages of boronizing.

d. Cr and Ni are partially incorporated into the FeB and Fe_2B.

e. FeB and Fe_2B can only absorb a limited amount of Cr and Ni, thus producing an increase in the concentration of both elements at the boride layer/metallic matrix interface. This behaviour has been demonstrated in previous works [23] and it is clear that the higher the Cr and Ni concentrations of the alloy substrate the greater in general will be the tendency to form distinct Ni, or Cr boride layers.

Naphthenic acid corrosion has been observed in oil refineries processing acidic crudes and amongst the factors influencing the corrosion behaviour of metals and alloys in these media are:

1. Crude neutralization number as defined by ASTM-D 664, or ASTM-D 974 standards.

2. Temperature.

3. Fluid velocity.

The principal products of naphthenic acid corrosion have been found to be hydrocarbon soluble metal naphthanates furthermore mechanical surface damage has been reported to play an important role in this type of corrosion.

In the case of 1020 steel, the boride layers showed few signs of attack, or corrosion produced formation (figure 4) and the original FeB/Fe_2B distribution and morphology remained unchanged.

These observations confirm the chemical inertness of iron borides to naphthenic acids and the real possibility of reducing this type of corrosion in oil refineries, processing highly acidic crudes. Furthermore, the extreme hardness [1600-1800 Hv] of these layers would also guarantee the effective protection of steel against the erosive attack reported with these acids under high fluid velocity conditions.

In the case of high temperature oxidation the thermogravimetric data and morphology of oxidized boronized steel are shown in figures 5 and 6 respectively.

The boronized steel was found to be very resistant during high temperature oxidation, forming a very thin protective oxide layer of about 10 μm, while the unprotected 1020 steel showed excessive exfoliation spalling and an almost linear rate of oxidation. It seems that the boride layer acts as a barrier to oxygen and/or Fe diffusion, thus reducing oxidation rate, furthermore the thermal expansion coefficients of the boride layers are almost the same as those of iron [$4.4-5.5X10^{-8}K^{-1}$], reducing thus the danger of cracking and spalling due to thermal stresses. Finally it was observed that the prolonged exposure [170 hours] at 650^oC, led to the elimination of area FeB layer. These processes seem to be the result of boron redistribution, within the boride layers and the formation of a more homogeneous layer, less prone to crack formation and flaking.

CONCLUSIONS

The boride layers provide substantial protection of low carbon steel against:

1. High temperature oxidation

2. Naphthenic acid corrosion in both liquid and vapour phases.

ACKNOWLEDGEMENTS

The authors thank Dr. O. Lingstuyl of INTEVEP for the continuous supply of materials and useful discussions throughout this project.

References:

1. S.LONG and G.E. McGUIRE. Thin Solid Films 64 [1979] 433
2. Y.SHIBUYA, Trans ISIJ [1981] 429.
3. G.SALOMON, Lubric, Engng [1981] 634.
4. S.C. SINGHAL, Thin Solid Films 45 [1977] 321
5. O. KNOTEK, E. LUYSCGEUDER and L. LEUSCHEN, ibid 45 [1971] 331.
6. H.C. CHILD, Met.Mater Technol 13 [1981] 303
7. W.FICHTL, Rev. Int. Hautes Temp.Refract.France 17 1980 33.
8. N. KOMATSU, M. OBAYASHI and J. ENDO, J. Jpn Inst. Metals 38 [1974] 379.
9. K. HOSOTAWA, H. NISHIMURA, M. UEDA and F. SELCI, ibid. 36 [1972] 418
10. O. KNOTEK, E. LUGSCHEIDER and H. REIMANN, Thin Solid Films 64 [1979] 365.
11. A.G. VON MATUSCHKA, "Boronizing" [Hanser and Heyden, Germany, 1980]
12. R.H. BIDDULPH, Thin Solid Filmrs 45 [1977] 341
13. C. LEA, Metals Sci. [1978] 301
14. R. ROLLS and R.D. SHAW, Corr. Sci. 14 [1974] 431
15. K. FUJII, T. KATAGIRI and S. ARAKI, J. Jpn Inst. Metals 38 [1974] 698
16. P. SURRY, Br. Corr. J. 13 [1978] 31
17. Z.M. GUAN, G.X. LIU, H. J. XU and S.R. PENG, J. Heat Treating 3 [1983] 16
18. J. RUS, C. LUIS DE LEAL and D.N. TSIPAS, J. Mater. Sci. Lett. 4 [1985] 558
19. G. PALOMBARINI, M. CARBUCICCHIO and L. CENTO, J. mater. Sci. 19 [1984] 3732
20. V.I. POKHMURSKII, V.G. PROTSIK and A.M. MOKRAVA, Sov. Mater. Sci. 10 [1980] 185
21. K. H. HABIG and R. CHATTERJEE-FISCHER, Tribology Int. [1981] 209
22. P. GOEURIOT, R. FILLIT, F. THEVENOT, J.H. DRIVER and H. BRUYAS, Mater. sci. Engng. 55 [1982] 9
23. G. PALOMBARINI et al J. Mater. Sci. 19 [1984] 3732

TABLE I General characteristics and thickness of boride layers obtained with mixture 1.

TYPE OF STEEL	BORIDE LAYERS THICKNESS (μ) AFTER 12 HOURS	OBSERVATIONS
1020	40	"TOOTH" SHAPED POROUS FeB/Fe_2B LAYERS.
5 Cr - 1 Mo	21	FLAT POROUS FeB/Fe_2B LAYERS.
304	17	FLAT POROUS FeB/Fe_2B LAYERS.
HADFIELD	32	FLAT POROUS FeB/Fe_2B LAYERS.

TABLE II General characteristics and thickness of boride layers obtained with mixture 2.

TYPE OF STEEL	BORIDE LAYERS THICKNESS (μ)			OBSERVATIONS
	6 HRS.	8 HRS.	12 HRS.	
1020	97	132	---	COMPACT "TOOTH" SHAPED FeB/Fe_2B LAYERS.
5 Cr - 1 Mo	47	57	71	SLIGHTLY "TOOTH" SHAPED FeB/Fe_2B LAYERS. THIN INTERFACE BETWEEN BORIDE LAYERS AND METAL MATRIX.
304	25	37	45	COMPACT FLAT FeB/Fe_2B LAYERS. PRONOUNCED POROUS INTERFACE BETWEEN BORIDE LAYERS AND METAL MATRIX.
HADFIELD	58	68	112	COMPACT FLAT FeB/Fe_2B LAYERS.

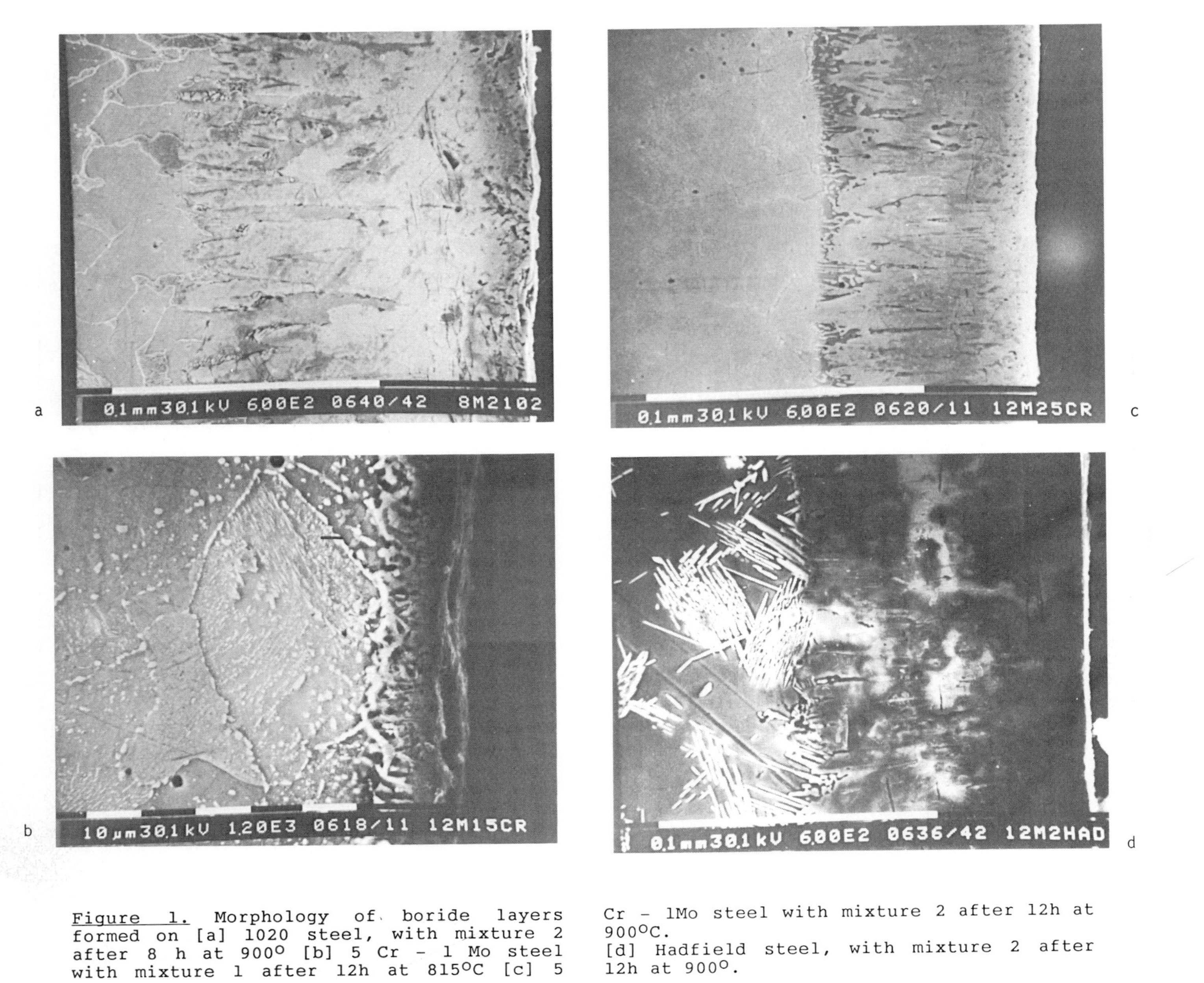

Figure 1. Morphology of boride layers formed on [a] 1020 steel, with mixture 2 after 8 h at 900° [b] 5 Cr - 1 Mo steel with mixture 1 after 12h at 815°C [c] 5 Cr - 1Mo steel with mixture 2 after 12h at 900°C.
[d] Hadfield steel, with mixture 2 after 12h at 900°.

<u>Figure 2.</u> Chromium concentration profile observed along indicated line, on a 5 Cr - 1Mo steel boronized with mixture 2 for 12h at 900°.

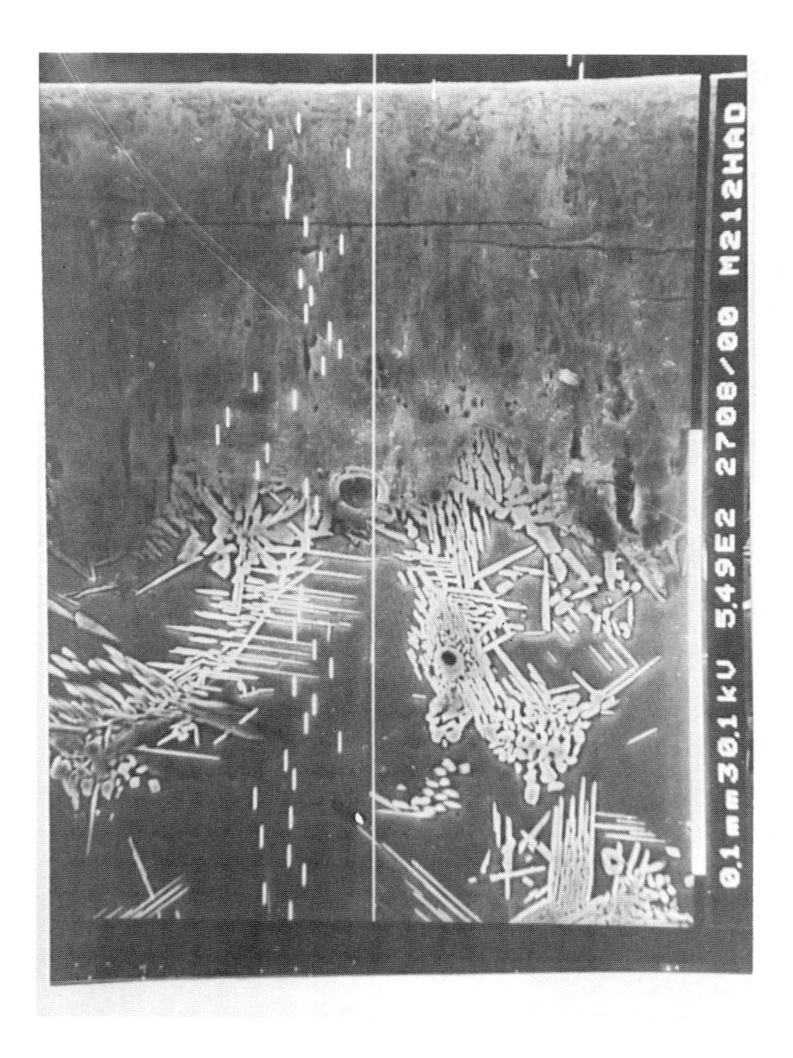

<u>Figure 3.</u> Manganese concentration profile observed along indicated line on Hadfield steel boronized with mixture a for 12 h at 900°.

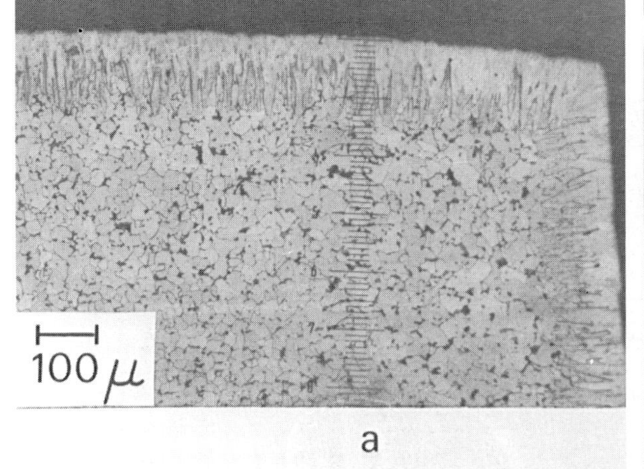

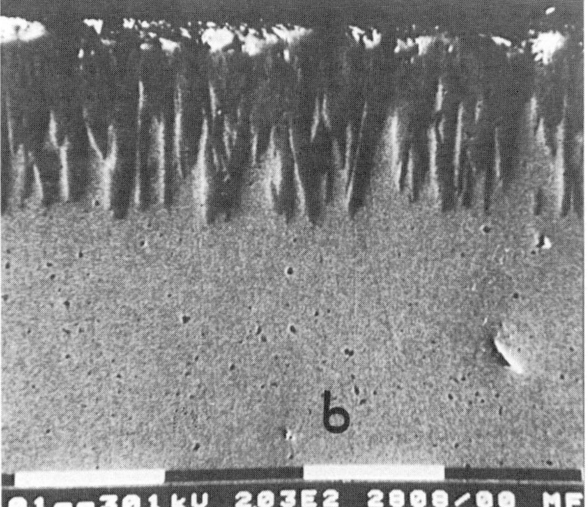

Figure 4. a] Typical boride layers obtained on 1020 steel. b] Morphology of boronized, 1020 steel subjected to liquid phase naphthenic acid corrosion, at 210° for 15 hours.

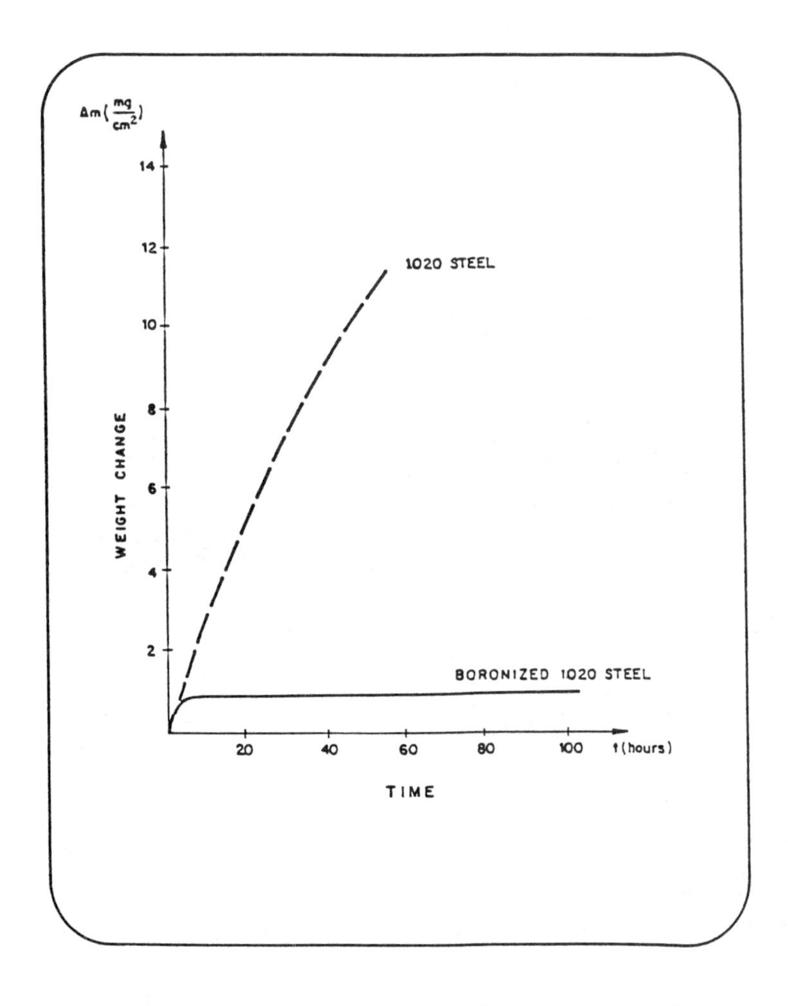

Figure 5. Thermogravimetric data for boronized, 1020 steel during high temperature oxidation in air at 650°.

Figure 6. Morphology of boronized 1020 steel after 170 hours of high temperature oxidation in air at 650°.

Boriding of Fe and Fe–C, Fe–Cr and Fe–Ni alloys: boride-layer growth kinetics

C M BRAKMAN, A W J GOMMERS and E J MITTEMEIJER

The authors are in the Laboratory of Metallurgy, Delft University of Technology, The Netherlands.

Synopsis

Specimens of pure Fe and of Fe-0.8 wt% C, Fe-0.5 wt% Cr, Fe-4.0 wt% Cr, Fe-4.0 wt% Ni and Fe-10.0 wt% Ni alloys were borided in commercially available "Ekabor 2" powder. In contrast to literature reports it was found that in all cases a surface-adjacent "FeB" sublayer was present on top of an "Fe_2B" sublayer in the compound layer. Layer-growth kinetics were analysed by measuring the extent of penetration of the "FeB" and "Fe_2B" sublayers as a function of boriding time and temperature in the range 1000-1250 K. Layer growth is dominated by B diffusion through "FeB/Fe_2B". This diffusion process is of strongly anisotropic nature. Consequently ragged interfaces occur between the substrate and the boride layers. The depths of the tips of the most deeply penetrated "FeB" and "Fe_2B" needles are taken as measures for diffusion in the fast diffusion directions. Assuming uni-directional B diffusion and parabolic layer growth a simple model of layer growth was applied. It accounts for the specific volume difference between "FeB" and "Fe_2B". As a result values were obtained for the activation energies of both "Fe_2B" *and* "FeB" layer growth.

1. Introduction

Boriding's most prominent technological advantage is the production of a very hard, wear-resistant surface layer on metallic substrates [1,2]. Industrial boriding is mainly applied to steel and ferrous alloys. For process control in automated installations, knowledge of kinetic parameters of the boriding process is essential. The separate and combined influences on boriding kinetic parameters of the many alloying elements in *steels* are not known.

This paper deals with a kinetic analysis of the boriding of pure iron and iron-carbon, iron-chromium and iron-nickel alloys. In earlier work devoted to the boriding of iron and carbon steels [3,4] no distinction was made between layer-growth rates of the (Fe,M)B and $(Fe,M)_2B$ sublayers. In the present analysis, use is made of boron-penetration depth obtained from needle-length measurements of both (Fe,M)B and $(Fe,M)_2B$ phases. A simple model for bilayer growth allowed *separate* determination of boron diffusion activation-energy values in both phases.

Results obtained indicate that alloying with chromium and nickel increases the activation energy values for both sublayers as compared to the pure Fe case. In particular, chromium and nickel have a strong influence on the activation energy for (Fe,M)B sublayer growth. Carbon has little effect.

2. Boride-layer growth kinetics

On boriding iron and ferrous alloys a boron-compound layer can develop on the surface of the specimen. In general the compound layer is composed of two sublayers: a (Fe,M)B sublayer on top of a $(Fe,M)_2B$ sublayer. An alloying element M originally present in the matrix can be incorporated in the borides. In the sequel the borides will be denoted by FeB and Fe_2B. The diffusion zone in the substrate beneath the compound layer can usually be ignored; the solubility of boron in Fe is very small [5].

It is assumed that:

(i) layer growth is rate-controlled by boron diffusion in the FeB and Fe_2B sublayers;

(ii) boride-layer growth occurs as a consequence of B diffusion perpendicular to the surface of the specimen.

The B concentration gradient within each sublayer is taken as constant (cf the small homogeneity ranges of about 1 at% of both borides reported in [6]). Further assuming that the boron surface and interface concentrations do not change during boriding, the kinetic model given in the Appendix can be adopted. Using $c_o \approx o$, it follows from eqs. (A-6) and (A-7):

$$C_{FeB} = D_{FeB}\,(c_s - c_a) = 2m\,(mc_a + nc_c) \quad (1a)$$

$$C_{Fe_2B} = D_{Fe_2B}\,(c_b - c_c) = 2n(\alpha m + n)c_c \quad (1b)$$

where m and n are the "parabolic" growth constants for the FeB and Fe_2B sublayers, respectively. The concentrations $c_s \approx c_a$ and $c_b \approx c_c$ (Appendix) and the iron atom volume ratio α (eq. (A-1)) can be assessed* for boriding of pure Fe from lattice parameter data [7]:

$c_s = 100.5 \times 10^3$ mol m^{-3}, $c_b = 59.8 \times 10^3$ mol m^{-3} and $\alpha = 0.84$.

---

* The orthorhombic unit cell of FeB has parameters a = 0.5506 nm, b = 0.2952 nm, c = 0.4061 nm; the tetragonal unit cell of Fe_2B has parameters: a = 0.5109 nm, c = 0.4249 nm. Both unit cells contain 4 molecules.

It is assumed that $c_s - c_a$ and $c_b - c_c$ do not depend significantly on temperature in the temperature range investigated. Then, the dependencies of C_{FeB} and C_{Fe_2B} on temperature are governed by the diffusion coefficients D_{FeB} and D_{Fe_2B}. Hence:

$$C_{FeB} = k_{FeB} \exp(-Q_{FeB}/RT) \qquad (2a)$$

$$C_{Fe_2B} = k_{Fe_2B} \exp(-Q_{Fe_2B}/RT) \qquad (2b)$$

where k_{FeB} and k_{Fe_2B} are constants. Consequently, the activation energies Q for B diffusion in FeB and Fe_2B can be derived from the slopes of the straight lines obtained by plotting $\ln C_{FeB}$ and $\ln C_{Fe_2B}$ as a function of the reciprocal boriding temperature.

Neither FeB nor Fe_2B has cubic crystal symmetry. Consequently, diffusion in these materials is of anisotropic nature and grains may exhibit needle-like growth. For both compounds [001] appears as the easiest diffusion direction [3,8]. Substitutional B diffusion can occur in both compounds as a consequence of small homogeneity ranges of about 1 at% [6] in the Fe-B phase diagram or unequilibrium amounts of vacancies at the boron sublattice.

Both sublayers have been shown to exhibit an 001 texture [3,8]. Further, it has been found that the Fe_2B 001 texture became increasingly sharp for increasing penetration depth [8]. A similar FeB behaviour is probable but difficult to prove (thin layer).

Initially, the direction of the steepest B concentration gradient is perpendicular to the specimen's surface where B atoms are supplied. Due to anisotropic diffusion, crystals with orientations allowing the largest B transport rate perpendicular to the surface grow out preferentially. Subsequently, the easy diffusion direction favours accelerated growth of these crystals although lateral B concentration gradients can occur.

In the present experiments it is expected that the needles most deeply penetrating into the substrate exhibit orientations close to or with [001] perpendicular to the surface.

The following interpretation is employed. On boriding first Fe_2B/FeB nuclei develop with more or less randomly distributed orientations. Thereafter, growth selection because of strong anisotropic diffusion leads to the 001 fibre texture becoming increasingly sharp for increasing penetration depths.

Hence, an unequivocal interpretation in terms of diffusion of boron in FeB or Fe_2B is only obtained if the respective needles penetrating most deeply are selected for indication of layer thickness*. The activation energies thus obtained correspond to diffusion in [001] directions for FeB and Fe_2B. This discussion implies that averaging over needles of various lengths to define an "average thickness" obstructs interpretation in terms of fundamental mechanisms.

* One also excludes then needles with the correct 001 orientations but: (i) not sectioned through their tips by the metallographic preparation technique, (ii) nucleated at a later stage in the process.

3. Results and discussion

3.1. Parabolic growth constants

Using measured data for the thickness parameters u and v (see Fig. A-1 and Eq. A-5)), plotting of $\overline{u^2}$ and $\overline{v^2}$ versus treatment time t yielded straight lines in general (but see below) indicating that parabolic layer growth occurred.

Usually these straight lines did not pass through the origin. The intercept of the time axis can result from inaccurate estimation of the time needed to heat up the specimen as well as from an (actual) incubation time for boride development. No indication was obtained for significant difference between intercepts of the time axis for $\overline{u^2}(t)$ and $\overline{v^2}(t)$ (apart from those differences due to experimental problems described below).

Part of the FeB sublayer can scale off easily during cooling after boriding and/or subsequent specimen preparation for examination, leading to too small values for u and v at variable times. Further, the boriding potential of the atmosphere may decrease as a result of exhaustion of the boriding powder, leading to too small values for u and v at long times and at high temperatures (Table 1). Because of the systematic nature of both errors, the following procedure was established:

(i) In the $\overline{v^2}$ vs t plot two points were selected yielding the *steepest* straight line. The intercept with the abscissa is taken as *the* (effective) incubation time.

(ii) The incubation time yields an additional point for the $\overline{u^2}$ vs t graph. Next, the (FeB) straight line is drawn through this point *and* the $\overline{u^2}$ data point leading to the *steepest* straight line;.

(iii) After determination of the slopes of the straight lines for both $\overline{v^2}$ and $\overline{u^2}$ vs. t, their square roots yield 2(n+m) and 2m, respectively, according to eq. (A-5).

Graphs of $\overline{u^2}$ and $\overline{v^2}$ vs. time for pure Fe specimen boriding are given in Figs. 1 to 5. The m and n values for the various materials and boriding temperatures are given in Table 2.

All values used for $\overline{u^2}$ and $\overline{v^2}$ are averages of 10 measurements. Errors for the slopes were assessed from maximal and minimal estimates for $\overline{u^2}$ and $\overline{v^2}$ according to $\{(\overline{u^2})^{1/2} \pm 2\sigma_u\}^2$ and $\{(\overline{v^2})^{1/2} \pm 2\sigma_v\}^2$ where σ_u and σ_v denote the standard deviations of u and v, respectively. Errors in C_I and C_{II} (cf eqs. (1a) and (1b)) as a consequence of these growth-constant errors are indicated in Fig. 6.

3.2. Activation energy for layer growth

Activation-energy values were determined on the basis of eqs. (1) and (2) using the parabolic sublayer- growth constants m and n gathered in Table 2:

For each material, plots of $\ln C_{FeB}$ and $\ln C_{Fe_2B}$ vs reciprocal boriding temperature were made (see Fig. 6 for the case of Fe). Straight lines were fitted to these data points employing a least-squares procedure. Activation-energy values were derived from the slopes of the straight lines (Table 3). These activation energies are interpreted here as representing B diffusion along [001] directions in FeB and Fe_2B respectively (see discussion in section 2).

Iron

Consider the projection of the crystal structures of both FeB and Fe_2B onto (001) (cf Figs. 22 and 24 in [3]). It follows that B atoms which would move along [001] are less hindered in Fe_2B than in FeB. Substitutional boron diffusion perpendicular to (001) appears possible in Fe_2B (there is a chain of B atoms along [001]), but in the case of Fe_2B a zig-zag path (corresponding to a zig-zag chain of B atoms) seems likely for net transport of B atoms along [001]. As a result it can be expected that diffusion of B is more constrained in FeB than in Fe_2B, which is consistent with a value for the activation energy larger for FeB than for Fe_2B (cf Table 3).

Activation-energy values were reported earlier for this case of boriding Fe; data are shown in Table 4. However, a comparison with the present results is difficult, because: (i) the FeB and Fe_2B sublayers were treated as one in Refs. [3 and 4] and the values reported for the activation energy reflect some average; (ii) it appears likely that the values for the activation energies reported for FeB and Fe_2B in Ref. [12] were derived just from the temperature dependence of alone m and alone n, respectively, thereby ignoring the interaction between m and n for bilayer growth as incorporated in eqs. (2a and b).

Iron-chromium and iron-nickel

Considerable amounts of Cr and Ni can be dissolved in FeB and Fe_2B [9].

The atomic radii of Cr and Ni are about the same and larger than that of Fe. Adopting the geometrical model for B diffusion, discussed above for boriding Fe, it can then be expected that Cr and Ni dissolved on the Fe sublattice of the borides will generally lead to activation energies for B diffusion larger than those for Fe (cf data for Fe, FeCr and FeNi in Table 3). In particular it may be anticipated that this geometrical constraint will be more pronounced for the FeB structure than for the Fe_2B structure. (Already for Fe the activation energy for FeB is larger than for Fe_2B). This effect is illustrated by the dramatic increase of the ratio Q_{FeB}/Q_{Fe_2B} on alloying with Cr and Ni (Table 3).

However, it should be recognized that on boriding Fe-Cr and Fe-Ni a redistribution of Cr and Ni can occur [10,11]. This could also influence the activation energies found.

Iron-carbon

Carbon does not dissolve significantly in FeB and Fe_2B [3]. On boriding carbon is driven ahead of the boride layer and, together with boron, it forms borocementite, $Fe_3(B,C)$ as a separate layer between Fe_2B and the matrix. (From lattice-parameter measurements it was found that the borocementite in the present case contained about 4 mass% B corresponding to $Fe_3(B_{0.67}C_{0.33})$ [8]). Thus, part of the boron supplied is used for the formation of borocementite.

Consequently, a bilayer-growth model for boriding Fe-C alloys does not hold in principle. However, a description based on an effective bilayer-growth model may yet be used if a number of simplifying assumptions are made:

(i) the kinetics for the three-layer system ($FeB + Fe_2B + Fe_3(B,C)$) are considered to be controlled only by B diffusion in FeB and Fe_2B, and

(ii) the fraction of all B atoms arriving at the $Fe_2B/Fe_3(B,C)$ interface that is used for $Fe_3(B,C)$ formation does not depend on time and temperature. Then, the fractional effect on the parabolic growth constant n is independent of temperature.

It can now be made likely that the changes of $\ln C_{FeB}$ and $\ln C_{Fe_2B}$ due to "B leakage" from Fe_2B do not depend on temperature in the temperature range investigated. Hence, minor changes of the Q values would occur (as compared with boriding Fe), which corresponds with the experimental findings (Table 3). Earlier work on boriding carbon steels also hinted at small effects of carbon on boriding kinetics [4].

4. Conclusions

(i) A simple model for bilayer-growth kinetics has been given allowing determination of the activation energy for diffusion-controlled growth of each sublayer.

(ii) Because of the anisotropic nature of boron diffusion, the (Fe,M)B and $(Fe,M)_2B$ sublayers produced on boriding iron-base alloys exhibit pronounced 001 fibre texture and needle-type morphology. Analysis of well-defined growth kinetics should be performed on the basis of penetration-depth measurements of those needles extending most deeply into the substrate. Activation energies obtained correspond with boron diffusion along [001] in both compounds.

(iii) Activation energies for growth of (Fe,M)B and $(Fe,M)_2B$ sublayers have been determined from parabolic growth constants. In all cases the activation energy of boron diffusion along [001] is larger for (Fe,M)B than for $(Fe,M)_2B$. This is ascribed to geometrical constraints more severe for boron diffusion in (Fe,M)B than in $(Fe,M)_2B$.

(iv) Activation energies increase on alloying with chromium and nickel; in particular the one for (Fe,M)B-sublayer growth.

(v) Boriding of iron-carbon (0.8 mass% C) leads to a surface layer composed of three compounds: FeB, Fe_2B and $Fe_3(B_{0.67}C_{0.33})$. If severe constraints are obeyed, boron diffusion in FeB and Fe_2B sublayers may still be analysed using the bilayer-growth model. The activation energies found are close to those determined for the case of boriding iron.

Appendix: Bilayer growth equations

The following case is analysed. A substrate of solvent element A has been saturated with solute element B (Fig. A-1). The chemical potential of B in the surrounding atmosphere is such that two layers, I and II, form on the substrate.Each layer corresponds to a certain compound of A and B. Layer/compound I is more rich in B than layer/compound II. It is assumed that:

(i) planar interfaces occur: inward diffusion of B is unidirectional (perpendicular to the surface);

(ii) differences in specific volume per solvent atom for the phases concerned (I, II and the substrate) are accommodated fully in the diffusion direction;

(iii) linear B-concentration profiles occur in both layers;

(iv) (inter)diffusion coefficients, D_I and D_{II} for layer I and II respectively, are constant. The same holds for the interface concentrations c_s, c_a, c_b, c_c and c_0. Local equilibrium is assumed at the interfaces.

Mass balance for growth of layer I

An increase du of the thickness of layer I in time dt occurs under simultaneous "consumption" of a layer of thickness αdu of layer II. The ratio of the specific volumes per solvent atom, V_{II} and V_I for compounds II and I, respectively, is given by: $\alpha = V_{II}/V_I$. For instance, if compounds I and II conform to AB and A_2B, respectively, it follows:

$$\alpha = V^m_{A_2B}/2V^m_{AB} \qquad (A-1)$$

where V^m denotes molar volume (in fact, α is equivalent to a so-called Pilling-Bedworth ratio).

The B mass-balance for the growth of layer I thus reads:

$$(c_a - \alpha c_b)\frac{du}{dt} = D_I\,(c_s - c_a)/u - D_{II}\,(c_b - c_c)/(v - u) \qquad (A-2)$$

where both u and v are distances to the (outer) surface.

Mass-balance for growth of layer II

Even if no reaction would occur at the layer II/ substrate interface, a shift of the interface would yet occur as a result of the reaction taking place at the I/II interface (if $\alpha \neq 1$). As follows from the above discussion, this interface shift amounts to $(1 - \alpha)du$. Therefore, in the analysis of intrinsic growth of layer II, a net shift dv_{net} has to be considered:

$$dv_{net} = dv - (1 - \alpha)du \qquad (A-3)$$

Then, proceeding analogously as for layer I, the B-mass balance for growth of layer II reads:

$$(c_c - \beta c_o)\frac{dv_{net}}{dt} = D_{II}\,(c_b - c_c)/(v - u) \qquad (A-4)$$

where β denotes the ratio of the specific volumes per solvent atom of substrate and compound II, respectively.

Note that the B amounts necessary to maintain the concentration gradients in the respective layers have been neglected in the left-hand members of eqs. (A-2) and (A-4).

The following general solutions [13] satisfy eqs. (A-2) and (A-4):

$$u = 2mt^{1/2} \quad \text{and} \quad v - u = 2nt^{1/2} \qquad (A-5)$$

After rearranging it follows:

$$C_I \equiv D_I\,(c_s - c_a) = 2m\{mc_a + nc_c - \alpha m(c_b - c_c) + -\beta c_o(\alpha m + n)\} \qquad (A-6)$$

and:

$$C_{II} \equiv D_{II}\,(c_b - c_c) = 2n(\alpha m + n)(c_c - \beta c_o) \qquad (A-7)$$

If the interface concentrations are not known the diffusion coefficients can not be determined from measurements of the "parabolic" growth constants m and n. However, the activation energies for diffusion can still be obtained from measurements of m and n provided the interface concentrations and in particular the differences $(c_s - c_a)$ and $(c_b - c_c)$ are approximately constant in the temperature range considered.

Acknowledgements

We are indebted to Ir. J.J. Smit for enthusiastic cooperation in an early stage of the project. We express our gratitude towards Ir. M.A.J. Somers and Professor B.M. Korevaar for stimulating discussions and towards Mr. P.F. Colijn for skilful metallographic analysis.

References

1. A. Graf von Matuschka: Boronising, 1980, Carl Hanser Verlag, Munich, FRG.
2. P.A. Dearnley and T. Bell: Surface Eng. 1 (1985) 203-217.
3. H. Kunst and O. Schaaber: Härterei-Tech. Mitt. 22 (1967) 1-25.
4. M.-J. Lu: Härterei-Tech. Mitt. 38 (1983) 156-169.
5. O. Kubaschewski: IRON-Binary Phase Diagrams, 1982, Springer Verlag, Berlin, FRG.
6. Binary Phase Diagrams, Vol. 1, T.B. Massalski et al. eds., ASM, Ohio, USA, 1986, p. 356.
7. R. Kiessling: Acta Chem. Scand. 4 (1950) 209-227.
8. J.J. Smit, Delft University of Technology, Laboratory of Metallurgy, 1984, unpublished research.
9. G. Hägg and R. Kiessling: J. Inst. Metals 81 (1952) 51.
10. C. Badini, C. Gianoglio and G. Pradelli: J. Mater. Sci. 21 (1986) 1721-1729.
11. Z.-S. Jiang, L.-X. Zhang, L.-G. Li, X.-R. Pei and T.-F. Li: J. Heat Treating 2 (1982) 337-343.
12. H. Plänitz, G. Treffer, H. König and G. Marx: Neue Hütte 27 (1982) 228-230.
13. W.E. Boyce and R.C. Diprima, Elementary Differential Equations and Boundary Value Problems, 2nd edition, New York, USA, 1969, p. 17.

Table 1

Examples of affected layer-thickness determinations due to scaling off and exhaustion of boriding powder

material and boriding temperature	boriding time (h)	FeB layer thickness (μm)	FeB+Fe_2B layer thickness (μm)
Fe-4Ni	1	nil	19.4
(1073 K)	4	11.0	53.7
Fe-4Cr	1	27.3	100.7
(1273 K)	21/2	nil	137.5
	4	nil	185.0
Fe	1/4	16.2	71.3
(1273 K)	1	25.0	133.5
	2	24.1	189.6

Table 2

Parabolic boride-layer growth constants m (upper value) and n (lower value). Constants m and n correspond to approximate layer growth in [001] directions (cf section 2)

	boriding temperatures (K)				
	1023	1073	1123	1173	1273
	FeB and Fe_2B layer growth constants m and n (eq. A-5 ($10^{-6}ms^{-1/2}$)				
Fe	0.0289	0.0570	0.0846	0.147	0.262
	0.154	0.212	0.313	0.517	0.842
Fe-0.8C	-	-	0.157	0.268	0.479
			0.231	0.347	0.552
Fe-0.5Cr	-	0.0642	0.113	0.159	-
	-	0.236	0.322	0.529	-
Fe-4Cr	-	0.0561	0.125	0.260	-
	-	0.102	0.147	0.242	-
Fe-4Ni	-	0.0479	0.141	0.289	-
	-	0.186	0.306	0.371	-
Fe-10Ni	-	0.0316	0.135	0.190	0.412
	-	0.196	0.262	0.358	0.482 (1223 K)

Table 3

Activation-energy values obtained using bilayer growth model (Appendix)

	Q_{FeB} (kJmol^{-1})	Q_{Fe_2B} (kJmol^{-1})	Q_{FeB}/Q_{Fe_2B}	correlation coefficient least squares fit Q_{FeB}	Q_{Fe_2B}
Fe	177	156	1.13	0.998	0.996
Fe-0.8C	164	143	1.15	0.991	0.993
Fe-0.5Cr	183	170	1.08	0.999	0.992
Fe-4Cr	290	208	1.39	0.999	0.996
Fe-4Ni	311	178	1.75	0.995	0.986
Fe-10Ni	286	159	1.80	0.983	0.999

Table 4

Activation-energy values for boriding iron reported in the literature

authors	activation energy reported (kJmol^{-1})	details
Kunst & Schaaber [3]	176; both layers treated as one	amorphous B powder boriding; B concentration measurements; armco specimens
Lu [4]	132; both layers treated as one; very little FeB	Ekabor 2 powder boriding; layer thickness measurements; armco specimens
Plänitz et al. [12]	FeB: 82.5 Fe_2B: 105.5	gas boriding; separate layer thickness measurements; armco specimens; $Q_{FeB}/Q_{Fe_2B} = 0.78$

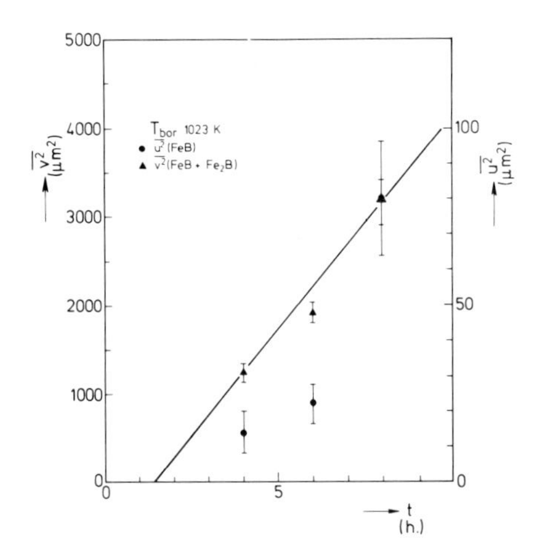

1. Averaged squares of layer thickness u and v (cf Fig. A-1) vs. boriding time for pure Fe specimen: 1023 K

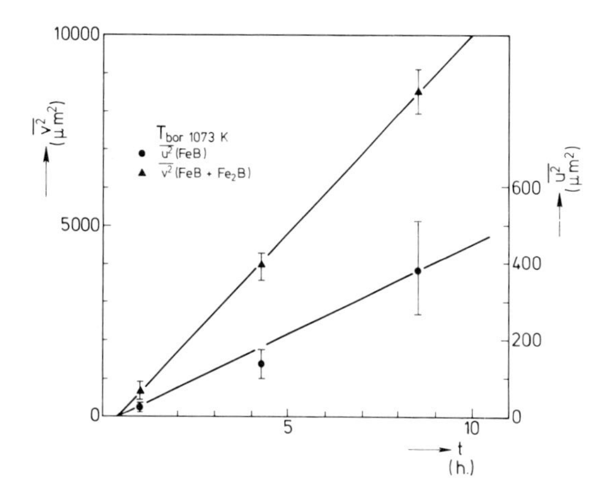

2. As Fig. 2, but boriding temperature: 1073 K

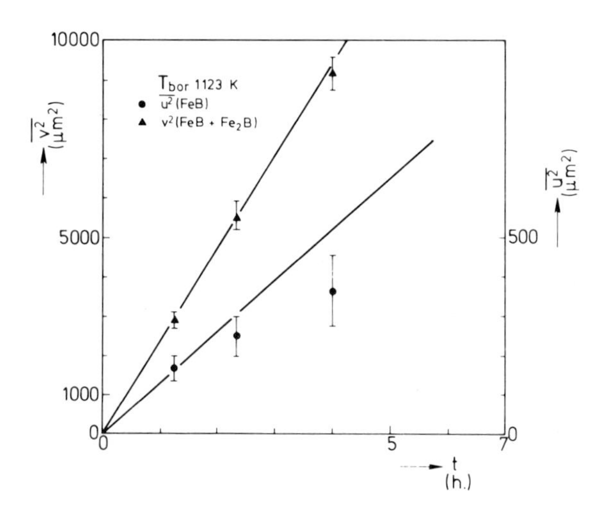

3. As Fig. 2, but boriding temperature: 1123 K

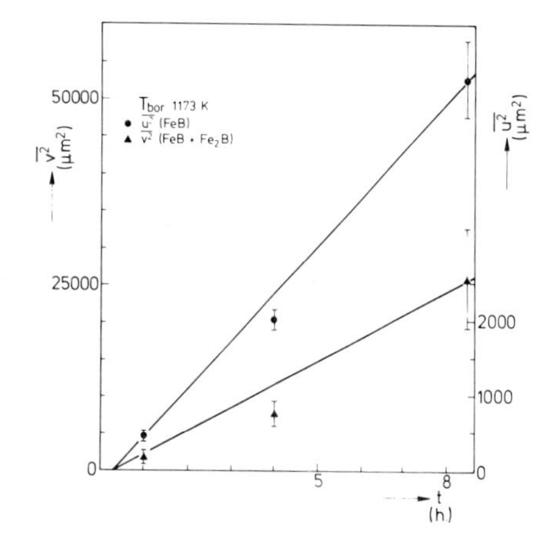

4. As Fig. 2, but boriding temperature: 1173 K

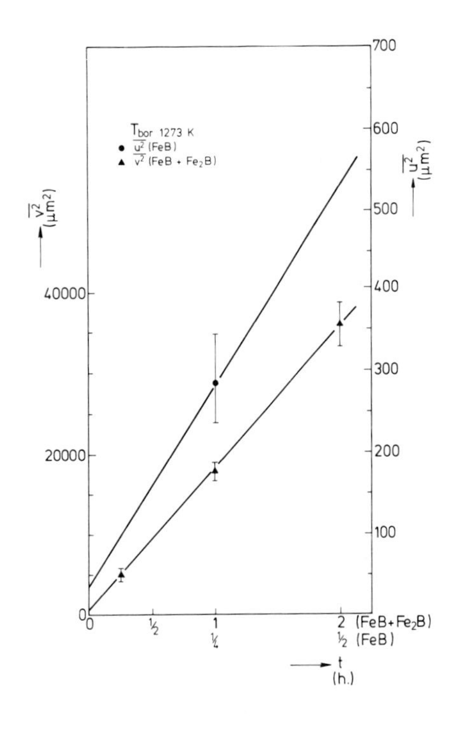

5. As Fig. 2, but boriding temperature: 1273 K

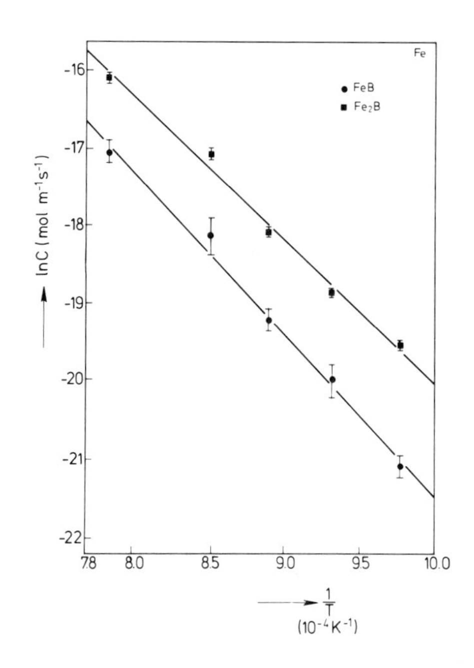

6. ln C_I and ln C_{II} vs. 1/T for pure Fe specimen (cf eqs. (1) and (2))

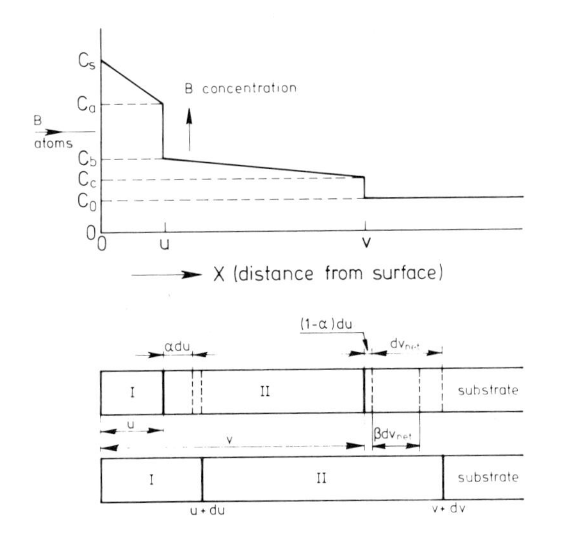

A-1 Solute-element concentration profile assumed in bilayer growth model

Computer-assisted thermochemistry and kinetics for heat treatment

T HOLM and J ÅGREN

TH is with the AGA Innovation, Lidingö, Sweden; JÅ is in the Department of Physical Metallurgy, Royal Institute of Technology, Stockholm, Sweden.

ABSTRACT

Thermochemistry and kinetics are important tools when predicting the outcome of a heat treatment. However, their usefulness is often restricted by the high complexity of a real industrial process compared to text-book problems. For example, a commercial alloy may typically contain more than 5 elements that all affect its equilibrium properties.

By the application of computer methods it is possible to overcome many of the difficulties and perform calculations of direct practical interest.

In this paper we will present some applications of computer operated methods for equilibrium calculations and simulation of surface and diffusional reactions. The equilibrium-calculation program and a thermochemical databank are the main components in an integrated system of computer programs called THERMO-CALC which have been developed over the last decade.

We have applied the THERMO-CALC system and a program for simulation of diffusional reactions to investigate:

1) Gas equilibria:
   - equilibrium composition of furnace atmospheres under different conditions
   - influence of gas composition and gas impurities on carbon activity

2) Dependence of alloy composition on carbon activity and carbide formation

3) Gas carburizing, kinetics of surface reactions related to gas composition and carbon profiles

4) Nitrogen equilibria in steel. Influence of alloying elements, chromium in particular, on solubility and nitride formation.

Experimental results are presented and discussed in relation to theoretical calculations.

1. INTRODUCTION

Thermochemical calculation is a necessary tool for proper control of heat treatment processes. Carbon-potential control in carburizing atmospheres is one important application. The use of computer-aided procedures makes it possible for the heat treater to apply thermochemistry and kinetics to complicated cases occurring in practice. From a practical point of view one may mention, for example,

1) influence of alloying elements upon carbon activity in a steel
2) prediction of equilibrium composition of atmospheres produced under different conditions
3) prediction of solubility of nitrogen in alloy steels
4) prediction of the progress of oxidation

Thermodynamics is strictly valid only for systems in equilibrium. In heat treatment this may be a realistic approximation in some cases, but often reactions are very slow and global equilibrium is far from being reached. In the latter case thermodynamics has to be combined with kinetic calculations. It may then be possible to simulate the reactions in a gas mixture, the reaction between gas and a metal at the surface of the metal and the diffusional reactions inside the metal.

In this paper some examples of thermodynamic and kinetic calculations related to heat treatment will be discussed. The THERMO-CALC system, developed at the Royal Institute of Technology (1), will be used for the thermodynamic calculations and a program for diffusional reactions (2) will be used for the kinetic calculations. The thermodynamic data for gases were taken from the SGTE substance data bank (3), the data for alloy steels from a compilation by Uhrenius (4), and the data for nitrogen-rich steels from the work by Hertzman (5). All these data banks are available on line in the THERMO-CALC system.

2. GAS EQUILIBRIA

The equilibrium composition of a gas is often estimated by regarding a main reaction such as the dissociation of methanol

$$CH_3OH \rightarrow CO + 2H_2 \qquad (1)$$

The atmosphere resulting from methanol will thus contain 33 vol % CO and 67 vol % H_2. However, several other reactions are possible, e.g.

$$2CH_3OH \rightarrow C + CO_2 + 4H_2 \qquad (2)$$

Using the THERMO-CALC system the equilibrium composition is calculated taking a large number of species into account. One example is shown in Table 1, the common 40/60 nitrogen/methanol mixture. As a comparison the compositions calculated for a number of atmospheres are given in Table 2. Only the concentrations of the major species have been included. The first example is the normal endogas, prepared from air/propane in the proportions 7.18/1. As can be seen there is an important difference: the inherent carbon activity is higher for the endogas. It should be emphasized that the calculations were based on the addition of a stoichiometric amount of O_2 (as air) in order to have a complete oxidation of propane. Already small amounts of extra oxygen, for example as water, will lower the carbon activity drastically. This is the reason why varying moisture content of the air affects the carbon potential of the endogas. The effect is demonstrated in the second line of Table 2 where the air/propane ratio is 7.23 rather than 7.18.

The rest of Table 2 shows the influence of impurities in methanol upon the resulting gas composition and carbon activity in the 40/60 mixture. It is shown that very low impurity levels of toluene, which is sometimes added to the methanol as a denaturant, drastically increase the carbon activity and thus the risk of soot formation whereas a low amount of H_2O decreases the carbon activity.

An interesting feature of mixtures between cracked methanol and N_2 is revealed in Fig. 1, which shows the carbon activity as a function of the N_2 content. The carbon activity is relatively insensitive to dilution with N_2.

3. INTERPLAY BETWEEN CARBON ACTIVITY AND ALLOY COMPOSITION

If a specimen of pure iron is equilibrated with a gas having a specified carbon activity the specimen will dissolve some carbon and the amount will depend on the temperature. In practice one often refers to this carbon content rather than the carbon activity of the gas, and it is called the "carbon potential" of the gas.

In order to heat treat an alloy steel the carbon activity in the atmosphere must be adjusted to the steel composition. It is well known that the alloy elements influence the carbon activity of a steel. In heat treating it is common to use simplified regression formulas, for example the ones originally presented by Gunnarson (6). However, these formulas have severe limitations. For instance, it is assumed that the steel is completely austenitic. In Table 3 a comparison is made between THERMO-CALC and the Gunnarson formula. Not only case-hardening steels but also high-alloy steels are considered. It is evident that for low-alloy steels the Gunnarson formula and the THERMO-CALC calculations give very similar results. For high-alloy steels, however, the Gunnarson formula does not give results that can be used in practice. The THERMO-CALC results, though, do agree well with practical results, see last column in Table 3.

A thermodynamic upper limit for the carbon activity in atmospheres is $a_C=1$, corresponding to the beginning of a driving force for the formation of soot in the gas or graphite on the surface. However, the limit for carbide precipitation may be lower. The carbon activity and the carbon potential at the carbide precipitation limit, calculated for some case-hardening steels, are shown in Table 4.

4. CARBURIZING

The kinetics of the carburizing process has been studied by many authors, for example by Collin et al (7) and by Grabke and Tauber (8). More recently Parrish and Harper (9) have presented an extensive review covering many aspects of the process.

The carburizing process can be divided into the following steps

- . molecular reactions in the gas
- . carbon transfer in the gas phase to the surface
- . surface reactions
- . diffusion of carbon in the steel

It is now well known that earlier simplifications assuming that diffusion of carbon in the steel is the rate-limiting step are not valid approximations except for very long heat treatments. A common opinion is that reactions and diffusion in the gas phase are very fast and do not need to be considered when calculating the carbon-transfer rate. On the other hand surface reactions are slow enough to be the rate-controlling step in the first part of the carburizing cycle.

In order to account for the limited rate of the surface reaction at the gas-metal interface it is common (7,9) to adopt the following empirical equation for the carbon flux entering the metal

$$J_C = \rho k(a_C^g - a_C^s) \quad (3)$$

where J_C is the carbon flux, i.e. the amount of carbon (grams or moles) per unit area and unit time, a_C^g is the carbon activity of the gas and a_C^s is the actual carbon activity at the steel very close to the surface of the steel; ρ is the density of the steel expressed as mole or gram per unit volume; k is the so-called mass-transfer coefficient and may depend on both temperature and gas composition. For low carbon contents (i.e. for steels) the above equation may be approximated as

$$J_C = k'(c_C^g - c_C^s) \quad (4)$$

where c_C^g is the carbon concentration, expressed as amount per unit volume, that the steel would have if equilibrated with the gas, c_C^s is the actual carbon concentration in the steel very close to the surface, and k' is k multiplied by the activity coefficient. This expression is more convenient for calculations and is usually applied in evaluation of experiments, see for example ref. 7. It should be emphasized though that the k value in eq. 3 is independent of the composition of the steel while k' is not, because it contains the activity coefficient. For low-alloy steels, like case hardening steels, the dependence is not very important and can be neglected but for high-alloy steels it must be taken into account.

Carburizing may occur through several reactions. The three main reactions are:

$$CH_4 \rightarrow C + 2H_2 \quad (5)$$

$$2CO \rightarrow C + CO_2 \quad (6)$$

$$CO + H_2 \rightarrow C + H_2O \quad (7)$$

In principle a mass-transfer coefficient may be evaluated for each reaction. This was done by Grabke and Tauber (8) and by Collin et al (7) who studied carburizing kinetics in CH_4/H_2, CO/CO_2 and CO/H_2 mixtures. They analyzed the concentration of different gas species and evaluated the rate constant for each reaction as a function of the concentration of the different species. Assuming that the different reactions proceed independently of each other the carburizing rate in a real atmosphere can be obtained as a sum of all the reactions. Slycke and Ericsson (10) and Collin et al (7) have presented calculations based on this approach. In atmospheres containing CO and H_2, i.e. endogas and cracked methanol, reaction 7 is the predominant one for carburizing (7,8) and the other reactions can be neglected.

The mass-transfer coefficient can also be evaluated from experiments with a carburizing atmosphere used in practice. In that case the single value obtained is some kind of average value which accounts for all the carburizing reactions and it is usually evaluated as a function of the over-all composition of the atmosphere. Such results have been reported by Wyss et al (11) who suggest that the mass-transfer coefficient is proportional to the product of CO and H_2 partial pressures.

Table 5 shows mass-transfer coefficients for different atmospheres, calculated from the parameters reported by various authors.

The carbon diffusion in a steel is most conveniently treated by means of numerical solutions of the diffusion equation. When such procedures are applied the influence of the limited rate of the surface reactions is easily accounted for. It is also possible to account for the concentration-dependent diffusion coefficient.

Such computer software has been developed recently (2) and applied to various cases, see for example refs. 12 and 13. It will now be applied to carburizing. The concentration-dependent diffusion coefficient of carbon was taken from ref. 14.

In Fig. 2 some carbon-concentration profiles (symbols) obtained experimentally by Sproge and Ågren (15) are compared with a calculated one (solid line). The atmosphere is the common 40/60 nitrogen/methanol mixture. The corresponding k' value can be found in Table 5. In the calculation the values according to Collin et al (7) were applied. As can be seen in Fig. 2 the agreement between calculated and measured profile is very good.

A common way of increasing the carburizing depth is to apply a two-step process, the so-called boost diffusion. In the first step the carbon potential is very high, typically above 1 wt% and a large amount of carbon is added to the surface layers of the steel. In the following step the potential is lowered to around 0.8 wt% and the carbon added in the previous step will now have time to diffuse into the steel. In the boost step pure methanol or pure endogas is used, i.e. no air is needed. Under these circumstances the k' value is twice as high in the methanol as in the endogas, see Table 5.

In Fig. 3 the calculated carburizing depth, defined as the depth to 0.30 wt% C, has been plotted as a function of carburizing time for different conditions. The lowest carburizing depth is obtained for the endogas and for pure cracked methanol in the one-step cycle. Air has been added to both gases in order to obtain a carbon potential of 0.8 wt%. The endogas will have the lowest k value, namely $1.9\ 10^{-7}\ m\ s^{-1}$ compared to $2.7\ 10^{-7}\ m\ s^{-1}$ for pure cracked methanol. As a comparison a curve calculated for complete diffusion control has been added

As can be seen the carburizing depth will increase considerably faster using a boost diffusion process, and in particular if it is based on pure methanol rather than, for example, a one-step process based on methanol or endogas.

5. SOLUBILITY OF NITROGEN IN ALLOY STEELS

The THERMO-CALC system has been applied to nitrogen/steel equilibria. Nitrogen is usually considered as an inert gas that can be used as a protection against oxidation of steels. However, it may instead give a nitriding effect because there is a strong tendency for nitrogen to dissolve in the steel.

The solubility of N_2 gas in pure iron is rather small although it is by no means negligible. In a stainless steel, which contains at least 13 percent chromium, the situation becomes radically different. Some calculated results for Cr-rich steels, based on the thermodynamic parameters from Hertzman (5), will now be presented. Chromium and nitrogen form some rather stable compounds and there is a large affinity between the two elements in the solid solution. A chromium rich steel equilibrated with N_2 gas at 1 atmosphere pressure will contain considerable amounts of nitrogen. This is shown in Table 6. It also shows that CrN starts to form above a certain Cr content. Furthermore Cr_2N may form at even higher Cr contents. These nitrides may cause the material to deteriorate.

In Fig. 4 the calculated equilibrium nitrogen content has been plotted as a function of the square root of pressure for an 18/8 stainless steel at 1050°C. As a comparison the results of some production-scale measurements (16), filled symbols, and some laboratory measurements from a recent work by Ekman (17), open symbols, have been included. In the production-scale measurements the time ranged from 1 to 10 minutes and the nitrogen content was obtained by chemical analysis of a 0.02 mm thick layer scraped from the surface. In Ekman's work the time was 10 minutes and the concentrations were evaluated from profiles obtained by microprobe measurements. It is evident that the measured nitrogen contents in both investigations are below the equilibrium contents.

It has been known for a long time (18) that if pure iron is surrounded by a purified, oxygen-free and dried mixture of N_2 and H_2 the rate of nitrogen pick-up is high. It is also known that small amounts of impurities in the gas, e.g. water, reduce the rate. The reason may be that a thin oxide film on the metallic surface inhibits the surface reaction. For an 18/8 stainless steel this effect was clearly demonstrated experimentally by Ekman. He found that by increasing the dew-point, and thus the oxygen activity, the nitrogen pick-up was reduced to undetectable levels whereas a reduced dew-point had the opposite effect, see Fig. 4.

SUMMARY AND CONCLUSIONS

Computer-assisted procedures have been applied to some practical problems in connection with industrial heat treating. These methods, as available in the THERMO-CALC system, make it feasible to predict various equilibrium properties of furnace atmospheres used in heat treating, e.g. the endogas from propane and air or cracked methanol. The methods are also capable of treating equilibrium properties of complex steels having many components and phases. It is thus possible to investigate, on the computer, different combinations of atmospheres and steels. By combining equilibrium calculations with simulation of diffusion and surface reactions one can simulate a whole heat treatment cycle, for example carburizing, and obtain valuable information of direct practical interest.

ACKNOWLEDGEMENTS

The authors are grateful to Professor Mats Hillert for many valuable suggestions. The calculations performed are parts of projects sponsored by The Swedish Board for Technical Development.

REFERENCES

1. B. Sundman, B. Jansson and J-O Andersson: CALPHAD 9 (1985) pp 153-190.
2. J. Ågren: J. Phys. Chem. Solids 43 (1982) pp 385-391.
3. The SGTE Databank 1985.
4. B. Uhrenius: Hardenability Concepts with Applications to Steel. (eds: D.V. Doane & J.S. Kirkaldy) TMS AIME 1977, pp 28-81.
5. S. Hertzman: Thermodynamic Aspects on Nitrogen in Steel, Thesis 1985, Royal Inst. of Techn. S-100 44 Stockholm, Sweden.
6. S. Gunnarson: Härterei-Techn. Mitt. 22 (1967) pp 293-295.
7. R. Collin, S. Gunnarson and D. Thulin: J. Iron and Steel Institute 210 (1972) pp 777-784, pp 785-789.
8. H.J. Grabke and G. Tauber: Arch. Eisenhüttenwes. 46 (1975) pp 215-222.
9. G. Parrish and G.S. Harper: Production Carburizing, Pergamon Press, Oxford 1985.

10. J. Slycke and T. Ericsson: J. Treating 2 (1981) 2 pp 3-19 and 2 (1981) 2 pp 97-112.

11. U. Wyss, R. Hoffman and F. Neumann: J. Heat Treating 1 (1980) pp 14-23.

12. J. Ågren: Acta Metall. 30 (1982) pp 841-851.

13. J. Ågren, H. Abe, T. Suzuki and Y. Sakuma: Metall. Trans. A 17A (1986) pp 617-620.

14. J. Ågren: Scripta Metall. 20 (1986) pp 1507-1510.

15. L. Sproge and J. Ågren: To be published.

16. T. Holm: unpublished research.

17. T. Ekman: Bachelor thesis 1987 (in Swedish) KTH, S-100 44 Stockholm, Sweden.

18. P. Grieveson and T. Turkdogan: TMS-AIME 230 (1964) pp 407-414 and pp 1604-1609.

TABLE 1: Calculated equilibrium properties of a mixture of 40% N_2 and 60% cracked CH_3OH at 930°C and 1 atm
a_c=1.05 (relative graphite)
Concentration of gas species

Species	Concentration	Species	Concentration
N_2	4.01999E-01	C_1H_2	4.56780E-15
H_2	3.95093E-01	C_4H_4	1.98331E-15
C_1O_1	1.97800E-01	C_4H_2	2.12056E-16
C_1H_4	2.50625E-03	H_1N_1	1.98008E-16
H_2O_1	1.86315E-03	$C_1N_1O_1$	1.72922E-16
C_1O_2	6.75663E-04	H_2N_2	1.25042E-16
$C_1H_1N_1$	3.19583E-05	C_2O_1	5.94425E-17
H_3N_1	2.99371E-05	$C_2H_4O_1$	3.07768E-17
C_2H_4	2.21985E-07	H_4N_2	7.11268E-18
H_1	1.31292E-07	$H_1N_1O_1$	4.37687E-18
$C_1H_2O_1$	8.09270E-08	N_1	2.99873E-18
C_2H_2	5.63379E-08	C_2N_1	2.68127E-18
C_1H_3	1.88765E-08	O_1	1.70235E-18
C_2H_6	1.49140E-08	N_2O_1	9.41055E-19
$C_1H_1N_1O_1$	1.11481E-08	C_4N_2	8.33631E-20
C_4H_8	1.19031E-09	C_1N_2	5.83348E-20
$C_1H_4O_1$	1.07802E-10	$C_1H_1O_1$	2.38428E-20
H_2N_1	1.58205E-11	C_1H_1	7.87020E-21
H_1O_1	5.20283E-12	O_2	4.06238E-21
C_3H_6	2.87462E-12	H_2O_2	2.87923E-21
C_2N_2	2.85974E-12	N_3	2.14406E-22
C_3H_8	2.01271E-13	C_8H_6	6.40031E-23
C_3O_2	5.60239E-14	C_1	1.19644E-23
N_1O_1	2.19394E-14	H_1O_2	8.92662E-24
C_1N_1	1.74696E-14	$H_1N_1O_2$	2.27676E-24
C_2H_1	8.60755E-15	C_3	6.54931E-25

TABLE 2: Calculated equilibrium properties of gas atmospheres at 1 atm.

Type of atmosphere	Temp °C	a_C (rel. graph)	Concentration of gas species (%) N_2	H_2	CO	CH_4	H_2O	CO_2
Endogas: C_3H_8+7.18 air	930	1.37	44.96	31.14	23.48	0.20	0.13	0.07
-"-	1050	1.01	44.84	31.41	23.60	0.07	0.05	0.02
Endogas: C_3H_8+7.23 air	930	0.88	44.95	31.14	23.46	0.21	0.13	0.11
40% N_2+60% CH_3OH	930	1.05	40.20	39.51	19.78	0.25	0.19	0.07
40% N_2+60% CH_3OH with 2.2% Toluene	930	15.8	41.05	35.69	20.07	3.12	-	-
40% N_2+60% CH_3OH with 1% H_2O	930	0.71	40.30	39.47	19.67	0.17	0.27	0.10

TABLE 3: Calculated carbon activity and carbon potential for various steels according to different sources

Steels									Calculated THERMO-CALC		Gunnarson formula	
SS/AISI	T°C	C	Si	Mn	Cr	Mo	Ni	V	W	a_C	carbon potential wt%	carbon potential wt%	Carbon potential practise wt%
Fe-C	850	0.42	-	-	-	-	-	-	-	0.35	0.42	0.42	-
2244/4140	850	0.42	0.27	0.75	1.05	0.22	-	-	-	0.28	0.38	0.38	-
2258/52100	850	1.0	0.25	0.35	1.5	-	-	-	-	0.72	0.81	0.89	0.61
2090/9255	850	0.56	1.75	0.75	0.15	-	-	-	-	0.66	0.76	0.67	0.72
2310/02	1000	1.5	0.3	0.45	12.0	0.8	-	0.9	-	0.12	0.29	0.50	0.23
M2	750	0.85	0.30	0.30	4.0	5.0	-	2.0	6.3	0.05	0.05	0.69	-

TABLE 4: Calculated maximum solubility of carbon in austenite at 930°C for various steels

Steels SS/AISI	C	Si	Mn	Cr	Ni	Mo	Equilibrium carbon content austenite/cementite wt%	Carbon potential wt%	Carbon activity relative graphite
FeC	-	-	-	-	-	-	1.32	1.32	1.03
2523/4317	0.20	0.3	0.9	1.0	1.2	0.12	1.15	1.12	0.81
2506/8620	0.20	0.3	0.8	0.5	0.55	0.20	1.20	1.19	0.88
2511/-	0.15	0.3	0.9	0.8	1.0	-	1.17	1.15	0.84
1370/1016	0.15	0.3	0.75	-	-	-	1.25	1.28	0.98

TABLE 5: Mass transfer coefficient k' for some gas compositions according to different sources

over all gas composition	k'(10^7 ms^{-1}) 900°C Collin et al*	900°C Wyss et al	930°C Collin et al*	930°C Grabke and Tauber*
cracked CH_3OH ($CO+2H_2$)	2.2	2.8	2.7	13
60% cracked CH_3OH+40%N_2	1.5	1.3	1.9	11
40% cracked CH_3OH 60%N_2	1.0	0.6	1.3	9
Endogas from C_3H_8	1.5	1.2	1.9	11.5
50% CO 50% H_2	2.4	3.1	3.0	14.4

*) The calculations are based on a carbon potential of 0.8 wt%

TABLE 6: Calculated equilibrium solubility of nitrogen in Fe-Cr alloys in pure nitrogen gas at 1050°C and 1 atm.

Cr content wt%	N content wt%	phases
0	0.0241	γ
5	0.0841	γ
10	0.268	γ+0.1 wt% CrN
20	3.22	γ+14.2 wt% CrN

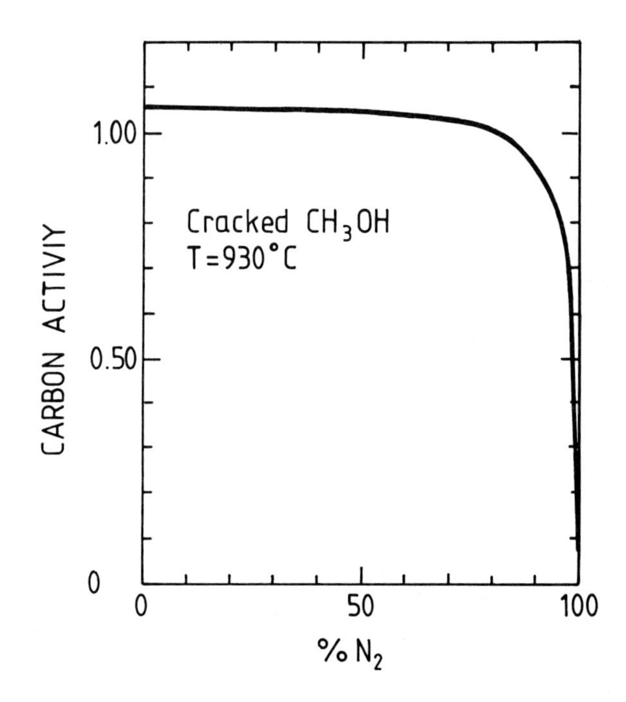

Fig. 1 Calculated carbon activity in atmospheres based on cracked methanol, at 930°C and 1 atm as a function of nitrogen dilution. The atmosphere is assumed to be at internal equilibrium.

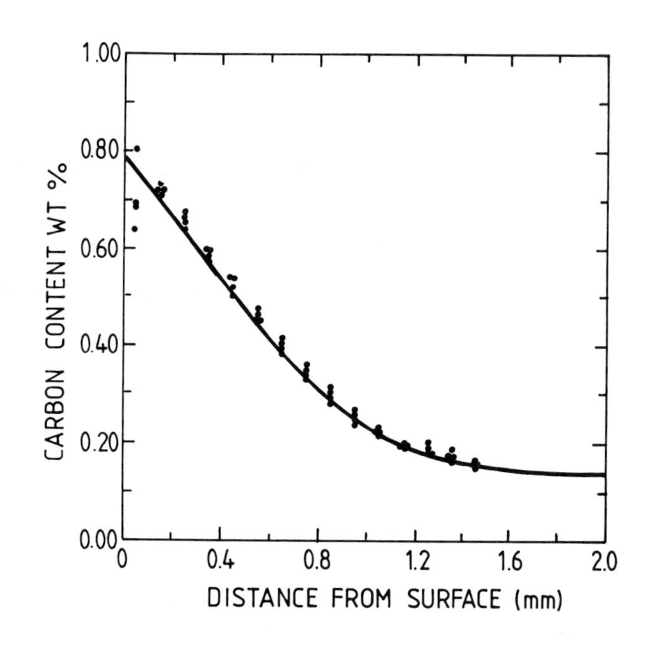

Fig. 2 Measured (symbols) and calculated (solid line) carbon concentration profiles. From ref. 15.

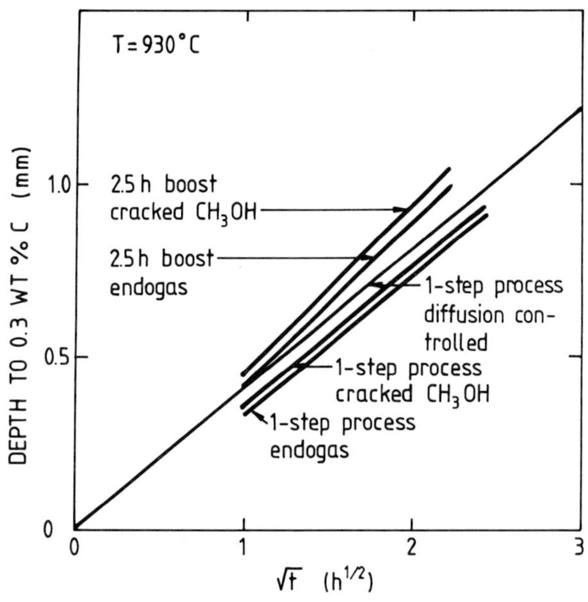

T=930°C
DEPTH TO 0.3 WT % C (mm)
1.0
0.5
0
2.5 h boost cracked CH_3OH
2.5 h boost endogas
1-step process diffusion controlled
1-step process cracked CH_3OH
1-step process endogas
0
1
2
3
$\sqrt{t}$ ($h^{1/2}$)

Fig. 3 Calculated carburizing depth to 0.30 wt% C as a function of the square root of time for different carburizing processes.

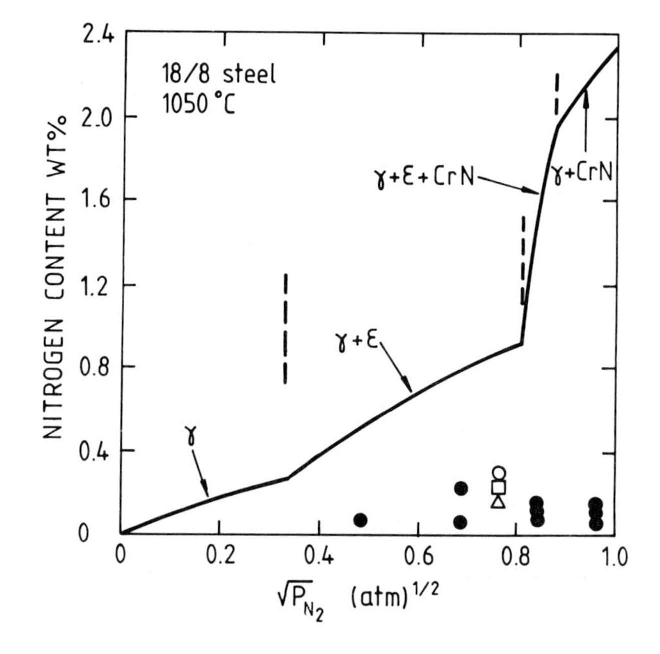

Fig. 4 Calculated solubility of nitrogen in an 18/8 stainless steel at 1050°C. Filled symbols: Nitrogen content of surface layer from production-scale measurements (ref. 16). Open symbols: Laboratory measurements at different dew-points (ref. 17), circles: DP=-46 C, squares: DP=-43 C, triangles: DP=-38 C text, section 5.

Utilization of databases of measured steel properties and of heat treatment technologies in practice

M GERGELY, T RÉTI, Gy BOBOK and Sz SOMOGYI

MG and SzS are in the Steel Advisory Centre for Industrial Technologies within the framework of Iron & Steel R & D Enterprise (VASKUT/Budapest/ Hungary); TR, formerly with VASKUT, is now with the Institute of Industrial Technologies; GyB, formerly with VASKUT, is now in the Ministry of Foreign Trade.

SYNOPSIS

The steel properties and heat treatment technologies are on computer files; the searchable fields are: size or type of workpiece, designation and composition of steels, heat number, tensile and yield strength, hardness, fracture toughness, employed heat treatment operation, producing enterprise, type of starting product, etc. The two presented databases are useful in steel selection, in determination of distribution of properties, mathematical connection between different characteristics, and in the selection of heat treatment technologies.

INTRODUCTION

Recognizing the need for exploitation of the potential reserves represented by steel characteristics, a project commenced in 1984 to establish a computerized data storage and retrieval system that can assist in forming rapid technical judgements, by supplying compositional, process and adaptational properties of steels. This paper briefly describes the interactive computerized system with special emphasis on the utilization of the two data bases mentioned in the title. Both the database of individual measured properties of machine components made of steel and the database of heat treatment technologies can be characterized by the fact that the central hardware is an extended personal computer IBM PC-XT and the software programs are based on dBASE III. Both the properties and technologies have been collected from the metalworking industry and refer to different engineering components. For the collection of information special data sheets were prepared and filled in by the experts in materials testing and heat treatment at the engineering and metallurgical plants. From each data sheet more than 100 items of numerical and textual information was fed into the data bases to derive the so-called "records".

DATABASE OF MEASURED STEEL PROPERTIES

The database of measured individual properties contains more than 9000 records. Each record contains the following 9 main data groups:

1/ Basic data
2/ Characteristics of starting product
3/ Technology of fabrication of the part
4/ Chemical composition of the part
5/ Results of tensile testing
6/ Impact testing results
7/ Results of hardness testing
8/ Results of Jominy test
9/ Other tests

The collection includes the following steel types:

- hardenable steels
- case-hardening steels
- structural steels
- free-cutting steels
- spring steels
- nitriding steels

The technique of searching in the data base is based on an interactive dialogue, it does not require any specially educated person. As a result of the search the computer furnishes the "found" records which satisfy the condition represented by the searching question. The data fields of the found records can be printed whole or in part and there are different possibilities to make statistical analysis on the numerical fields. The utilization of the data base is explained by the following types of examples:

Steel selection

As the first application-example let us search in the database of measured properties for records satisfying the conditions, formulated as follows:

Characteristic dimension of part:	60-90mm
Yield strength:	min. 650 MPa
Surface hardness:	min. 240 HB
Elongation (A5)	min. 14 %
Reduction in area:	min. 60 %
Impact energy:	min. 50 J

It should be noted that no steel can be found in the standards providing these mechanical properties in the specified dimension range.
Fig. 1 shows the searching. At the top of the figure the searchable fields and the keys (codes) can be seen. At the bottom the defined property ranges and the number of obtained records (70) satisfying the conditions given in the searching question can be read.
One of the found 70 records is printed out in Fig. 2.
Some of the properties of 6 records - selected from the 70 pieces displayed one after the other - are summarized in Table I and Table II.
There is a possibility of making a statistical evaluation of the fields from the 70 chosen records: distribution, mean and standard deviation of C, Si, Cr, ... contents or hardness at a certain distance from the end of the Jominy specimen, etc.
As an example Fig. 3 presents the histogram of impact energy values on the 70 records. According to this figure the mean value is 102 J, the standard deviation is 15.8 J.

Determination of the probable properties

In this case the searcher specifies the section size, the steelgrade or the composition range or the Jominy band and the tempering temperature range and wants to know the tensile strength distribution, the mean value of yieldpoints, elongation, reduction in area etc.
As the second application-example of this database let us compare the statistical characteristics of the standard 39Cr4 steel with the heat satisfying a narrower compositional range, namely:

C:	0.35-0.40		- 0.42
Cr:	0.85-1.05	instead of	- 1.20
Mn:	0.45-0.70		- 0.90

In Fig. 4 a the yield strength histogram can be seen for the 471 heats of the steelgrade 39Cr4. Fig. 4 b shows the yield strength histogram of the heats with narrower composition limits. On the basis of the two diagrams it can be stated that there is no use in introducing stricter specifications because the probable yield strength remains the same.
The results of the statistical analysis of the properties are summarized in Table III to see the comparison. It is necessary to mention that all of the 471 heats were used for the same product (crank shaft) and the heat treatment of these shafts was also the same, namely after normalizing from the austenitization temperature (850°C), quenching in water, and tempering at 600°C.

Determination of the mathematical connections between different characteristics

As the third application-example of the database of measured properties let us determine the dependence of surface hardness, yield strength, and elongation upon the tensile strength on the basis of the records originated from a Hungarian factory exporting forged products for the automotive industry. This means that the records contain data for products made of hardenable steels and the quality control tests were executed after the heat treatment (quenching, tempering).
The user program of the database provides the facility to display in the form of diagrams the required connections and to fit polinoms with least squared methods to the points.
Fig. 5 serves as an example showing the connection between the surface hardness measured in Brinell (HB) and the tensile strength (TN) measured on testpieces cut from the inside of the forgings. According to this diagram the connnection can be written in the form of a linear equation:

$$HB = 5.524 + 0.291\ TN$$

With similar analysis the equation for the elongation (EL) is:

$$EL = 27.512 - 0.0114\ TN$$

and the yield strength (YP) can be estimated with the formula:

$$YP = -290.51 + 1.175\ TN$$

DATABASE OF HEAT TREATMENT TECHNOLOGY

The detailed methods of heat treatment of machine parts are stored in the Database of Heat Treatment Processes. Before making their own decisions, the technologists can inquire about the different applied methods for similar parts.

The records for the actual workpieces contain the following 4 main data groups:

1. Basic data
2. Specified properties of the part and of the starting product
3. Instructions for heat treating operations
4. General instructions for the method of production and qualification.

The database hasn't got an English version, and because of the numerous textual (string type) fields there is no use in showing hardcopies of the records in Hungarian.

The following examples may be used to illustrate the searching methods:

a. Searching question:
   - type of part: not gear
   - max. dimension of part: min.400 mm
   - employed heat treatment operation: case hardening, or salt bath nitriding.

b. Searching question:
   - producing enterprise: C.K.V.
   - type of part: shaft
   - steel type: case hardening steel
   - type of starting product: rolled bar
   - characteristic dimension of the part: 45-60 mm
   - specified tensile strength of the part: 850-1000 MPa

This database may serve as a starting point for the computerized technological designing of methods of heat treatment and also may render assistance in developing novel, more up-to-date references, guidelines and standards for material selection, design, dimensioning and production technology.

SUMMARY

The two databases demonstrated by some examples fit into the information system recently developed in VASKUT, and consists of five main blocks:

- Database of technical literature
- Database of domestic and foreign steel standards
- Database of heat treatment technologies
- Database of measured steel properties
- Database of derived data

By using the system, technical decisions can be made more quickly, and on a firmer basis.
The processing and statistical analysis of the stock of data may result in concrete proposals for rationalizing the methods of heat treatment in use at the enterprises, e.g. for eliminating unjustified and superfluous heat treating operations, for reducing material and power costs.

ACKNOWLEDGMENT

The authors wish to express their profound indebtedness to the following research workers for their valuable advice in proceeding with this study and for their cooperation in data collection: Dr E. G. Fuchs, Dr G. Buza (staff of VASKUT), Dr T. Konkoly (Technical University, Budapest), I. Szerényi, L. Varga (staff of NOVOTRADE).

Table I Some characteristics of 6 records selected from the 70 obtained in Fig.1

Steel grade	Name of part	C	Si	Mn	S	P	Cr	Ni	Mo	Cu	Al
		Chemical composition of the parts in vol.%									
39Cr4	crank shaft	0.38	0.26	0.62	0.014	0.013	0.97	0.17	-	0.15	0.05
38X2H2MA	coupling bar	0.36	0.25	0.38	0.020	0.014	1.47	1.40	-	0.13	-
35XMA	gear	0.37	0.23	0.62	0.013	0.020	1.00	0.12	0.18	0.20	-
40X	drive hub	0.42	0.27	0.63	0.020	0.022	0.91	0.11	-	0.12	-
20XGC	shank stud	0.20	1.01	0.95	0.010	0.012	1.00	0.09	-	0.14	-
38XC	round bar	0.39	1.12	0.38	0.008	0.011	1.30	0.18	-	0.14	-

Table II Some characteristics of 6 records selected from the 70 obtained in Fig.1

Steel grade	Name of part	Charact. dim. mm	Yield str. MPa	Tensile str. MPa	Hardness on surf. HB	Elonga- tion %	Reduction in area %	Impact energy J	Heat treat- ment
39Cr4	crank shaft	78	662.1	818.1	247.0	18.2	60.2	103.5	WQ 850 T 600
38X2H2MA	coupling bar	70	951.3	1038.8	302.0	16.5	60.9	92.3	OQ 850 T 600
35XMA	gear	77	754.6	887.5	262.0	18.2	65.2	118.6	OQ 850 T 600
40X	drive hub	78	714.4	887.8	244.5	15.5	60.0	62.6	OQ 860 T 600
20XGC	shank stud	80	687.6	799.5	241.0	20.4	65.2	108.0	WQ 880 T 450
38XC	round bar	70	888.0	1056.5	317.0	19.5	60.9	110.8	OQ 900 T 630

Table III Comparison between the statistical values of mechanical properties of the standard 39Cr4 steel and the values obtained by analysing heats within a narrower compositional range

Steel grade/ Number of records		Yield str. MPa	Elongation %	Impact energy J	Hardness at 15mm on Jominy bar, HRC
39Cr4	Mean	673.3	17.6	77.8	39.5
N=517	Std. dev.	49.3	2.1	28.3	3.6
Narrower 39Cr4	Mean	673.0	18.0	87.1	38.6
N=247	Std. dev.	52.1	2.1	27.4	3.3

```
  V A S K U T                  SEARCHING IN THE DATABASE               87.04.22

IHHHHHHHHHHHHKHHHHHHHHHHHHKHHHHHHHHHHHHKHHHHHHHHHHHHKHHHHHHHHHHHHKHHHHHHHHHHHH;
: MAIN DATA  : CHEMICAL   :  TENSIL    :  IMPACT    : HARDNESS   :  JOMINY    :
LHHHHHHHHHHHHNHHHHHHHHHHHHNHHHHHHHHHHHHNHHHHHHHHHHHHNHHHHHHHHHHHHNHHHHHHHHHHHH9
:AN=ACESS NO :C  CR  CU  B:M1=TYP SIZE : M2=TYP SIZE: H5=HEAT TEM: JL=JOM.LENG:
:LO=LOCATION :SI NI  AL  O:H1=1.HEATTEM: H3=HEATTEM.: T5=HEAT TIM: AT=AU. TEMP:
:SC=ALLOY SP :MN MO  W   H:T1=1.HEATTIM: T3=HEATTIM : H6=HEAT TEM: AI=AU. TIME:
:TY=CLAS.PRT :P  V   CO  N:H2=2.HEATTEM: H4=HEATTEM.: T6=HEAT TIM:            :
:            :S           :T2=2.HEATTIM: T4=HEATTIM : HA=HARDNESSLHHHHHHHHHHHH9
:            :            :EL=ELONG.   : CP=IMP.ENER:            :   OTHER    :
:            :            :RA=REDUCTION: FR=IMP.STR.:            LHHHHHHHHHHHH9
:            :            :TN=TENSIL   :            :            : CM=TEST COD:
:            :            :YP=YIELD PO.:            :            :            :
HHHHHHHHHHHHHJHHHHHHHHHHHHJHHHHHHHHHHHHJHHHHHHHHHHHHJHHHHHHHHHHHHJHHHHHHHHHHHH<
  CONTINUE ? Y/N                                                          _
IHHHHHHHHHHHHHHHHHHHKHHHHHHHHHHHHHHHHHHHHHHHHHHHHHHHHHHHHHHHHHHHHHHHHHHHHHHHHH;
: KEY :  CP         :                                                         :
:                   :                                                         :
: RECORDS :     70  :                                                         :
HHHHHHHHHHHHHHHHHHHHJHHHHHHHHHHHHHHHHHHHHHHHHHHHHHHHHHHHHHHHHHHHHHHHHHHHHHHHHH<
  M1: 60- 90 YP: 650.0-1000.0 HA: 240.0- 320.0 EL:14.0-24.0 RA:60.0-70.0
  CP: 50.0-120.0
```

Fig.1 Searching example from the database of measured properties

```
V A S K U T                    PAGE :    1                    DATE : 87.04.22
-------------               -------------              ------------------
                             MAIN FILE OF DATABASE
+------------------------------------------------------------------------------+
! ACCESS.NUM: 8511077   LOCATION :                                             !
!                                                                              !
!   DRAWING NUMBER           NAMING OF THE PART        NAMING OF THE PRODUCT   !
!                          FORGATTYUS TENGELY                                  !
!                                                                              !
!     CLASSIFICATIONS OF THE PART         ALLOY SPECIF.  HEAT NUM    MASS      !
! TENGELY                                 39CR4          403921      120.00    !
!                                                                              !
!  ENCLOSE SIZES      MAXIMAL SIZE      TYPICAL SIZE    SERIES  [PR/YR]        !
! 1024  192  138         1024               105                                !
+------------------------------------------------------------------------------+
                          RESULTS OF CHEMICAL TESTING
+------------------------------------------------------------------------------+
! LOCATION OF SAMPLING:ADAG          MODE OF ANALYSIS:KVANTO                   !
!   C: 0.38  SI: 0.26  MN:   0.62  S: 0.014  P: 0.013  CR:   0.970  NI:   0.170 !
!   MO:   0.000  V: 0.000  CU: 0.150  AL: 0.050  W:   0.000  CO:   0.000       !
!   TI: 0.000  B: 0.000  O:    0  H:    0  N:    0  PB: 0.000                  !
+------------------------------------------------------------------------------+
                          RESULTS OF TENSIL TESTING
+------------------------------------------------------------------------------+
!TYP.  SAMPL. SPECIMEN    TST      HEAT TREATMENT          ELON REDUCTENSIL  YIELD P!
!SIZE  LOC DI TYP-SIZE    TEM  CD TEM  TIM CD TEM TIM  [%]  [%] [N/QMM] [N/QMM]!
!                                                                              !
!  78  R/3 H  10,5D          20 VE  850 300 MV 600 475 18.2 60.2  818.1    662.1!
+------------------------------------------------------------------------------+
                          RESULTS OF IMPACT TESTING
+------------------------------------------------------------------------------+
! TYPIC  SAMPLING  SPECIMEN    TST      HEAT TREATMENT       IMP.EN.  IMP.STR.  !
! SIZE  LOC. DIREC TYP-SIZE    TEM  CD TEM  TIM CD TEM TIM [JOULE]  [J/QCM]    !
!                                                                              !
!  78  R/3  H     10*10*55,2   20 VE  850 300 MV 600 475  103.5     129.4      !
+------------------------------------------------------------------------------+
                          RESULTS OF HARDNESS TESTING
+------------------------------------------------------------------------------+
! LOC.OF TEST  TEST PARAMET. TST      HEAT TREATMENT        METH. RESULT       !
!                            TEM  CD TEM  TIM CD TEM TIM                       !
!                                                                              !
! D78 FELULETE 10/3000/30       20 VE  850 300 MV 600 475  HB     247.0        !
+------------------------------------------------------------------------------+
                          RESULTS OF JOMINY TESTINGS
+------------------------------------------------------------------------------+
!  SAMPLING      AUSTEN.        H A R D N E S S   R E S U L T S      [HRC]     !
!  LOCATION      TEM  TIM 1.5.  3.   5.   7.   9.   11.  13.  15.  20.  25.  30.!
! BUGA SARKA     860   30 54.0 54.0 52.0 48.0 46.0 41.0 37.0 35.0 33.0 30.0 30.0!
!                          35.  40.  45.  50.                                  !
!                          0.0  0.0  0.0  0.0                                  !
!                                                                              !
+------------------------------------------------------------------------------+
                          RESULTS OF OTHER TESTINGS
+------------------------------------------------------------------------------+
!  TESTING       TYPIC      HEAT TREATMENT              TESTING                !
! CODE  TEMP     SIZE    CD TEM  TIM CD TEM TIM         RESULT                 !
!                                                                              !
!                          NON EXISTING RECORD !                               !
!                                                                              !
+------------------------------------------------------------------------------+
```

Fig.2 One of the found 70 records of the searching shown in Fig.1

```
M1: 60- 90 YP: 650.0-1000.0 HA: 240.0- 320.0 EL:14.0-24.0 RA:60.0-70.0
CP: 50.0-120.0

                              STATISTICS

IMMMMMMMMMMMMKMMMMMMMMMMMMMKMMMMMMMMMMMMMKMMMMMMMMMMMMMKMMMMMMMMMMMMM;
:  CHEMICAL  :    TENSIL   :    IMPACT   :  HARDNESS   :    JOMINY   :
LMMMMMMMMMMMMMMMMMMMMMMMMMMMMMMMMMMMMMMMMMMMMMMMMMMMMMMMMMMMMMMMMMMMM9
:C  CR  CU  B:EL=ELONG    :CP=IMP.ENER.:HA =VALUE   :J1,J3,J5,J7,:
:SI NI  AL  O:RA=REDUCTION:FR=IMP.STR. LMMMMMMMMMMMM9J9,J11,J13, :
:MN MO  W   H:TN=TENSIL   :            :HB =BRINELL :J15,J20,J25,:
:P  V   CO  N:YP=YIELD PO.:            :HV =VICKERS :J30,J35,J40,:
:S          :             :            :HRC=ROCKWELL:J45,J50     :
HMMMMMMMMMMMMJMMMMMMMMMMMMMJMMMMMMMMMMMMMJMMMMMMMMMMMMMJMMMMMMMMMMMMM<

                         VARIABLE  CP
                         PARTS        70
                         NO.OF MES    81
                         MINIMUM     54.000
                         MAXIMUM    147.000
                         RANGE       93.000
                         MEAN       102.099

                      HISTOGRAM ? (Y/N)


M1: 60- 90 YP: 650.0-1000.0 HA: 240.0- 320.0 EL:14.0-24.0 RA:60.0-70.0
CP: 50.0-120.0
                               CP STATISTICS
IMMMMMMMMMMMMMMMMMMMMMMMMMMMMMMMMMMMMMMMMMMMMMMMMMMMMMMMMMMMMMMMMMMMMMMMMMMMMMMMM;
:PARTS:    70  NUM.OF MEASURE:    81  MIN: 54.000   MAX:147.000   RANGE: 93.000  :
:LOWER  50.000  UPPER 147.500 CELL:     7.500 MEAN :102.099  ST.DEV: 15.794      :
LMMMMMMMMMMMMMMMMKMMMMMMMMMMMKMMMMMMMMMMKMMMMMMMMMMMMMMMMMMMMMMMMMMMMMMMMMMMMMMMM9
:      RANGE      :FREQUENCY :REL.FREQ.:             H I S T O G R A M           :
: 50.000   57.500:       1   :   0.01  :DD                                       :
: 57.500   65.000:       1   :   0.01  :DD                                       :
: 65.000   72.500:       1   :   0.01  :DD                                       :
: 72.500   80.000:       2   :   0.02  :DDDDD                                    :
: 80.000   87.500:       8   :   0.10  :DDDDDDDDDDDDDDDDDDDDD                    :
: 87.500   95.000:      14   :   0.17  :DDDDDDDDDDDDDDDDDDDDDDDDDDDDDDDDDDDDD    :
: 95.000  102.500:      14   :   0.17  :DDDDDDDDDDDDDDDDDDDDDDDDDDDDDDDDDDDDD    :
:102.500  110.000:      14   :   0.17  :DDDDDDDDDDDDDDDDDDDDDDDDDDDDDDDDDDDDD    :
:110.000  117.500:      16   :   0.20  :DDDDDDDDDDDDDDDDDDDDDDDDDDDDDDDDDDDDDDDDD:
:117.500  125.000:       6   :   0.07  :DDDDDDDDDDDDDDDD                         :
:125.000  132.500:       2   :   0.02  :DDDDD                                    :
:132.500  140.000:       0   :   0.00  :                                         :
:140.000  147.500:       2   :   0.02  :DDDDD                                    :
:                :           :         :                                         :
:                :           :         :                                         :
HMMMMMMMMMMMMMMMMJMMMMMMMMMMMJMMMMMMMMMMJMMMMMMMMMMMMMMMMMMMMMMMMMMMMMMMMMMMMMMMMM<
```

Fig.3 Statistical evaluation of impact energy values of the 70 records in Fig.1

SC:39CR4 C :0.35-0.45 CR:0.850-1.250 MN:0.45-0.95

YP STATISTICS

PARTS: 471 NUM.OF MEASURE: 517 MIN:375.000 MAX:830.900 RANGE:455.900
LOWER 450.000 UPPER 825.000 CELL: 25.000 MEAN :673.348 ST.DEV: 49.308

RANGE		FREQUENCY	REL.FREQ.	H I S T O G R A M
450.000	475.000	0	0.00	
475.000	500.000	0	0.00	
500.000	525.000	1	0.00	
525.000	550.000	1	0.00	
550.000	575.000	4	0.01	D
575.000	600.000	22	0.04	DDDDDDD
600.000	625.000	45	0.09	DDDDDDDDDDDDDDD
625.000	650.000	92	0.18	DDDDDDDDDDDDDDDDDDDDDDDDDDDDDD
650.000	675.000	107	0.21	DDDDDDDDDDDDDDDDDDDDDDDDDDDDDDDDDDD
675.000	700.000	119	0.23	DDDDDDDDDDDDDDDDDDDDDDDDDDDDDDDDDDDDDDDD
700.000	725.000	73	0.14	DDDDDDDDDDDDDDDDDDDDDDDD
725.000	750.000	22	0.04	DDDDDDD
750.000	775.000	14	0.03	DDDD
775.000	800.000	9	0.02	DDD
800.000	825.000	6	0.01	DD

SC:39CR4 C :0.35-0.40 CR:0.850-1.050 MN:0.45-0.70

YP STATISTICS

PARTS: 228 NUM.OF MEASURE: 247 MIN:518.000 MAX:825.000 RANGE:307.000
LOWER 500.000 UPPER 825.000 CELL: 25.000 MEAN :673.080 ST.DEV: 52.112

RANGE		FREQUENCY	REL.FREQ.	H I S T O G R A M
500.000	525.000	1	0.00	
525.000	550.000	0	0.00	
550.000	575.000	3	0.01	DD
575.000	600.000	13	0.05	DDDDDDDDDD
600.000	625.000	27	0.11	DDDDDDDDDDDDDDDDDDDD
625.000	650.000	45	0.18	DDDDDDDDDDDDDDDDDDDDDDDDDDDDDDDDDD
650.000	675.000	52	0.21	DDDDDDDDDDDDDDDDDDDDDDDDDDDDDDDDDDDDDDDD
675.000	700.000	46	0.19	DDDDDDDDDDDDDDDDDDDDDDDDDDDDDDDDDDD
700.000	725.000	26	0.11	DDDDDDDDDDDDDDDDDDDD
725.000	750.000	15	0.06	DDDDDDDDDDD
750.000	775.000	10	0.04	DDDDDDD
775.000	800.000	6	0.02	DDDD
800.000	825.000	2	0.01	D

Fig.4 Comparison of yield strength distribution of standard 39Cr4 steel and heats with narrower compositional range

HA - TN STATISTICS NO.OF POINTS :1082

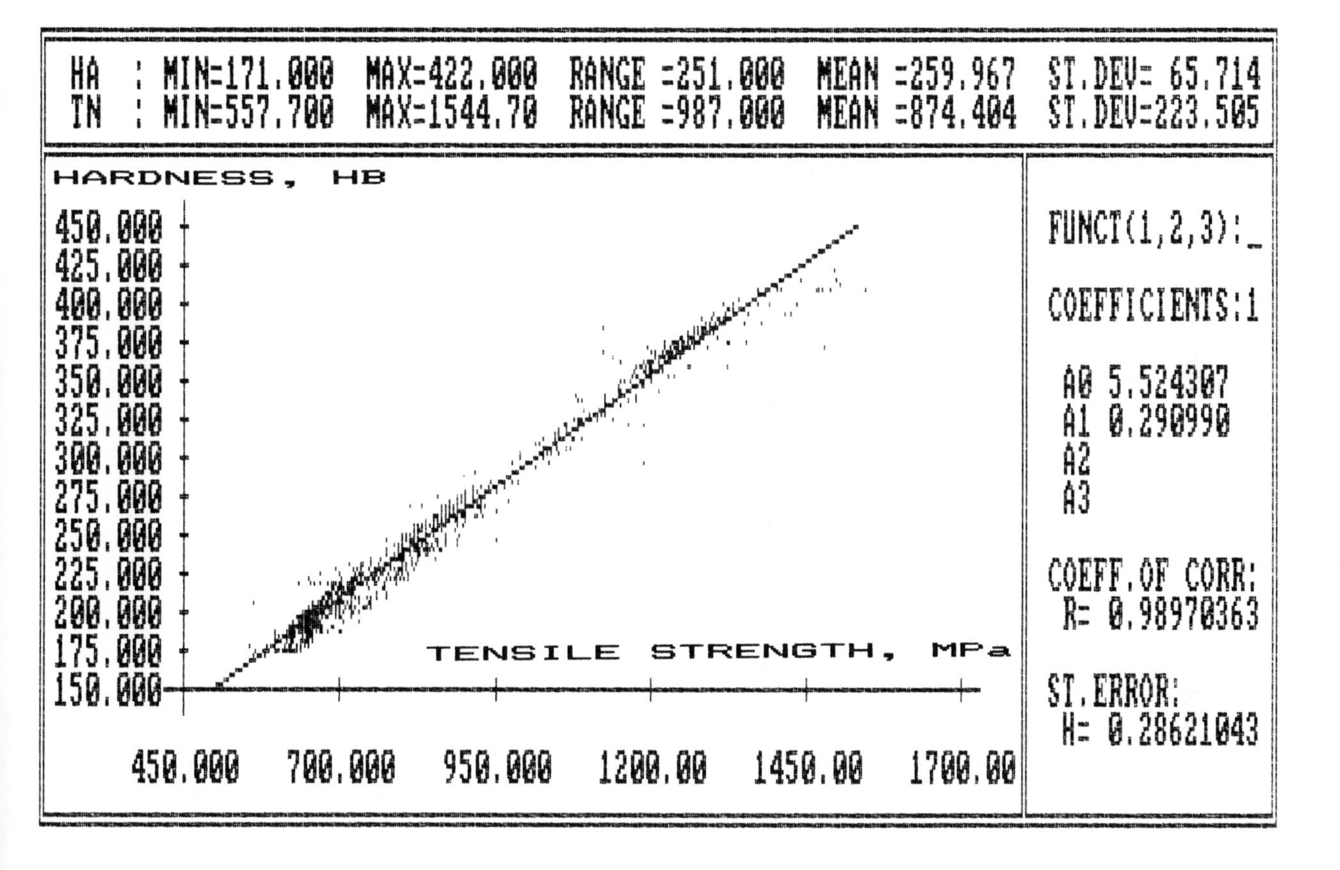

Fig.5 Connection between the surface hardness and tensile strength of specimens cut out of forged shafts

□

Application of microprocessors for property monitoring and control in modern heat treatment practice

T RÉTI, M GERGELY and Gy BOBOK

TR, formerly with VASKUT, is now with the Institute of Industrial Technologies; MG is in the Steel Advisory Centre for Industrial Technologies within the framework of Iron & Steel R & D Enterprise (VASKUT/Budapest/Hungary); GyB, formerly with VASKUT, is now in the Ministry of Foreign Trade.

SYNOPSIS

A continuous property-predicting computer is described which may be used with any traditional furnace to monitor the heat treating process and indicate the property changes. The predicting computer is characterized by the fact that utilizing suitable input data it simulates the transformation processes. The model on which the software is based has phenomenological character. It includes differential equations the parameters of which depend in part on the individual chemical composition and microstructure of the alloy to be heat treated and in part on the momentary temperature value. The concrete industrial application is presented in two examples. One of the examples is the monitoring of the graphitization process during annealing of malleable cast iron in a batch-type furnace, the other concerns the prediction of the hardness after quenching and tempering of shaft-type parts in a continuous furnace. Experiments up to date have shown that the relatively cheap microprocessor-based unit is an effective tool for reducing the time and energy demand of heat treatment and for improving the reproducibility of production.

INTRODUCTION

Since the beginning of the energy crisis the modernization of the control of heat treating processes has become of especial importance. Research of the optimalization and automation of heat treating processes has been substantially accelerated by the recent rapid development in microelectronics and computer techniques. Novel results were achieved in the microprocessor control of heat treating processes. The computer-assisted units developed for control of gas carburizing processes deserve especial mention /1,2,3 /. It seems, however, that the development has just begun. The manifold possibilities of microprocessor in this respect are far from being exhausted.

This paper describes a novel type of property monitoring and control unit based on micro-computer. It is an effective tool for reducing the time and energy demand of heat treating processes. The unit connected to the furnace indicates continuously during the course of heat treatment the property changes occurring due to the transformations and informs the user about the momentary state of the process and its eventual completion.

After describing the construction of the property monitoring and control unit the basic characteristics of the physical-metallurgical model, on which the software is based, are sketched. Industrial application will be presented with concrete examples.

GENERAL DESCRIPTION OF THE PROPERTY MONITORING COMPUTER

Figure 1 shows the theoretical schema of the property-predicting computer. The computer uses data entered directly through the keyboard at the beginning and data furnished continuously by the furnace through a data logger. Utilizing the data the micro-computer follows the microstructural transformations occurring in the alloy during heat treatment and the processes causing property changes.

The input data entered through the keyboard are mostly as follows:

- geometric and dimensional characteristics of the part to be heat treated,
- individual chemical composition of the alloy, characterizing the charge in question,
- data concerning the initial state and micro-structure of the alloy (e.g. grain size, volume of phases, specific surface, etc.),
- mode of heat treatment, technical parameters of heat treating furnace,
- value of the property to be realized by heat treatment (e.g. hardness, mechanical or micro-structural characteristics).

The continuously measured furnace signals form another important part of the input data. These are the so-called process variables: temperature of the furnace and the part, atmospheric pressure, composition etc.

The output data are numerical values of a definite property of the heat treated part (e.g. hardness) which readily characterizes the progress of the heat treating process as a function of the time.

Heat treatment is completed when the continuously indicated property reaches a pre-determined value.

The property monitoring computer can efficaciously reduce the time and energy demand of heat treatment because the input data supply excess information which is usually not available when using traditional heat treating processes.

KINETIC MODEL OF THE PROPERTY MONITORING SYSTEM

Earlier investigations have proved that most of the thermally activated processes occurring during heat treatment may be readily described phenomenologically by common differential equations of the following form:

$$\frac{dP}{dt} = f(P, T, b_1, b_2 \dots b_N) \quad (1)$$

where P is a given property of the alloy, characterizing the microstructure,
T is the temperature and
b_1, b_2, b_N are parameters depending on the alloy composition, the initial microstructure and temperature, etc.

From this recognition the physical metallurgical model forming the core of the monitoring software was constructed as a differential equation representing a special case of equation (1).

$$\frac{dP}{dt} = K\, y^{b_1} (1-y)^{b_2} \left(\ln \frac{1}{1-y}\right)^{b_3} \quad (2)$$

Using differential equation (2) furnishes two important advantages. First, by suitably selecting parameters K, b_1, b_2 and b_3 transformation processes of various types may be readily described. Second, it is relatively easy to determine the unknown model parameters after logarithmization of equation (2) with the aid of multivariant linear regression.

According to our experiences with this simple model and software based on equation (2) computerized property monitoring may be employed in the following areas /4-7/:
- non-isothermal grain size growth in the austenitization of steels,
- assessment of hardness in tempering quenched steels,
- cementite dissolution in the austenitization of unalloyed tool steels,
- determination of hardness reduction in recrystallization annealing of cold worked alloys,
- calculation of austenite decomposition products during continuous cooling,
- dissolution of the non-equilibrium second phase during homogenization of aluminium alloys /8/.

The solution of some simpler property monitoring tasks in the course of heat treatment has been presented earlier /4,6/. The possibilities of using the property monitoring computer shall be presented below on the example of two heat treating operations.

CONTROLLED ANNEALING OF MALLEABLE CAST IRON

Annealing of malleable cast iron is intended - as we know - to produce a ferritic matrix with graphite nodules from a microstructure containing initially eutectic cementite. The schema of heat treatment process is shown in Figure 2. The first annealing stage - producing austenite and graphite out of eutectic carbides - is rather long; its duration depends on the chemical composition and microstructure of the cast iron and may take t_A = 40-70 hours.

The time of graphitization may be reduced effectively by treating in the furnace at the same time castings from the same charge only - i.e. with the same composition - and using a microcomputer control unit for predicting the graphitization process. In this case prediction is based on the fact that the graphitization process in non-isothermal conditions may be described with sufficient precision with the following generalized Avrami-type kinetic function /9/:

$$y(t) = 1 - \exp\left\{-\int_0^t k\ (T)dt\right\}^2 \quad (3)$$

where y is the extent of graphitization and
k(T) is the well-known Arrhenius expression, characterizing the process rate.

Among the two parameters of the Arrhenius expression, Q - the apparent energy of activation of the process - may be regarded as constant according to the measurements, with a value of Q = 250 kJ/mole. Factor k_o, the so-called pre-exponential factor, however, depends significantly on the composition of white cast iron and its value characterizes the tendency to graphitization of the latter. Parameter k_o should be determined for each charge composition previous to malleablizing: dilatometric tests have proved the most suitable for this aim, due to their rapidity. When heat treatment starts parameter k_o is also entered through the keyboard as input information.

Although the controlled and indicated period is a relatively small fraction of the total cycle of heat treatment, the computerized prediction of graphitization may lead to significant savings in time and energy.

QUENCHING AND TEMPERING IN A CONTINUOUS PUSHER-TYPE FURNACE

The property-monitoring computer described below has been developed for monitoring heat treatment processes comprising several steps.

The computer unit predicts after quenching and during tempering the Brinell hardness of the part made of the hardenable steel. This hardness is continuously indicated and permits suitable management and rationalization of the heat treating process. The computer hardware is described in ref. /10/. The model on which the predictive software is based is illustrated in Fig. 3.
The program is composed of 12 modules. The input data, shown at the right of the block scheme, are as follows: chemical composition of the workpiece to be hardened, its initial state (annealed, normalized, quenched, tempered), the heating medium (gas atmosphere), diameter of the workpiece, distance from the cooled surface of the point where we wish to know the probable microstructure and properties, austenitizing temperature, cooling medium (water, oil), tempering temperature. The short description of the individual modules is as follows:
Module 1 checks whether the chemical composition of the workpiece corresponds with the specified composition of the steel type for which the

predicator program has been developed, and then calculates the A_1, A_3, B_s, M_s temperatures.
Module 2 computes the heating curve, taking into account also the furnace features. Calculations are based on the application of an approximate method developed to solve heat-conduction problems for simple geometries.
Module 3 calculates the Ac_3 temperature, the time required to reach the Ac_3 temperature and to approach the austenitizing temperature to the nearest 20°C. Ac_3 is determined - apart from the composition - by the initial state and the heating rate.
Module 4 computes the austenite grain size, taking into consideration also the non-isothermal conditions, according to the published method /5,6/.
Module 5 calculates the characteristics to the TTT chart as a function of the chemical composition and austenite grain size, taking into account also the temperatures A_1, Ac_3, B_s, M_s.

The information content of the traditional TTT charts is built into the program in the form of kinetic differential equations /4-6/.
Module 6 computes the cooling curve in the given point of the workpiece. The principle of the calculation is identical with the method applied in the module 2.
Module 7 plays a central role in the system, calculating the progress of the transformation process during continuous cooling from the TTT characteristics. For the calculation of the amount of martensite a novel formula is used /11/ instead of the well-known equation proposed by Koistinen and Marburger /12/. Estimation of the ferritic, pearlitic, bainitic fraction during each step is based on the methods detailed in the references /4-6/.
Module 8 provides information about transformation temperatures during continuous cooling. In reality there are no transformation temperatures but definitions for the start of the ferrite-pearlite reaction and for the temperature range of the bainite reaction. In our experience the start is defined as either 1% or 5%.
Module 9 gives us the microstructure, namely the amount of ferrite-pearlite, bainite, martensite and retained austenite. For this kind of information the program uses the above-mentioned definitions, e.g. bainite is the transformation product obtained from austenite between the temperatures B_s and M_s.
Module 10 figures out the probable hardness after quenching on the basis of the microstructure and the carbon content.
Module 11 computes the heating curve to reach the tempering temperature.
Module 12 calculates the hardness during tempering according to the method /13/.

Figure 4 presents the time-temperature chart of the industrial heat treatment process comprising several stages. In the course of the treatment the part passes through the aggregate continuously according to a predetermined schedule. The progress of heat treatment is unequivocally defined by the position of the charging car (Figure 4). In this facility shaft-type parts made of unalloyed and low-alloy hardenable steels are treated: the parts are first normalized, then cooled below the transformation temperature A_1 and finally quenched and tempered.

The use of the property-monitoring computer improved the reproducibility of heat treatment, the quality of the heat treated parts, the scatter of the mechanical properties and the percentage of rejection reduced.

CONCLUSIONS

Experience to date has shown that the use of property-monitoring computers furnishes novel possibilities of rationalizing and optimalizing heat treatment processes. The advantages of using this device are as follows:

1/ The time and energy demand of heat treatment is reduced.
2/ The reproducibility of production is improved, the rate of rejection is reduced.
3/ The heat treated product is more uniform in quality, the scatter in mechanical characteristics is accordingly reduced.
4/ The modernization of control results in labour savings.

REFERENCES

1 J. Wünning: Zeitschrift für Wirtschaftliche Fertigung, 1982, 9, p. 424-246.

2 A. Knieren, H. Pfau: in "Heat Treatment '84", p. 13.1-13.8, 1984, London, The Metals Society

3 P. Sommer: Heat Treatment of Metals, 1987, 1, p. 7-10.

4 M. Gergely, T. Réti, P. Tardy, G. Buza: in "Heat Treatment '84", p. 20.1-20.7, 1984, London, The Metals Society

5 T. Réti, G. Bobok, M. Gergely: in "Heat Treatment '81", p. 91-96, 1983, London, The Metals Society

6 M. Gergely, Sz. Somogyi, G. Buza: Materials Science and Technology, 1985, 1, p. 893-898.

7 M. Gergely, Sz. Somogyi, T. Réti: in Proc. 5th International Congress on "Heat Treatment of Materials", Budapest, October 1986, Scientific Society of Mechanical Engineers, Vol. 1, p. 186-195.

8 A. Roósz, Z. Gácsi, G. Fuchs: Acta Metall. 1984, 10, p. 1745-1754.

9 E. Fuchs, M. Gergely: Bányász. Kohász. Lapok (Kohász.), 1971, 8, p. 185-187.

10 B. Borossay, Gy. Bobok, K. Kovács: in Proc. 5th International Congress on "Heat Treatment of Materials", Budapest, October 1986, Scientific Society of Mechanical Engineers, Vol. 1, p. 137-140.

11 E. Füredi, M. Gergely: in Proc. 4th International Congress on Heat Treatment of Materials, Berlin, June 1985, Arbeitsgemeinschaft Wärmebehandlung und Werkstofftechnik, Vol. 1, p. 291-301.

12 D. P. Koistinen, R. E. Marburger: Acta Metall., 1959, 7, p. 59-60.

13 Sz. Somogyi, M. Gergely: in Proc. 4th International Congress on Heat Treatment of Materials, Berlin, June 1985, Arbeitsgemeinschaft Wärmebehandlung und Werkstofftechnik, Vol. 1, p. 84-90.

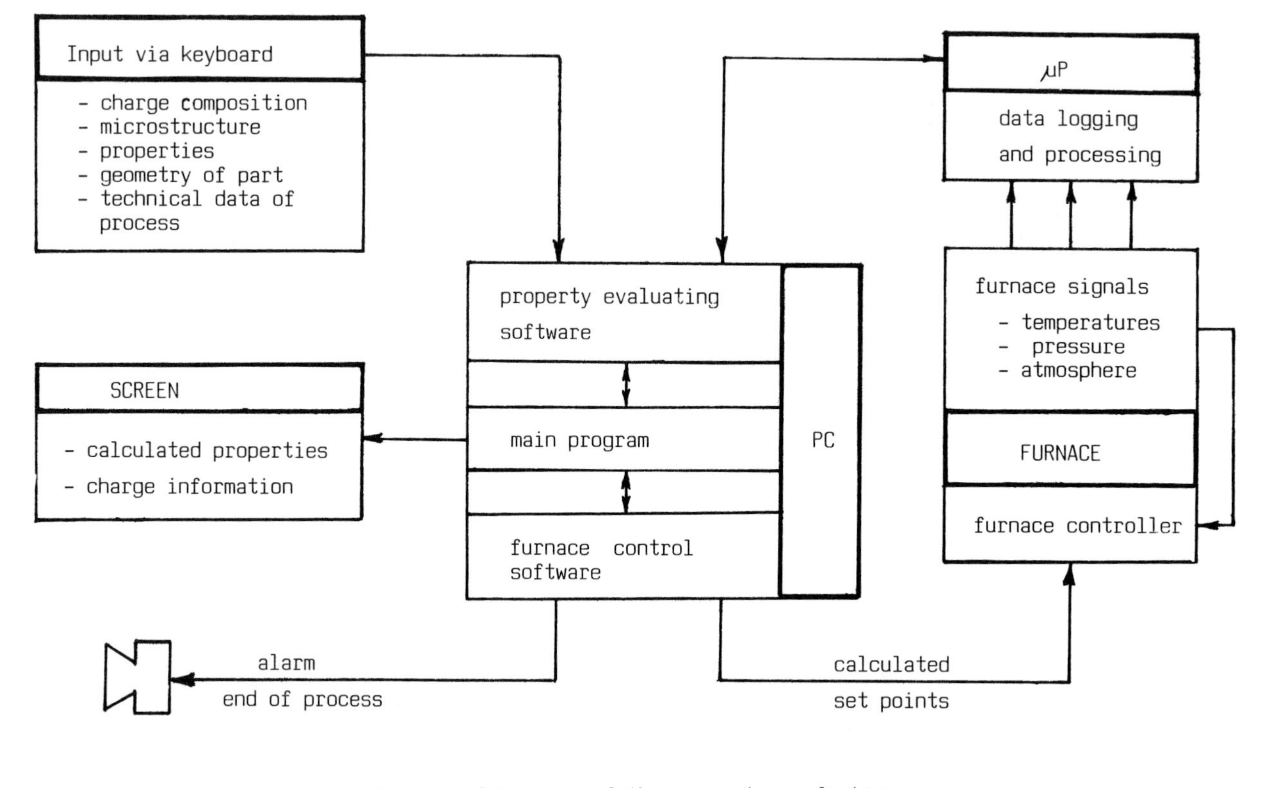

Fig. 1 Schematic of the property-predicting computer

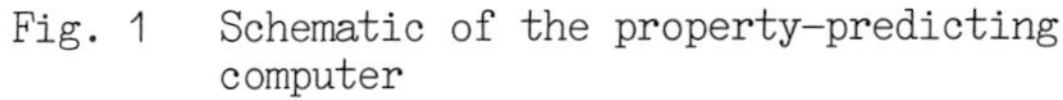

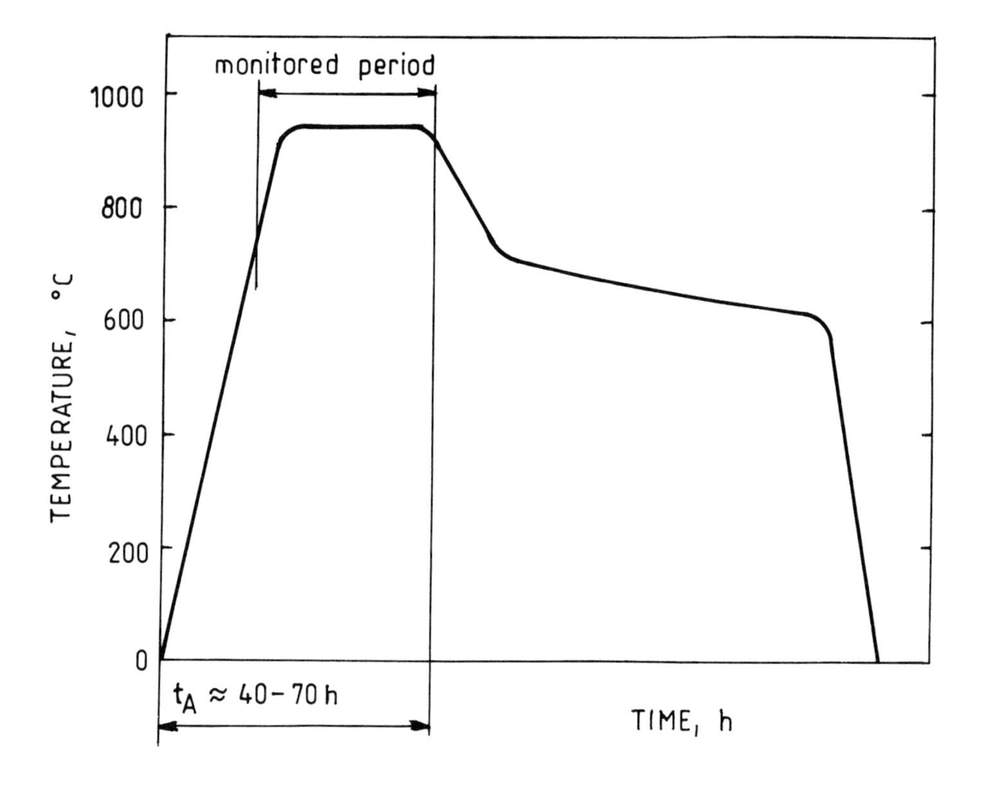

Fig. 2 Controlled annealing cycle for producing a blackheart malleable iron

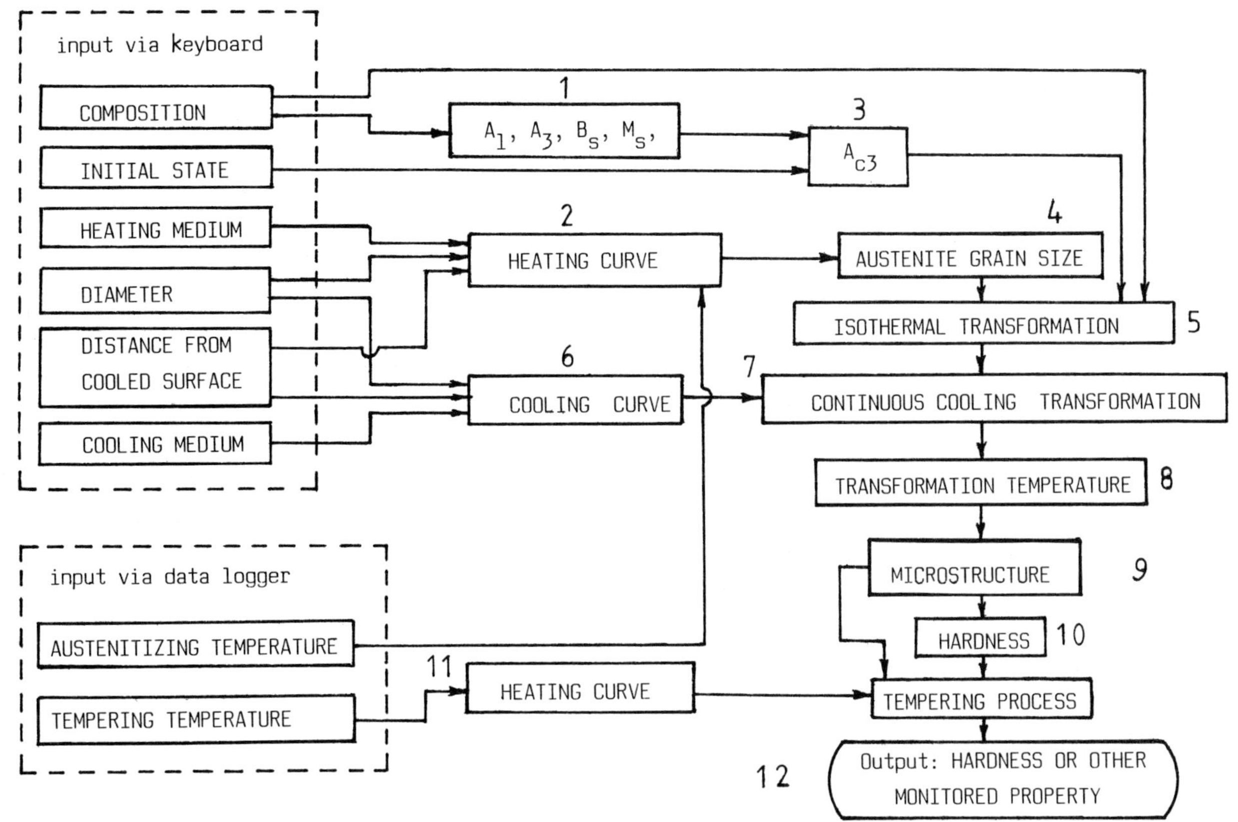

Fig. 3 Principle of the computerized property-monitoring system

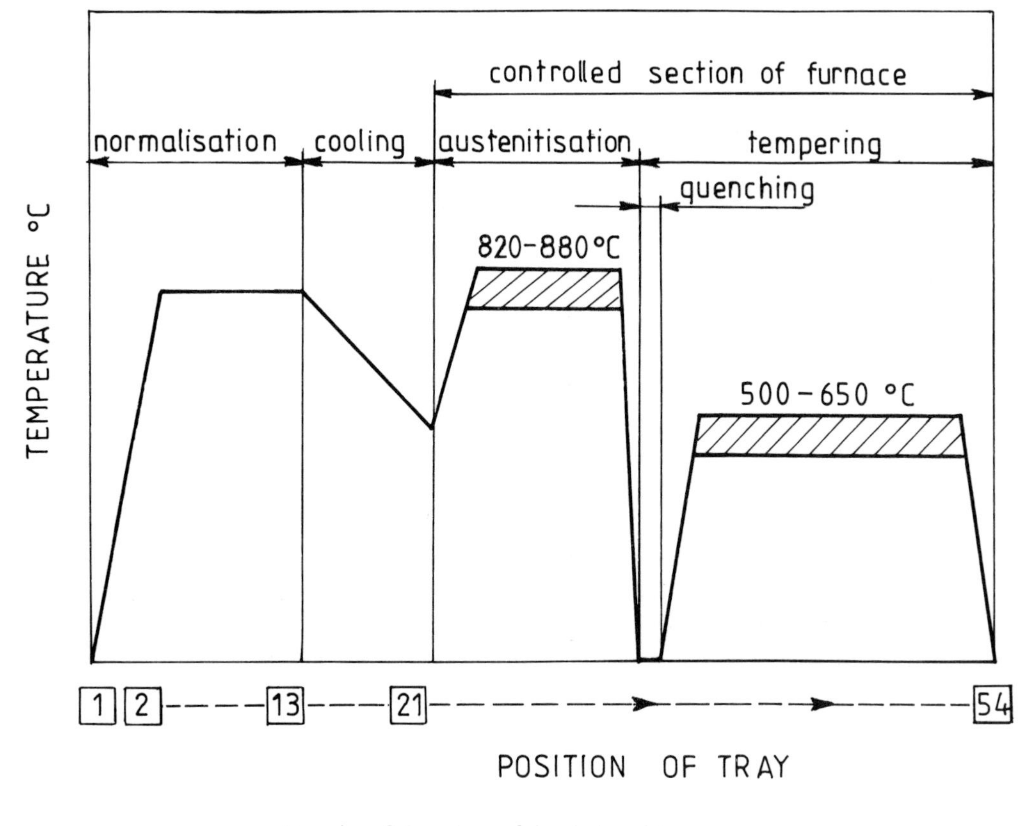

Fig. 4 Schematic of heat treating in continuous pusher furnace

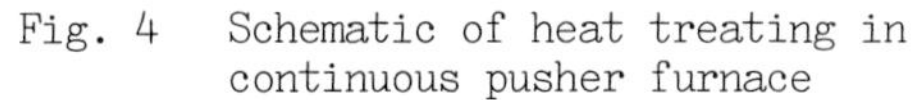

Commercial heat treatment: finance, quality and technology

T L ELLIOTT

The author is Managing Director of Senior Heat Treatment Ltd., Watford, UK.

SYNOPSIS

Despite the decline of our traditional engineering industries, e.g. Automobiles, Commercial Vehicles, Tractors, Machine Tools, Bearings, Fasteners, etc., there has been a tremendous growth in the number of Commercial Heat Treatment Companies and people employed by that industry. This is because Commercial Heat Treatment has three important advantages to offer the United Kingdom's engineering industry, viz. FINANCE, QUALITY and NEW TECHNOLOGY.

For a variety of reasons many engineering organisations have elected to sub-contract their heat treatment requirements to a specialist company and concentrate their efforts on operating plant for which they are more qualified.

It is this growth in the Sub-Contract Heat Treatment Market that has enabled modern Commercial Heat Treatment Companies to invest in furnaces employing the latest technology so that a quality heat treatment service is offered at a competitive price.

FINANCE

Unfortunately most engineering companies have had to curtail their activities in recent times and, faced with reduced loads for their own "in house" furnaces, many companies have found considerable savings could be made by closing down this equipment and sub-contracting the work out to a specialist Commercial Heat Treatment Company. The author's own company has some 60 customers who closed their facility and now send out their heat treatment.

Quite often those companies who are able to show a marginal case for an adequate return on an investment in heat treatment plant have decided, after considering the need to hire the specialist skills for necessary process control and maintenance and the loss of valuable floor space, in favour of sub-contracting out their heat treatment. Limited funds can then be used for investment in plant and machinery that they are more suitably qualified to operate.

Clearly there will always be those large companies who have sufficient volume to purchase plant employing the latest technology and be able to operate this equipment efficiently. Nevertheless, these companies will still maintain contact with their local Commercial Heat Treatment Company, using them to even out peaks in their production schedules, cater for breakdowns and for processing small quantity experimental work.

Finally, there is the multitude of small and large companies who have insufficient work to justify installation of their own plant, who provide the backbone of the Commercial Heat Treatment Market.

QUALITY

"The largest growing market in the world today is for goods of high quality and good value, not just cheapness" - Lord Sieff, who to many is synonymous with Marks & Spencer.

If we are to compete with increased international competition then quality standards for heat treatment have to be continually monitored and improved.

Quite often old furnaces are unable to meet modern quality control standards and only by employing the latest furnace materials, technology and micro-processor controls can these standards be achieved.

Modern stress relieving and normalizing cycles will now specify both uniformity of temperature during the climb to the soak temperature as well as uniformity at temperature.

By fitting micro-processor control units to each of the four burners of the Low Thermal Mass furnace shown in fig. 1, which is used for stress relieving and normalizing fabrications and castings weighing up to 20 tonnes in weight at our Blackburn heat treatment plant, the type of control demonstrated by fig. 2 is obtained.

Micro-processor Control units have been in operation for controlling Vacuum furnaces for some time and Fig. 3 shows a typical unit which is used in controlling complete heat treatment cycles for five Pressure Quench Vacuum units installed at various centres of Senior Heat Treatment Ltd.

Vacuum heat treatment is particularly useful in the heat treatment of tool steels where excellent surface finish and minimised distortion are obtained. They are also used for the heat treatment (both annealing and hardening) of stainless steel as well as the heat treatment and brazing of the various exotic nickel alloy materials used in the aircraft industry.

It goes without saying that good temperature control is vital for all heat treatment processes. However, of equal importance is the control of furnace atmospheres in gas carburising cycles, to ensure the optimum distribution of carbon within the case structure of gas carburised gears, if the optimum residual stress patterns and hardness gradients, which are so vital for maximum fatigue strength, are to be obtained.

The relative benefits of Infra Red CO_2 analysers versus Oxygen probe are still being debated for the control of carburising atmospheres. The Infra Red CO_2 unit shown in Fig. 4 is used to control the furnace atmospheres in five Sealed Quench furnaces at one of our Midland heat treatment centres, Fig. 5. These Sealed Quench furnaces are used for hardening tool steel parts as well as the heat treatment of conventional alloy wrought steels and the gas carburising/ carbonitriding of various gears and machine tool/ automobile component parts.

The engineering industry at present is probably paying more attention to quality than it has for a long time. The traditional approach to inspect the final product and screen out those component parts not meeting specification was very wasteful in terms of the time and material invested in unusable products. A more sensible approach is to control the process to ensure minimum variability, minimum waste (scrap and rework) and hence maximise the use of resources. When in control, a programme of continuous improvement can then be embarked upon. This is the basis of SPC which is now being used by commercial heat treatment companies to control their heat treatment cycles. The benefits of increased throughput due to the elimination of waste and rectification, together with better control of energy, far outweigh the initial cost of implementing SPC, which is generally only a cost relating to training.

The advent of B.S.5750, which many Commercial Heat Treatment Companies have or are in the process of obtaining, has brought a common unified approach to quality within industry. Again the concept of inspect and correct after heat treatment has been discarded in favour of quality assurance, which is applied prior to and during heat treatment through a co-ordinated documented system. Its continuity within a company is guaranteed by systematic reviews and audits, both intrinsic and extrinsic.

The implementation of such a scheme, where each operator becomes his own inspector, requires a high degree of commitment at ALL levels within the Company. It is particularly important that Management show LEADERSHIP of this commitment.

Commercial Heat Treatment Companies will also have the Approval of the multi-national engineering conglomerates as well as in certain cases the necessary Quality Approvals to carry out heat treatment for Ministry of Defence and Aerospace Contractors.

TECHNOLOGY

One's immediate reaction might be that there has not been much change in heat treatment procedures in recent times, but there have been substantial changes, e.g. the growth of vacuum heat treatment. Fifteen years ago there were only about ten vacuum furnaces being operated by Commercial Heat Treatment Companies, today there are at least sixty such units in operation.

Improvement in furnace materials, technology and controls have made the modern furnace both easier and cheaper to operate as well as being capable of meeting the latest stringent metallurgical specifications. The growth that has occurred in the Contract Heat Treatment Market has enabled the Commercial Heat Treater to invest in furnaces incorporating these developments.

Before considering some of the newer heat treatment processes which are now in regular use, some of the well established furnaces should not be overlooked.

Sealed Quench Furnaces

Probably the most popular furnace used by the Commercial Heat Treater for the heat treatment of machined ferrous parts. Heat treatment cycles carried out in these furnaces include Annealing (Bright and Sub Critical), Carbonitriding, Carburising, Carbon Restoration Hardening and Ferritic Nitrocarburising. Recent developments include the use of microprocessors for complete automation of heat treatment cycles and an increase in the use of ceramics.

Salt Bath Furnaces

These are still being used and are particularly useful for the heat treatment of long slender parts and the heat treatment of high speed steel cutting tools where the precipitation of proeutectoid carbides is minimised.

Molten salt is still useful in overcoming cracking and distortion problems by hot quenching alloy steels. Fig. 6 shows a 1½ tonne, 1.5 m. long cold forming punch being quenched into hot nitrate salt at 260°C, after austenitizing, under the protection of Endothermic gas at 830°C.

Whilst there will always be those unique requirements that suit salt bath furnaces, in general the long term future of salt baths has to be questioned in the light of recent salt price increases, difficulties in disposing of salt and the advent of the pressure quench vacuum furnace.

Induction

In Induction Hardening a high frequency alternating current is used to induce eddy currents into the surface of the steel component to be surface hardened. The lower the frequency the greater is the depth of penetration. Heat penetration will also be proportional to the time the current is applied. After heating the component part will be hardened by quenching into a suitable cooling medium, which will generally be water or a polymer, depending on the alloy content of the steel being used.

Power generation was generally by Thermionic Valves for high frequency, and for lower frequencies by Motor Generators, but more recently Solid State Inverters have made great progress in Induction Hardening techniques. Microprocessors are again used to control the process and Fig. 7 shows a simple fixture, with microprocessor control, used for selectively hardening the heads of ball studs.

The advantages of Induction Hardening are :-

(i) restricted hardening is possible with minimal power consumption and distortion;
(ii) short process times;
(iii) generation of compressive stresses within the surface; and
(iv) complete automation allows the use of unskilled labour.

Its disadvantage is its rather specialised nature and relatively high capital cost, which makes it only suitable for long production runs. It is in this respect that the Commercial Heat Treater is able to assist the engineer, who may require selective hardening on small production runs. The Commercial Heat Treater is able to design and manufacture inductors suitable for these unique applications and carry out the hardening on one of his generators at a respectable price.

Some of the "new" heat treatment processes now being offered by Commercial Heat Treatment Companies include :-

Plasma Nitriding

There are now several impressive units being operated by Commercial Heat Treatment Companies and the process, which involves subjecting parts to be nitrided to a negative potential of about 1000 V., and then introducing a mixture of nitrogen and hydrogen at a pressure of 0.1 to 10 mbar, is well covered in other Papers presented at the Conference. The major advantages of this process over the conventional gaseous process is the freedom from white layer, the possibility of using lower nitriding temperatures and shorter process times. The process is used extensively on the Continent but the market in the U.K. is still being slow to change to this process.

Titanium Nitride Coating

Coating hardened tool steel parts with a layer of Titanium Nitride produces a surface hardness of 2500 - 3000 HV, compared to 900 HV as obtained with conventionally hardened and double tempered High Speed Steel. The deposition of the coating is either by Chemical Vapour Deposition (CVD) or Physical Vapour Deposition (PVD).

For various reasons the Commercial Heat Treater seems to prefer the PVD process, which is carried at a much lower temperature, 350/450°C, compared to the higher temperature, 800/1000°C, used for CVD. The lower temperature of PVD means distortion is virtually zero.

The first part of the PVD involves subjecting the surface of the tool or part being treated, to a glow discharge in Argon, in order to clean and prepare the surface. Titanium within the chamber is then evaporated by an electron beam and nitrogen is fed into the chamber to replace the Argon. Nitrogen and Titanium are attracted to the workpiece because it still has a negative pressure. After the required time, when the desired thickness of Titanium Nitride is obtained the part is cooled under Argon.

Because this coating has a high hardness, good abrasion resistance, low coefficient of friction, high melting point and thermal conductivity it improves cutting tool life phenomenally. The process is particularly useful for cutting tools, i.e. drills, gear cutting hobs, taps and milling cutters, etc.

Fig. 8 shows a selection of Gear Shapers and Hobs, Titanium Nitride Coated by Holt Brothers (Halifax) Ltd.

Laser Heat Treatment

Is no longer a laboratory plaything and Inductoheat (Tewkesbury) have had a unit in production for some time. Whilst Laser Heat Treatment cannot compete commercially with Induction, its advantages are in hardening complex profiles, which would require expensive tooling and the heat treatment of finished ground parts. The elimination of liquid quenching prevents cracking and distortion.

In conclusion, it can be seen that the Commercial Heat Treatment Industry has invested in plant employing the latest technology and with a high degree of professional dedication in its management is able to offer the United Kingdom's engineering industry a quality heat treatment service which will enable it to compete internationally.

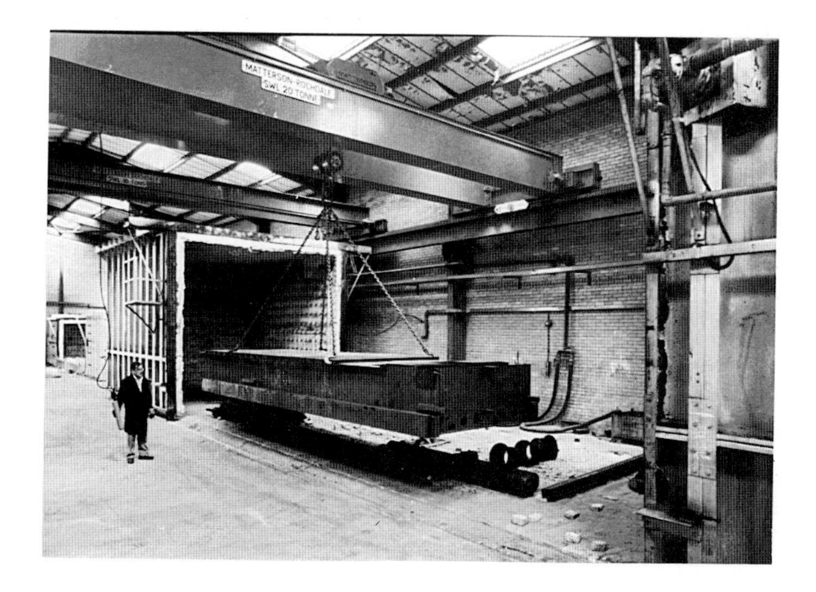

Fig. 1   Low Thermal Mass, Gas Fired Stress Relieving/Normalizing Furnace.

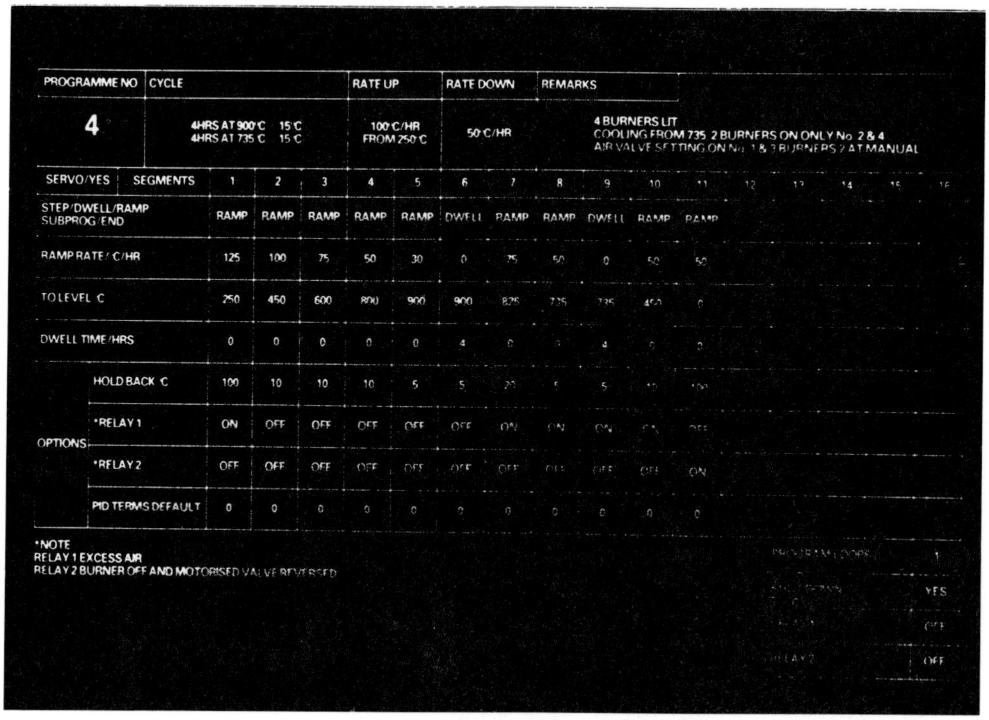

PROGRAMME NO	CYCLE	RATE UP	RATE DOWN	REMARKS
4	4HRS AT 900°C 15°C 4HRS AT 735°C 15°C	100°C/HR FROM 250°C	50°C/HR	4 BURNERS LIT COOLING FROM 735 2 BURNERS ON ONLY No 2 & 4 AIR VALVE SETTING ON No 1 & 3 BURNERS 2 AT MANUAL

SERVO/YES	SEGMENTS	1	2	3	4	5	6	7	8	9	10	11	12	13	14	15	16
STEP/DWELL/RAMP SUBPROG/END		RAMP	RAMP	RAMP	RAMP	RAMP	DWELL	RAMP	RAMP	DWELL	RAMP	RAMP					
RAMP RATE °C/HR		125	100	75	50	30	0	75	50	0	50	50					
TO LEVEL °C		250	450	600	800	900	900	825	735	735	450	0					
DWELL TIME/HRS		0	0	0	0	0	4	0	0	4	0	0					
OPTIONS	HOLD BACK °C	100	10	10	10	5	5	20	5	5	10	100					
OPTIONS	*RELAY 1	ON	OFF	OFF	OFF	OFF	OFF	ON	ON	ON	ON	OFF					
OPTIONS	*RELAY 2	OFF	OFF	OFF	OFF	OFF	OFF	OFF	OFF	OFF	OFF	ON					
OPTIONS	PID TERMS DEFAULT	0	0	0	0	0	0	0	0	0	0	0					

*NOTE
RELAY 1 EXCESS AIR
RELAY 2 BURNER OFF AND MOTORISED VALVE REVERSED

PROGRAM LOOPS	1
[illegible]	YES
[illegible]	OFF
RELAY 2	OFF

Fig. 2   Typical Normalizing Cycle, as obtained with microprocessor control.

Fig. 3 Control Panel for complete automation of Vacuum Heat Treatment Cycles.

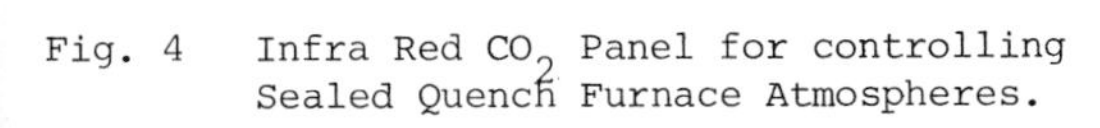

Fig. 4 Infra Red CO_2 Panel for controlling Sealed Quench Furnace Atmospheres.

Fig. 5 Sealed Quench Furnaces for hardening and carburising various component parts.

Fig. 6

Marquenching into nitrate salt at 260°C a $1\frac{1}{2}$ tonne, 1.5 metre long, cold forming die, machined from Nickel Chromium Tool Steel, after austenitizing at 830°C, under the protection of Endothermic Gas.

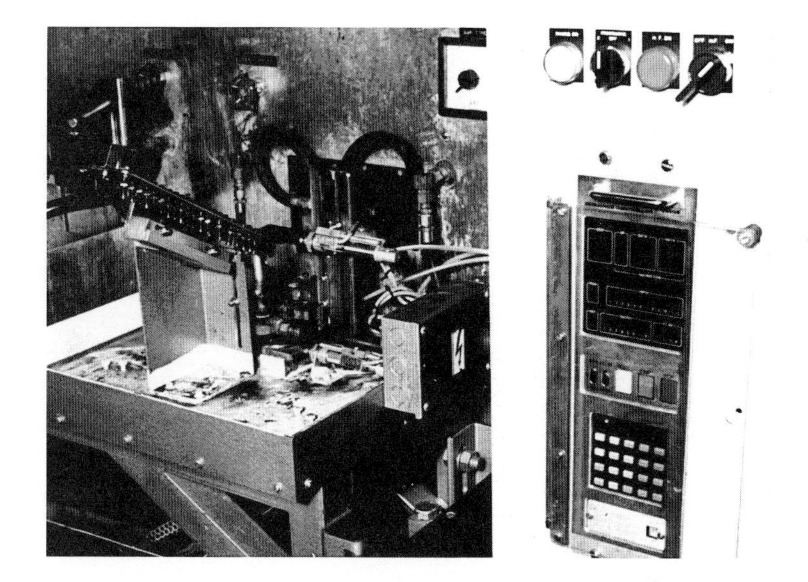

Fig. 7

Microprocessor Controls, fitted to the handling fixture, for Induction Hardening Ball Studs.

Fig. 8

Titanium Nitride Coated Gear Shapers and Hobs, Courtesy Holt Bros, Halifax.

□

Vacuum heat treatment of low alloy steels with 5 bar rapid gas quenching

P LISTEMANN

The author is Head of the Processing Division and Production Department, Schmetz GmbH, Menden, Germany.

1. Synopsis

Besides all the advantages of vacuum heat treatment, modern furnaces offer the possibility of high quenching rates and special quenching programmes such as martempering and austempering.

By virtue of those high quenching rates some low alloy steels can be fully through-hardened in a modern vacuum furnace showing very high strength and excellent toughness without any tempering ranges.

A newly developed furnace with a dual heating system as most important feature represents indeed a further contribution to a more economic heat treatment under vacuum even in lower temperature ranges.

2. Introduction

In earlier times, at the beginning of the vacuum furnace era, they were mainly used for bright annealing and for such processes where high quenching rates were not necessary. For those - mostly - long-term processes factors such as good vacuum results and an excellent temperature uniformity were the most important requirements.

The break-through for vacuum furnaces can be traced back to the introduction of 'overpressure gas quenching' which made possible fully through-hardening of a wide range of tool steels and high speed steels.

The next step in its development was to minimize the distortion rate of parts to be treated by changing the gas flow direction according to a time or temperature sequence.

Nevertheless, a still unsatisfied demand for economic heat treatment under vacuum could increasingly be observed.

Heating rates of vacuum furnaces in the lower temperature range showed unsatisfactory results due to insufficient radiation.

This challenge was entirely met by the development of the new SCHMETZ Furnace Type I 2R "FUTUR" featuring a dual heating system "convection/radiation" in the temperature range up to 850 °C (see Fig. 1 and 2).

3. Cooling Rates in Modern Vacuum Furnaces

Fig. 3 shows the cooling rate in the center of a 500 kg load of 20 mm bar cuts. Curves were taken from the core of a 20 mm bar cut which was placed in the center of the batch.

Cooling rates as a function of the parts' diameter are shown in Fig. 4.

These values were taken from the core of test specimens having different diameters and the position in the center at a practical 500 kg.

Knowing the quenching capacity of the furnace (Fig. 4) and the actual TTT-diagram of the material to be heat treated, the heat treater is able to calculate quite easily whether or not a part is through-hardenable.

4. Vacuum Heat Treatment of Low Alloy Steels

Comparing the quenching rate obtainable in modern vacuum furnaces with TTT-diagrams of certain low alloy steels, fully through-hardening of diameters of more than 80 mm must be feasible (see Fig. 5).

Another important fact is that there is indeed no other quenching medium where the cooling intensity can be varied in such wide ranges than in the gas stream itself by controlling the pressure and the gas quantity led through the load.

Tensile test specimens and notched bars for impact test were vacuum heat treated and quenched with different cooling rates. Chemistry of the different materials is listed in Fig. 6.

The quenching rate was determined by thermocouples mounted in the core of the test specimen; in addition to that, the actual TTT-diagrams of the test material were issued. For the test material 42CrMo4 it was really impossible - even by using the fastest quenching rate - to harden this type of material without any bainitic transformation (see Fig. 7).

4.1. Vacuum Heat Treatment of the Material 34 CrNiMo 6

Containing 1,5% each of nickel and chromium, the material 34 CrNiMo 6 is through-hardenable in vacuum furnaces up to 80 mm. The actual TTT-diagram as well as the cooling curves of the test specimen are clearly shown in Fig. 8.

For comparison purposes, specimens of the same material were heat treated under a protective atmosphere (CP = 0,35%) and tempered to the same hardness as the vacuum treated specimens. (Please bear in mind the fact that the vacuum treated specimens were not tempered!).

4.2. Mechanical Properties of Vacuum Heat Treated 34CrNiMo6

The mechanical properties of 34CrNiMo6 after heat treatment are listed in Fig. 9, thanks to which it becomes obvious that elongation as well as notched bar impact strength of not tempered vacuum heat treated specimens are at least equal to those values of material treated in the conventional way. The yield strength is significantly lower (see Fig. 10).

The advantage of vacuum treated parts can clearly be seen in the fact that - with the same yield strength and toughness - the ultimate tensile strength is substantially higher, which indeed leads to a higher safety rate against breakage of vacuum heat treated parts of the material in question (see Fig. 11) in comparison with conventional heat treatment.

5. Saving Costs with the Dual Heating System

In the lower temperature range the efficiency rate of heat transfer by radiation is known to be poor; and it is without any doubt that heating-up under convection is much faster (see Fig. 12).

It is the newly developed furnace SCHMETZ Type I 2R FUTUR which renders possible heating of the load under convection up to 850°C, the furnace then evacuated. In addition to all those advantages the heat treater can benefit from a conventional vacuum furnace such as overpressure gas cooling and reversal of the gas stream.

To illustrate to you the time saving effect, the whole cycle of a high speed steel batch for vacuum heat treatment is shown in Fig. 13. One tempering cycle included, the new furnace type does save about 50 % of the bottom and the bottom time compared with conventional furnace types.

The temperature uniformity in the lower temperature range within the load is indeed much better in those furnaces with revolved protective gas (nitrogen).

It is without any question - as clearly revealed in Fig. 14 - that there is an extremely high uniformity in the time range in which parts show the desired tempering temperature (outer layer/ center of the batch) under protective gas.

Considering all those results and advantages as mentioned above you can come to the conclusion that the latest innovation SCHMETZ FURNACE Type I 2R FUTUR is indeed an interesting and fruitful contribution to a more economic and time-saving heat treatment under vacuum.

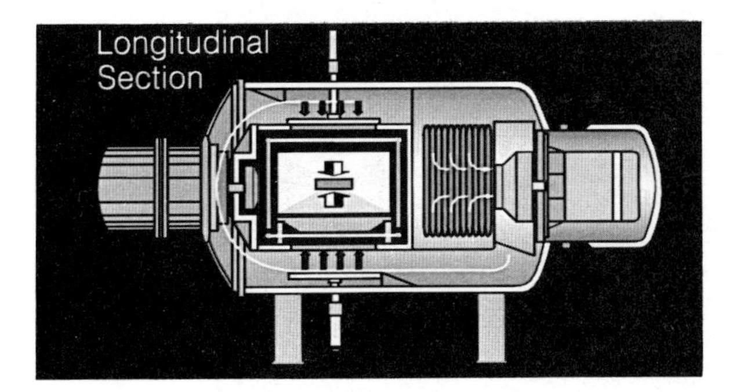

Fig. 1

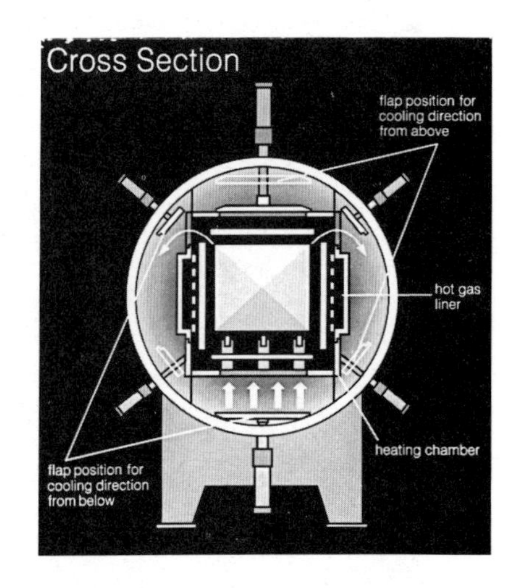

Fig. 2

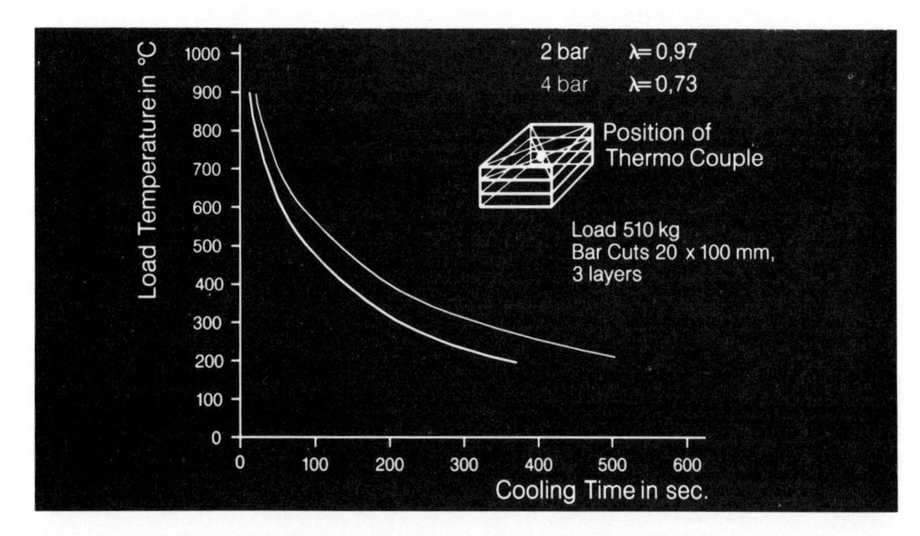

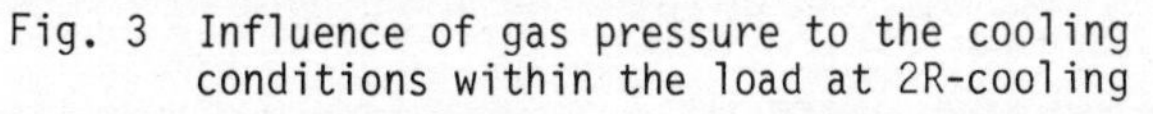
Fig. 3 Influence of gas pressure to the cooling conditions within the load at 2R-cooling

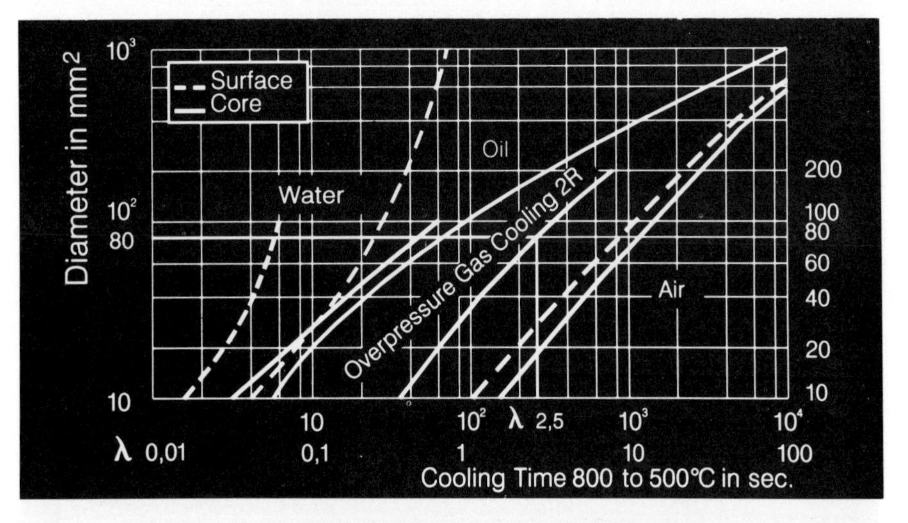

Fig. 4 Quenching intensity in different media (a. Hougardy)

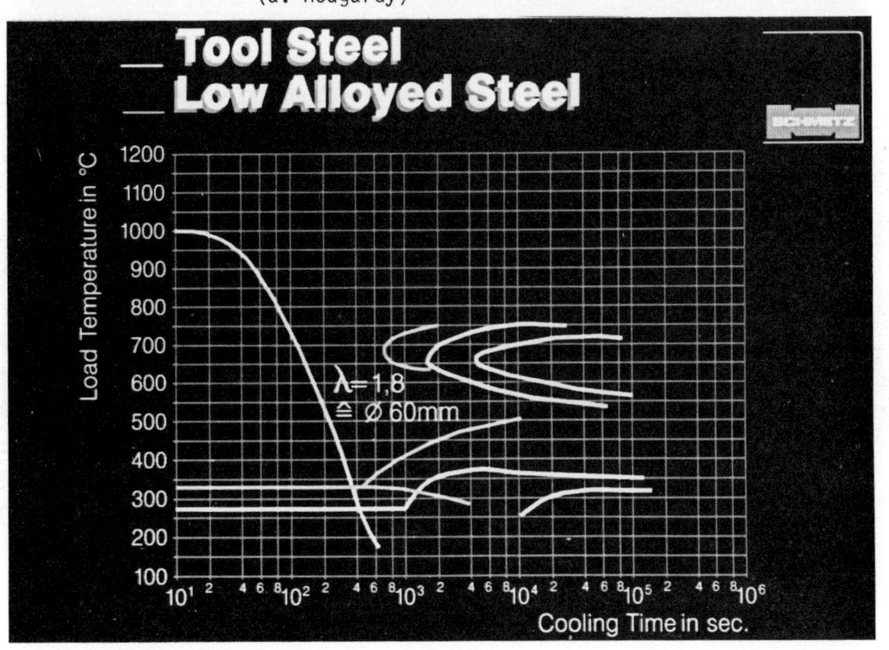

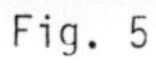
Fig. 5

Fig. 6

Material \ Analysis Weight %	C	Si	Mn	Cr	Ni	Mo	S
30CrNiMo6	0,31	0,26	0,45	1,96	1,83	0,30	0,035
34CrNiMo6	0,36	0,24	0,54	1,54	1,43	0,20	0,023
42CrMo4	0,41	0,23	0,71	1,14		0,16	0,026
SAE 9260	0,56	1,88	0,76	0,17	0,14	0,20	0,021

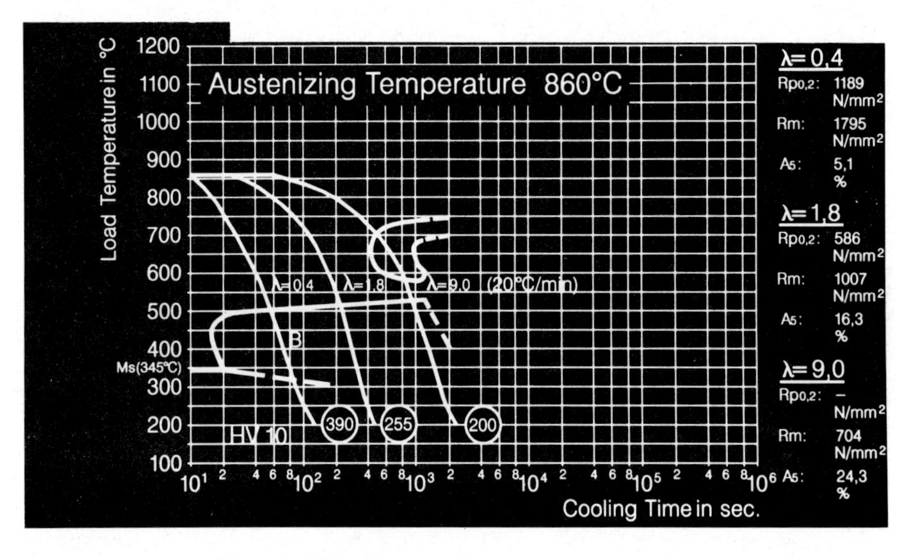

Fig. 7 TTT diagram for continuous cooling: material 42CrMo4

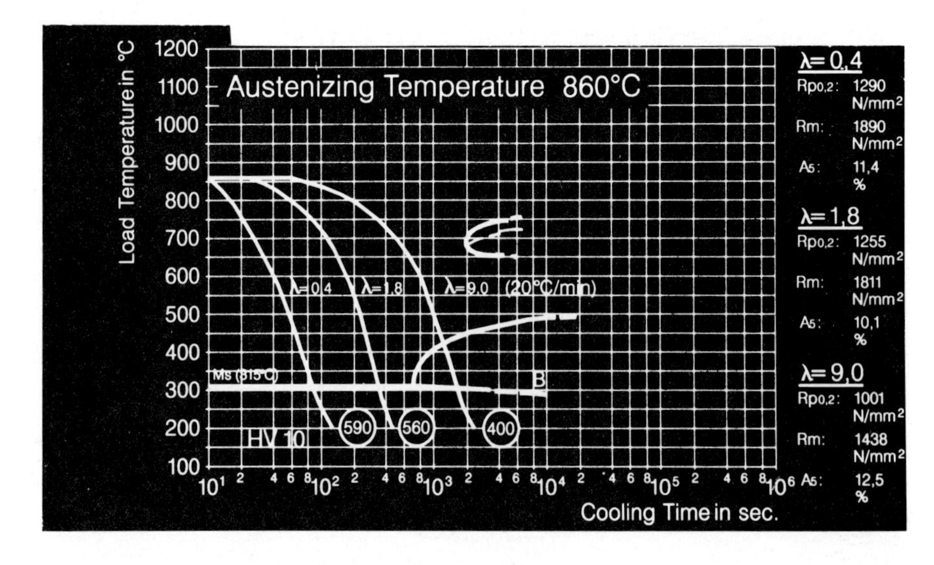

Fig. 8 TTT diagram for continuous cooling: material 34CrNiMo6

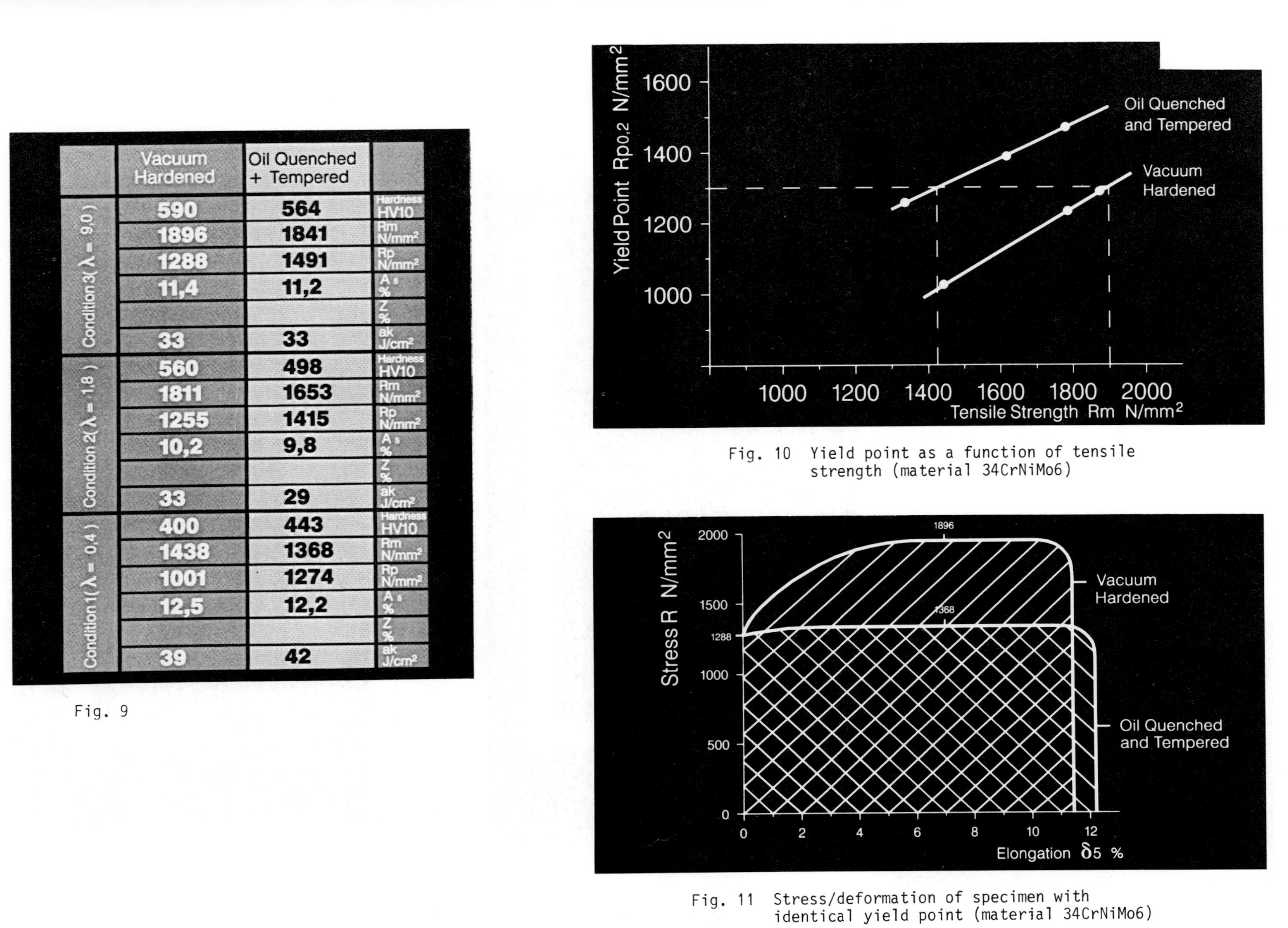

	Vacuum Hardened	Oil Quenched + Tempered	
Condition 3 (λ = 9,0)	590	564	Hardness HV10
	1896	1841	Rm N/mm²
	1288	1491	Rp N/mm²
	11,4	11,2	A s %
			Z %
	33	33	ak J/cm²
Condition 2 (λ = 1,8)	560	498	Hardness HV10
	1811	1653	Rm N/mm²
	1255	1415	Rp N/mm²
	10,2	9,8	A s %
			Z %
	33	29	ak J/cm²
Condition 1 (λ = 0,4)	400	443	Hardness HV10
	1438	1368	Rm N/mm²
	1001	1274	Rp N/mm²
	12,5	12,2	A s %
			Z %
	39	42	ak J/cm²

Fig. 9

Fig. 10 Yield point as a function of tensile strength (material 34CrNiMo6)

Fig. 11 Stress/deformation of specimen with identical yield point (material 34CrNiMo6)

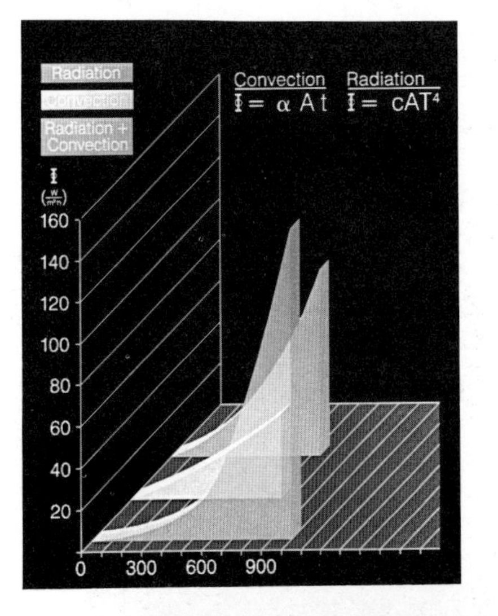

Fig. 12 Heat transfer with convection and radiation as a function of temperature

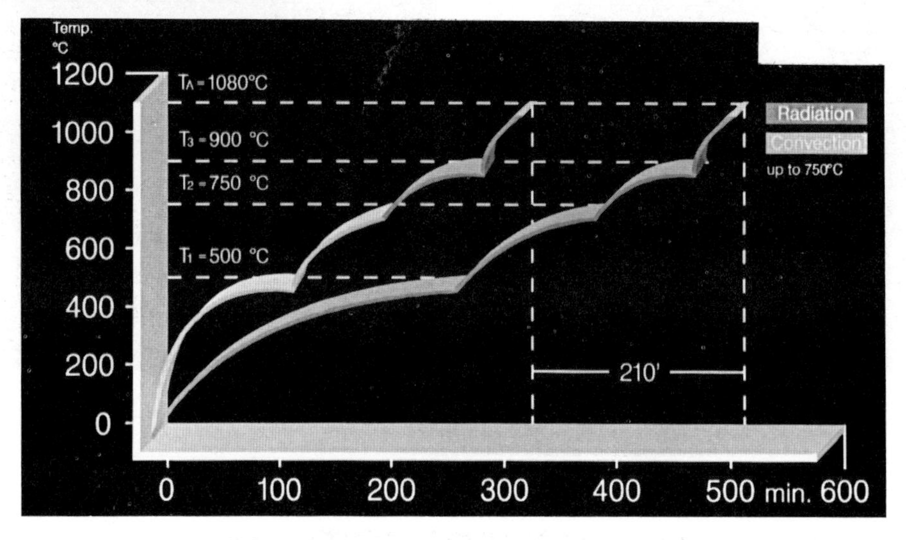

Fig. 13 Heating up of a sensitive batch load with three soaking steps

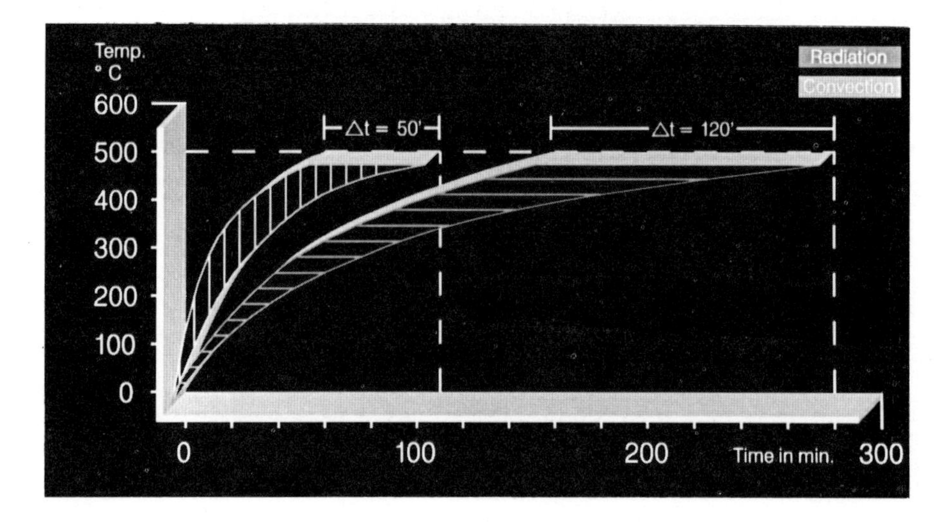

Fig. 14 Temperature difference within the load

Opportunities for combined HIP/heat treatment

L E TIDBURY and B A RICKINSON

LET is Project Engineer and BAR Technical Director of H.I.P. Ltd., Chesterfield, UK.

1. INTRODUCTION

The successful application of Hot Isostatic Pressing (HIP) to cast material was first demonstrated over twenty years ago. It was shown that the density and mechanical properties of cast alloys could be considerably improved after HIP treatment, as a direct consequence of the removal of internal defects during such processing (ref 1). Initially, for economic reasons, commercial work was restricted to the field of nickel base superalloy gas turbine components. At this early stage of its industrial development, HIPping was regarded as a means of recovering castings rejected due to high porosity levels, and clearly this was most cost effective when high value parts (Ni,Ti base) were affected.

Despite the demonstration that porosity could be removed from a range of other cast alloys by HIP, it remained unattractive for some time, purely on economic grounds. This situation has changed considerably over the last five years for two primary reasons.

Firstly the advancement in equipment design and operating technique, have combined to improve overall operating efficiency. Secondly, a considerable reduction in overall operating costs per unit volume, has accompanied the trend toward increased HIP unit size.

Equally, there has been a shift in emphasis away from scrap recovery, and towards property enhancement. It has been shown that the mechanical properties of castings which have achieved all normal N.D.I. requirements, can still be increased by HIP treatment (ref 2).

Specific improvements in properties achieved after HIP treatment varies to some extent with alloy type and casting technique, but in general terms the following can be expected:

Tensile strength	-	5-15% increase
Yield strength	-	5-10% increase
Ductility (El%)	-	50-100% increase
Fatigue life	-	3-10 fold increase

Other benefits such as improved stress rupture performance and corrosion resistance have also been reported, (ref 2,3).

The reduction of data scatter, which is a further characteristic of HIPped castings, is probably more important than any of the above individual improvements. This increase in product consistency is crucial from a designers point of view. HIPped castings can be used with far greater confidence, and the full performance potential of the material can be approached. It is therefore possible to consider modification of the design factor which designers apply to cast components.

2. HIP CONDITIONS

The operating conditions which are used for HIP treatment can be expressed in terms of three basic parameters, namely temperature, pressure and time. This is not of itself surprising since defect removal is largely diffusion controlled. From experience, however, it is fair to say that for most alloys there is a reasonably large temperature window within which pore closure can be achieved. The lower limit can be considered as the minimum temperature at which plastic yielding and pore closure can be guaranteed, and the upper limit as the alloy solidus. The preferred temperature is clearly one which will also impart an acceptable as-HIP microstructure.

In addition to the primary HIP conditions the further variable involving post HIP sustain cooling rate is a crucial parameter relating to as-HIPped microstructure.

From the foregoing it can be appreciated that since HIPping is invariably a relatively high temperature procedure, by careful modification of the sustain temperature and correct selection of the cooling rate, some elements of the standard heat treatment for many alloys can be provided within the HIP cycle.

As a consequence it might be suggested that;

i) Direct HIP costs could be reduced by modifying the severity of sustain temperature conditions.

ii) Subsequent heat treatment stages would no longer be required, thereby reducing overall energy costs.

With this background in mind, specific results from recent work can be reported which indicate the feasibility of a combined HIP/heat treatment approach.

3. EXPERIMENTAL WORK

3.1 Alloy Steel

Experimental work was undertaken with a view to optimise process conditions for HIPping of a chromium/molybdenum steel (EN40B - ref 4). The primary objective was to isolate the minimum process temperature and to this end HIP trials were conducted using temperatures between 500°C and 1205°C. Subsequent metallographic examination and tensile testing confirmed that material processed at 500°C and 800°C had porosity remaining after treatment, (figure 1). HIPping at temperatures in excess of 800°C resulted in a fully dense microstructure, (See Figure 2). Properties typical of the alloy in the as hardened condition could be obtained in as-HIPped material, cooled from a sustain temperature greater than 800°C at a rate approaching 10°C per minute (figure 3).

In retrospect this latter point is not surprising since, although the hardening specification calls for an oil quench from the austenitizing temperature, the alloy is virtually air hardenable; However, this observation is crucially important with respect to combining HIP and heat treatment. It demonstrates that the cooling regime specified for heat treatment purposes is the most convenient method of exceeding the critical cooling rate for the alloy, rather than that which is absolutely necessary to achieve the specified mechanical properties.

An attempt was made to HIP at the recognised austenitizing temperature, such that the as HIPped sample could be directly tempered to produce the specified heat treated mechanical property levels. One can observe (fig 4) that the measured ductility provides a clear indication as to the presence of porosity, and all samples HIPped at 900°C or higher have a consistent ductility within the specification. Although further trials need to be conducted to establish the treatment temperature window at or near 900°C to guarantee pore closure, the potential clearly exists to combine HIP with the hardening treatment for steels of this type. Despite the fact that relatively low cooling rates were successful in imparting a typical hardened microstructure to small test bars, higher cooling rates available in those HIP units equipped with rapid cool facilities (to be described later) would probably be required to ensure satisfactory treatment of larger section sizes.

3.2 Niobium Stabilized Stainless Steel

Similar trials to those noted earlier were conducted with cast bars of a niobium stabilized stainless steel (CF8-C) to assess the feasibility of HIPping at the solution treatment temperature, in this case 1050°C. No evidence of porosity was found in material HIPped at 980°C or higher (figures 5a and 5b). A later cycle, involving HIP treatment at 1050°C, with a cooling rate typical of that achieved after vacuum heat treatment, resulted in a microstructure similar to that of traditionally heat treated material. Given that a sufficiently high cooling rate (~ 60°C per minute) could be generated at the conclusion of the sustain part of the HIP cycle it would appear that solution treatment and HIPping of stainless steels could be combined.

Previously reported work has claimed that the delta ferrite content of stainless steel weldments can be modified by the influence of pressure during HIP treatment (ref 5). A comparison of several grades of cast austenitic stainless steel, solution treated both at atmospheric pressure, and under typical HIP pressure, did not substantiate this earlier work. Optical metallography did not reveal any significant variation in microstructure with pressure when applied at the solution treatment temperature, (see table 1).

Consideration of the Clausius Clapeyron relationship suggests that pressure differences of the order of 1000 atmospheres would be equivalent to modifying reaction temperatures by approximately 20°C. Such a change should not result in a significantly modified delta ferrite content, and this supports our experimental evidence from conventional cast parts.

Having in mind the accuracy of HIP temperature control (+/- 10°C), it is suggested that the effect of pressure should only be considered as significant when thin walled castings or weldments are processed. The refinement of ferrite size and spacing achieved by the rapid liquid to solid transition accompanying such process routes could explain those differences in experimental results noted in this and earlier work, (ref. 5).

3.3 Precipitation Hardening Stainless Steel

Considerable work has been carried out over the last three to four years to optimise the total production route for high strength stainless steel (17-4PH) castings (ref 6). In addition to the refinement of melting and casting practice, it has been shown that subsequent HIPping is essential if cast parts are to challenge forged components in fatigue sensitive applications. Recent work within this study (ref 7) has proposed the possibility of combining HIP with solution treatment for this and similar compositions. Trial work to date indicates that while there may be a very slight fall in fatigue limit compared to the separately HIPped and heat treated material, this is not considered significant.

In the above case homogenization and solution treatment practices are progressed sequentially within the HIP vessel. Porosity removal can be considered as complete after the homogenization stage, and thereafter the gas pressure available at the conclusion of the solution treatment phase is desirable to effect rapid quenching.

As a point of clarification it is important to recognise that the development of argon gas pressures of the order of 1000 atmospheres yields a fluid medium having a density approaching that of water. The gas state is clearly retained, but such a viscous fluid provides the perfect medium to transfer heat in a uniform and efficient manner.

At HIP treatment temperatures of less than 1300°C then heat transfer is achieved largely by convection. Radiative heat transfer only increases in importance as furnace and charge temperatures exceed 1500°C. Conversely during charge cooling the promotion of strong convection loops between hot and cold gas within different zones of the pressure vessel provides the opportunity for rapid heat removal. At the conclusion of the combined cycle cast products are unloaded from the HIP vessel to receive the lower temperature ageing treatment. Arguably this operation could also be embodied within the HIP practice, but in view of the duration of this treatment such a procedure is unlikely to be cost effective.

3.4 Aluminium

The HIP treatment of aluminium alloys, particularly cast materials, has seen a steady increase in importance from the original work conducted in the mid 1960's. Under normal circumstances such processing is conducted within the temperature range 480°C-520°C and as such the treatment is well below solidus and solution treatment temperatures.

Despite this fact, a study involving L99 (A356) alloy attempted to demonstrate that a HIP treatment of 530°C, combined with post sustain cooling rates approaching 60°C/minute, was effective in providing a measure of first stage heat treatment. Results noted in table 2 were taken from cast test bars exposed to the HIP and precipitation treatments only. Although the minimum proof stress has not been achieved both the UTS and ductility results are promising. It would seem likely that if temperature control could be enhanced by the use of high output base metal thermocouples, and the cooling rate improved, the possibility of effecting a combined HIP/solution treatment might be approached.

4. HIP EQUIPMENT

Having outlined the metallurgical potential of combined HIP and heat treatment it is useful to consider the demands which these requirements place on HIP equipment design. Current HIP furnace designs offer excellent process (sustain) temperature uniformity within the charge volume (normally better than plus or minus 10°C) which would be considered acceptable for most higher temperature heat treatment operations. Much attention has been focussed upon the problem of post sustain charge cooling, a subject which can be basically divided into two components:

a) Absolute cooling rate. (°C per minute)

b) Uniformity of cooling rate throughout the charge volume.

4.1 Cooling Rate

In consideration of the absolute cooling rate, it might be expected that cooling rates typical of gas pressure quenching furnaces should be achievable. Equally and by contrast, rates in excess of 100°C per second, typical of quenching into liquids, would not appear to be a practical proposition. Such operational limits are indeed correct. However, it is only in recent times that more elegant design engineering has provided the means by which rates in excess of 50°C/min could be developed in a HIP unit containing a charge of 500 kg.

4.2 Uniformity of rate

The uniformity of cooling rate achieved throughout the charge is of significant practical interest. The manifestation of a cooling rate gradient through the charge was a characteristic typical of early rapid cool systems. Uniformity of cooling rate could be improved to a limited extent by varying the component packing density within the charge, but such an approach was empirical and in any event resulted in a reduced operating efficiency.

4.3 State of the Art

During 1986 a large HIP unit has been installed in the UK which has a rapid cool system designed to overcome these traditional drawbacks. This 1.1 metre diameter unit (see fig. 6) represents a major advance in HIP furnace technology. The successful integration of advanced argon manipulation systems within the pressure vessel with highly sophisticated temperature control has resulted in the most advanced rapid cool facilities currently available.

Cooling rates of 100°C/min for a total charge weight (including furniture) of 2.5 tonnes, have been demonstrated, and the uniformity of rate as recorded from thermocouples disposed throughout the charge is shown in figure 7a.

As a consequence of the extensive and controllable thermal absorption capability available within the press, interruption of the cooling sequence at any point confirms that the charge lags behind gas temperature by no more than 40°C. Cooling rate data collated from charge thermocouples is therefore representative of that rate experienced by the charge.

It should be emphasized that although a maximum cooling rate in excess of 100° per minute has been successfully demonstrated, the same degree of cooling rate control and uniformity can be applied to any cooling rate within the equipment capability (fig 7b).

HIP equipment of a size appropriate to the economic processing of ferrous and other components is therefore available with cooling capabilities necessary to enable combined HIP and heat treatment to be considered on a routine basis. Experimental work to date has indicated that those properties typically obtained by HIP and first stage heat treatment as separate operations can be achieved in a combined pressure cycle. It is nevertheless fair to state that the novel effect of HIP pressure upon phase transformation sequences requires further detailed investigation.

5. CONCLUSIONS

1. For most alloys there is considerable scope for modifying HIP conditions without compromising the effectiveness of the process for porosity closure.

2. HIP equipment design has advanced to a stage where uniform cooling rates of 100°C per minute from sustain temperature can be achieved.

3. Combining HIP and heat treatment is technically feasible for many alloys, and should result in economic benefits when compared to HIP and heat treatment as separate operations.

REFERENCES

1. H. Hanes et al : 'Hot Isostatic Processing' Metals and Ceramics Information Centre, Columbus, Ohio. Report MClC - 77-34.

2. P. D. Holmes : Proceedings '2nd Int Isostatic Pressing Conference' - Stratford, England - Sept 1982 (16).

3. P. G. Bailey & W. H. Schweikert : Superalloys - Metallurgy & Manufacture 1976, 451-465.

4. L. E. Tidbury : Un-published work.

5. R. K. Malik : Metals Progress, 119 (3) P 86.

6. R. McCallum & W. Lang : 'Influence of Processing route on the Fatigue Behaviour of Investment Cast High-Strength Steels : Metallurgia, March 1985.

7. R McCallum et al : Proceedings '20th E.I.C.F. Conference on Investment Castings' - Brussels, June 1986 - P 5.01-5.18

TABLE 1 The independent effects of HIP and heat treatment upon the residual delta ferrite content within two sand cast and one investment cast stainless steels

COMPOSITION	AS CAST	CONVENTIONAL HEAT TREATMENT 1050°C	H.I.P. 1050°C	CONVENTIONAL HEAT TREATMENT 1160°C	H.I.P. 1160°C
% DELTA FERRITE					
SAND CAST COMPOSITION A	20.5	20.4	20.3	15.6	14.5
SAND CAST COMPOSITION B	10.5	10.8	10.2	9.9	8.6
INVESTMENT CAST COMPOSITION C	7.8	5.8	6.0	3.6	3.9

TABLE 2 The effect of HIP + ageing treatment only on the properties of sand cast LM25 (L99)

	MINIMUM SPECIFIED UTS (MN/m²)	MINIMUM ELONGATION %
SAND CAST + FULL HEAT TREATMENT	230	2
HIP AND RAPID COOL+ PRECIPITATION TREATMENT	211	4.3

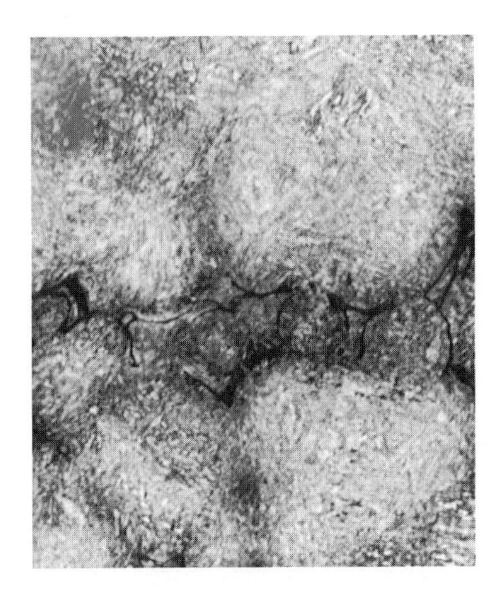

1. The effect of HIP treatment on cast EN40B temperatures within the range 500-800°C. Note residual interdendritic porosity.

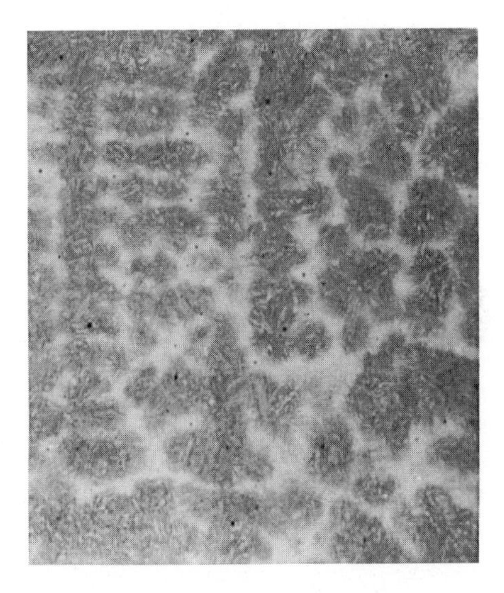

2. Typical microstructure of cast EN40B after HIP treatment at 1000°C.

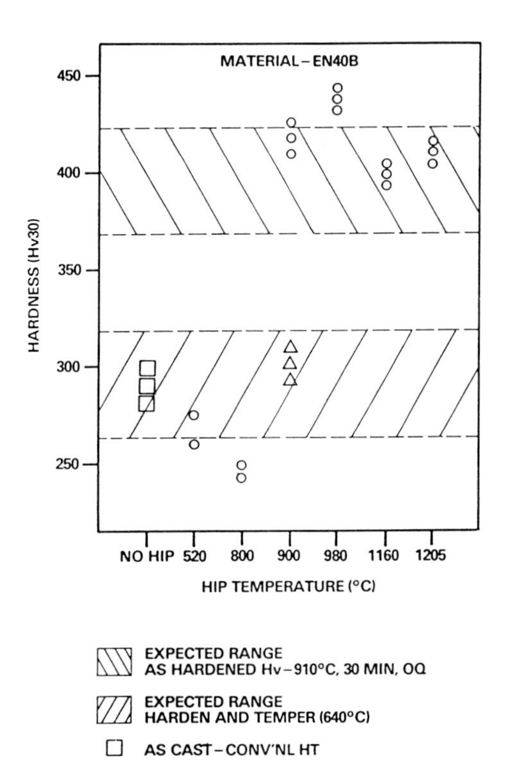

3. As HIPped hardness as a function of sustain temperature. Conventional heat treatment ranges are noted. - Material EN40B.

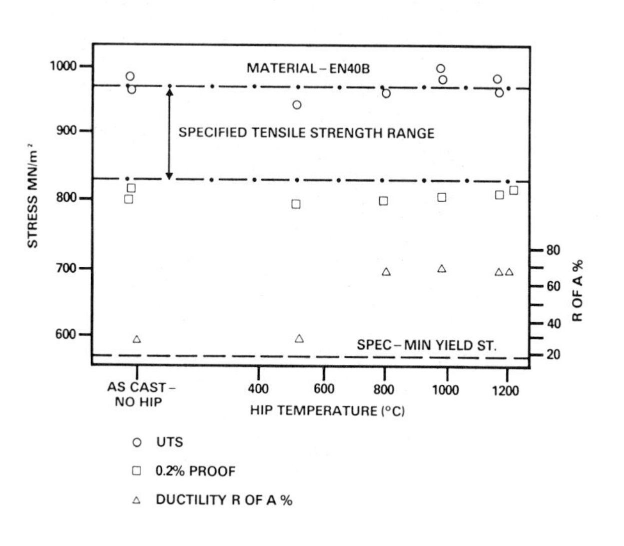

4. The effect of HIP temperature and porosity removal on the mechanical properties of cast EN40B following tempering at 640°C.

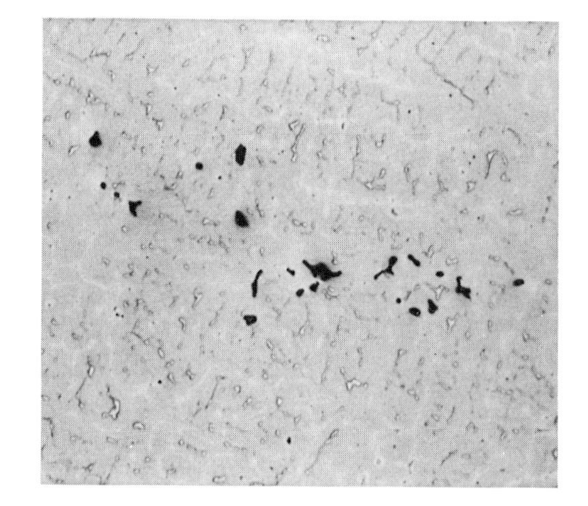

5a. Cast and solution treated Niobium stabilised stainless steel. Note colonies of residual microporosity.

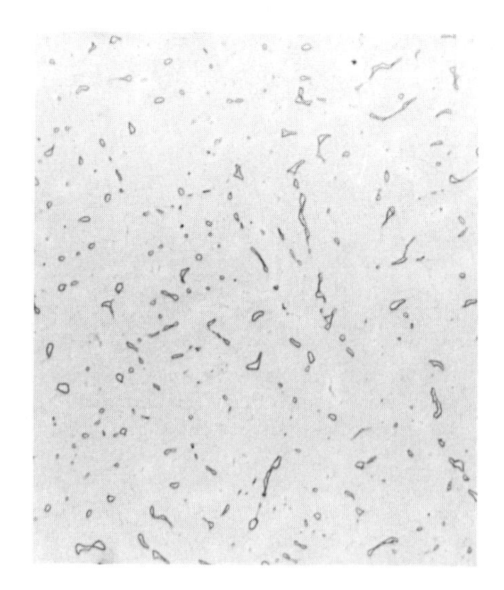

5b. Material as in 5a, after HIP treatment at 980°C.

6. MegaHIP the largest Hot Isostatic Press in Europe and equipped with sophisticated rapid cool capability.

7a. Typical cooling trace from thermocouples placed within a 2.5 tonne steel charge after programming for cooling at a rate of 100°C/min.

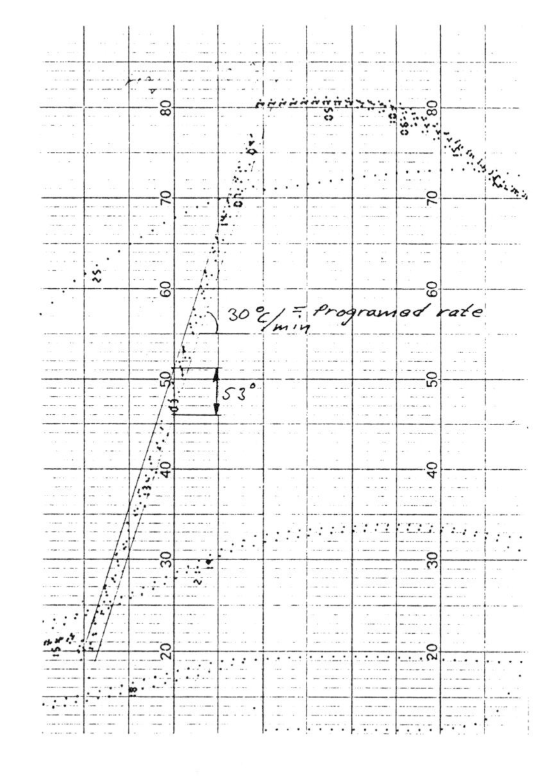

7b. Similar charge to that described in 7a, programmed to cool at a rate of 30°C/min.

Impact of ion implantation and titanium nitride coating technologies

R A SANDERSON

The author is with Tecvac Ltd., Stow-Cum-Quy.

The subject of coatings for surface engineering has been receiving increasing attention from research workers, designers and production engineers. There is a growing awareness of the benefits of using known and proven materials and enhancing their performance by the application of the new surface techniques. The best known and greatest successes have been recorded in the tooling applications.

The paper today will concentrate on two processes, the titanium nitride coating deposited by the PVD process, and nitrogen ion implantation. These are two tried and proven treatments which have become commercially available over recent years, finding ever increasing applications in industry. More importantly they are probably the forerunners of a family of new treatments that will evolve from these processes in the next ten years.

The techniques selected for this paper are quite different in their approach but impart very similar properties to the component. The titanium nitride coating by value of its own properties and nitrogen ion implantation by the conversion of the surface of the component.

In both cases the enhanced properties are improvements in adhesive wear, abrasive wear and a reduced co-efficient of friction. In titanium nitride there are the further beneficial properties of very high surface hardness and also chemical inertness. The nitrogen ion implantation would be expected to impart an improved surface hardness and also improve other properties. However, the implanted layer is in the order of 0.1 micro thick and special techniques are required to measure the hardness, although in use nitrogen and its benefits greatly exceed this depth due to a migration characteristic.

Titanium Nitride Coating (PVD)

There are basically two techniques used to apply titanium nitride coatings; chemical vapour deposition (CVD) and physical vapour deposition (PVD). The CVD process became available in 1969 and uses gaseous reaction in a chamber with a slight vacuum. The operating temperature is in the region of 900 - 1000°C

The TiN (PVD) process became commercially available in 1981. The major advance in the technology was to evaporate titanium from the solid and cause it to react with the nitrogen gas. The PVD process is carried out between 350 - 500°C which allows it to be used on a wide range of steels in use in the engineering industry either in the form of tools, or components.

There are a number of PVD techniques employed. **TECVAC** uses the electron beam method which produced a fine TiN vapour which offers the benefits of a very smooth uniform coating.

The Titanium Nitride Process

The equipment used to deposit titanium nitride is based on a vacuum chamber able to achieve levels of 10-6 torr before the coating process is undertaken.

There is perhaps a less high technology but crucial part of the process and that is the precleaning of the components. As a good guideline the deposition of the titanium nitride requires at least as good a surface as that required for good quality electroplate. In practice that will mean solvent and alkali degreasing with ultrasonic assistance.

This is followed by a rinse in distilled water and drying. In the case of components with significant and tenacious oxide the most successful preparation is the mechanical abrasion of the surface by a vapour blasting technique.

The components when mounted in the chamber will rotate during the process to ensure a uniformity of coating. The chamber is evacuated and a quantity of argon gas introduced.

Applications of Titanium Nitride

The initial impact of titanium nitride on the engineering industry was on carbide inserts coated using the CVD technique. These have been highly successful and have achieved a much improved tool life with higher machining feeds and speeds. The considerable successes achieved in the tooling applications using the CVD technique led to the concentration on that market when the more versatile PVD process became available. The early successes with the PVD process were found on high speed cutting tools, drills, taps, reamers and hobs. The typical claim of 400% improvement is commonly found in cutting applications of many types.

In the case of titanium nitride coated drills the standard required by the drill manufacturers is inevitably rather higher than the claims made in their sales literature. The exact test details vary slightly from one manufacturer to another but are based on a comparison with uncoated drills. In these tests an uncoated drill will burn out in one hole. A steam oxide treated drill will last for perhaps five holes and a titanium nitride coated drill must last for greater than 100 holes. The requirements set by the drill industry is therefore a 20 times life improvement. This is a tough requirement.

Another successful metal cutting operation is in machine taps where there is a tendency to cut unevenly at the thread start. Using coated taps it was found that the threads were cut very much more cleanly and that the bell mouthing at the start of the cut was very much reduced. The very clean cut found using titanium nitride is a feature of the process and on some applications it was found that the cutters were apparently cutting undersize.

The benefits found in cutting tools have been shown to be equally attractive in press tools. In one example a company blanking and forming stainless steel components was using the traditional tools made from D2. They achieved 4000 components from a set of tools. The tools were then manufactured from ASP 23 and the life was increased to 10,000 components and with titanium nitride coating this was further improved to 92,000.

In a metal powder compacting operation coated punches were found to give an improvement of 400%. In a subsequent tooling procurement for a new job it was estimated that the requirement would be for 400 punches because of the high abrasive wear rate. By using coated punches only 100 were required at a cost of £100 each thus showing a saving of £30,000, less the cost of coating.

The non tooling applications are perhaps a better guide to the range of properties that can be exploited by titanium nitride coatings. There are many examples which demonstrate the improvements in reducing the abrasive wear and the adhesive wear. It is also possible to benefit from the reduction in the coefficient of friction and the inert characteristics of the coating.

In the reduction of adhesive wear a particularly good example has been in the coating of titanium. Titanium and titanium alloys have very attractive strength to weight ratios as well as excellent corrosion resistance. They do however have relatively poor wear properties. In titanium to steel contact, the titanium very quickly demonstrates galling. The titanium, coated with titanium nitride, on contact with steel shows a low coefficient of friction and very little wear. There has been a growth of titanium and its alloys and the area responding most enthusiastically to this work is the motor racing industry.

This perhaps is one area where the testing of the coated components is achieved very quickly and therefore a successful result is rapidly incorporated into the design.

An example of severe abrasive wear is the rubbing of a metal pen on the paper of a chart record. In this case the pen made from a nickel-chrome wire was titanium nitride coated and the life of the pen was considerably extended.

The reduction of the coefficient of friction has been most effectively demonstrated in an aircraft oxygen safety release mechanism. The unit was made from stainless steel and involved a pin acting on a plate. The mechanism was activated when a given pressure level was reached and the pin pushed clear of the plate by an actuator. The coefficient of friction between the pin and the plate was sufficiently high to cause the mechanism to function. The pin and the plate were both titanium nitride coated. The coating reduced the pressure at which the mechanism would operate and made the design more effective.

In the use of the inert properties of the coating the application being given is perhaps more akin to tooling than to a product. It is as a solder mask for flow soldering of printed circuit boards on a production line. A material was required that could be used as a mask for solder which would not be wetted and thus solder coated.

The mask that was used with considerable success was made from stainless steel and titanium nitride coated.

Guidelines for Designers

The coating deposited is only 2 - 3 micros thick. The performance will therefore very much depend on the surface onto which it is deposited and the material of the component. The main areas of importance are:

The surface finish of the component

For a thin coating on a surface with large asperities the loading on the coating will be such to cause premature failure. In general the better the finish the better the results from the titanium nitride coating.

Substrate Strength

The coating has a very high hardness with values of approximately 2600 VPN. However, it is a thin coating and must rely on the strength of the substrate to support it in high load applications. Titanium nitride cannot turn a low strength material substrate into one of high strength. If the substrate undergoes plastic deformation the coating may crack and subsequently fail. There are examples where components have been nitrided to give a component surface strength and then titanium nitride coated for wear resistance.

Substrate Material

The coating process is carried out at between 350 & 500°C. The temperature at which the coating takes place is important to ensure a good coating structure and adhesion. The material being coated must therefore be capable of withstanding this temperature. Successful results have been achieved with a wide range of materials including high speed steel, carbide, tool steels, maraging steels, and stainless steel.

Components Containing Joints

The coating process is a vacuum operation and great care must be taken with any brazed or soldered parts. Low melting point, solder or braze materials can evaporate and thus coat the chamber and component. Similarly care must be taken to avoid organic adhesives or pressed inserts. These substances will prevent the titanium nitride coating from adhering to the component.

ION IMPLANTATION

The Ion Implantation process involves the implantation of the ions of one atomic species into the surface of the chosen material. The process has been used extensively since the 1970's to implant dopant elements into semi-conductors and this is now very big business.

In the engineering field extensive work has been carried by the Nuclear Physics Division of the Atomic Energy Research establishment at Harwell. Considerable benefits have been demonstrated using a wide range of different implanted species. There have been improvements in abrasive and adhesive wear and also in the reduction of the coefficient of friction and in the improvement in oxidation and corrosion resistance.

The most thoroughly investigated technique has been the implantation of nitrogen. This is attractive as a commercial project in that the equipment to implant gaseous species is less complex and less expensive and therefore more cost effective than that to implant the non gaseous species.

Nitrogen Ion Implantation

In this process a beam of nitrogen ions is accelerated towards and collides with the surface of the material being implanted. The nitrogen is embedded in the surface ofthe component.

The ion source uses an intense self excited plasma in a tubular chamber with a narrow slit down one side. Nitrogen ions which emerge from the slit are accelerated across a high voltage gap and travel through the intervening space to strike the target. The ions travel in straight lines so the process is 'line of sight' and affect only those surfaces seen by the beam. To achieve wider coverage or to treat components with uneven or re-entrant surfaces manipulation of the component is required.

There is a release of kinetic energy as the ionised gas strikes the components. This results in a heating effect of some one to two watts per square centimetre which is generally a low order effect and unimportant except for thin wall or very light components. For low alloy and carbon steels it may be desirable to keep the temperature below 200°C to avoid tempering them. For other materials there may be a lower limit to the temperature, below which the effect is diminished.

The typical concentration of nitrogen ions required to produce the improved properties is in the order of 1-5 x 10.17 nitrogen ions/cm2 with an optimum of 2-3 x 10.17 nitrogen ions/cm2. The latter will give a nitrogen concentration of approximately 10 atomic percent with a typical depth of penetration of 0.1 micro.

Wear tests on samples have shown that the reduced wear rates have continued far beyond the depth of the initial ion implantation.

Analysis carried out after wear of 12 micros showed that over 30% of the nitrogen was still present i.e. the nitrogen was therefore still offering improvement at greater than 100 times the depth of original implantation. An explanation for this phenomenon has been offered by Dearnaley. The nitrogen can act in two ways to harded the substrate and to reduce abrasive wear.

In plain carbon and low alloy steels the nitrogen forms a solid solution. When the metal surface is plastically deformed a dense network of dislocations is created. There is a strong attraction between the interstitial nitrogen atoms and the dislocations which causes a resistance to further dislocation movement. The rate of deformation and hence the rate of wear are reduced.

In alloys containing strong nitride forming elements such as chromium, aluminium, titanium, the nitrogen creates a very fine dispersion of hard nitride precipitates such as CrN. The hardening effects cannot be entirely explained by the nitrides present. It has been shown that a component plasma nitrided and then nitrogen implanted shows an increased resistance to wear when compared with the non implanted components.

Applications of Nitrogen Ion Implantation

The range of applications where ion implantation can be used is broadly similar to that of the titanium nitride coating.

The areas where there is a positive advantage in using nitrogen ion implantation are:

a) When the material substrate will soften if heated to greater than 350 degrees C.

b) When components are so precise that no change in dimension can be tolerated.

c) For applications in which there is a natural resistance to using a coating that could come off e.g. food, medical.

Medical Prostheses

An area of particular interest is that of the surgical implant of joint replacements. But the design of the prosthesis and the choice of materials presents the medical engineer with a difficult problem and many alternatives have been put forward. One now gaining favour is based on a titanium alloy which is almost inert, is light and strong but has suffered somewhat from uncertain wear characteristics. Nitrogen implantation of the spherical surface of this type of implant now offers a radical improvement. Trials are expected to confirm a working life greatly exceeding the life expectancy of even the youngest recipient. The nitrogen implantation of the metal ball was found to reduce the wear to negligible levels.

The approval of this technique by the medical authorities would allow hip joints to be provided for much younger people. The limiting factor would become the durability of the metal to bone bond.

Injection Moulding Tools

Typical examples of successful implantation in the plastics industry have been injection moulding screws. The nitrogen ion implantation of the injection moulding screw increased the life of this component by a factor of 4 times.

Similarly in the injection nozzles, gate areas and cavity blocks suffering from wear experienced when working with filled plastics.

E.D.M. or textured steel surfaces can also benefit from the reduction in wear.

Chrome plated tooling can also benefit from the implantation of nitrogen. The already chromed surface is improved by creating chromium carbides, reducing wear and forming a protection against corrosive gases and fluids.

Ceramics

There is an ever increasing move to harness the attractive high temperature properties of ceramics. The low ductility however tends to limit the considerable potential applications for these materials. There has been some very successful work completed recently which has shown that nitrogen ion implantation of ceramic components can give improved tensile strength and improved fatigue properties.

Guidelines for Designers

Material

The substrate material to be implanted can be any solid material. Successes have been reported using metals, ceramics and plastics.

Geometry

The process is strictly line of sight and therefore the ability to present the surface to be implanted to the beam is essential. For example it is hardly possible at this stage to implant the bore of a tube to a depth to width ratio of much more than 2 to 1. It is possible to manipulate a component under the beam to increase the number of faces exposed for implantation.

Temperature

The greatest number of applications of nitrogen ion implantation have been on tool steels. In this case the major benefit on the lower temperature tempering tool steels e.g. AB01 & D2. These steels are extensively used in engineering and in the majority of applications, ion implantation would offer a considerable improvement in performance.

Many components are manufactured from carburised and hardened material and are therefore low temperature tempered. These components could be finished machined and ion implanted on the wear surfaces to give a improvement of 2-10 times the life without any reduction in mechanical properties.

CONCLUSIONS

This paper has only given a glimpse of this technology but in summary it is clear that there is enormous potential to enhance the properties of engineering materials.

They are still relatively new and are gaining acceptance every day.

They are only the first of a large family of treatments that will become available within the next five to ten years.

It makes good commercial sense for those companies offering these processes to extend their market as wide as possible. It will also be beneficial to industry to take advantage of the considerable savings offered by these techniques.

Improvements have been and are being made in new applications every day, it is important to ensure that the solution to the problem is a cost effective solution. □

AUTHOR INDEX

SUBJECT INDEX